General Chemistry Workbook

FOURTH EDITION

General Chemistry WORKBOOK

HOW TO SOLVE CHEMISTRY PROBLEMS

Conway PIERCE

Professor Emeritus
University of California, Riverside

R. Nelson SMITH

Pomona College

W. H. FREEMAN AND COMPANY
San Francisco

Printed in the United States of America

International Standard Book Number: 0-7167-0157-X

1   2   3   4   5   6   7   8   9

Preface

This revision of the Workbook reflects several changes in our own teaching, and incorporates several terms and conventions that have become widely accepted since the publication of the third edition. Because most students are poorly prepared to handle experimental data in graphical form, and to derive important data from graphs, we have added Chapter 6, which is on graphical representation. Because of the increased emphasis on thermo-dynamics in first-year courses, Chapters 15 and 23, which deal with this subject, have been expanded and more detail has been added to them. The expansion of the material on synthesis and on the prediction of the course of chemical reactions (Chapter 25), including some helpful drawings, stems from the continued minimization and disorganization of this subject in most first-year texts. The problems involving gases (Chapter 11) are treated more analytically than they were in the earlier editions, and all solutions are shown to result from the direct application of the ideal gas law. As before, many chapters will be useful in courses that include much quantitative analysis in the laboratory work.

Since the third edition, one of us (Smith) has published a general chemistry textbook entitled *Chemistry: A Quantitative Approach* (New York: Ronald Press, 1969). It is natural that the writing of such a textbook should have contributed to the improvement of this Workbook, in the same way that its earlier editions aided in the writing of the text.

We wish to thank the many teachers and students who have given us criticisms and comments on the material of the preceding editions, and we hope that those who use this edition will be equally frank in calling to our attention any errors or omissions they may note.

CONWAY PIERCE
R. NELSON SMITH

Contents

General Chemistry Workbook

Chapter 1

How to Study Problems

For many of you the first course in chemistry will be a new experience and perhaps a difficult one. To understand chemistry, you will need to work hundreds of problems; for many students the mathematical side of the course is just too much, and the casualties are high.

If you are serious about wanting to learn chemistry, a few suggestions now may help make the way much easier.

1. Learn each assignment before going on to a new one. Chemistry has a vertical structure; that is, each new concept depends on previous material. Don't pass over anything, expecting to make up later. And don't postpone study until exam time. Keep up to date.

2. Understand problems before you try to work them. Chemistry problems use terms that may be new to you, and too often students fail to understand what these terms mean.

When you read a problem, try to analyze it before solving it. Every problem can be broken down into two parts that can be thought of as questions: (a) What information is given in the data of the problem? (b) What are you asked to find in working the problem?

To illustrate (without using chemical terms), suppose we are given this problem: "A boy rows a boat at the rate of 4 miles per hour. He heads upstream in a current of 5 miles per hour. Where will he be in 1.5 hours?"

First, what do the data tell us? If the boat were in still water, it would go 6 miles in 1.5 hours. But, while the boat is moving against the current, that current is carrying it downstream at the rate of 5 miles per hour; in 1.5 hours the boat, if it were not being rowed, would be carried 7.5 miles downstream.

Now, what is asked in the problem? We want to calculate the position of the boat after 1.5 hours. Using the information, we see that the boat drifts downstream 7.5 miles while being rowed upstream 6 miles. At the end of the given time interval the boat is $7.5-6 = 1.5$ miles below the starting point.

This looks, and really is, easy. Most of the chemistry problems you will encounter in this course will also be easy if you make sure that you understand them before trying to solve them. Don't become frightened if a problem is stated in moles, liters, or other new terms. They are just as simple as miles and hours once you understand them.

Another practice that often helps translate word problems into the mathematical equations needed to solve them is to draw simple sketches and label them to show how the different parts are related.

3. Know how to perform the *mathematical* operations you need in solving problems. The mathematics used in general chemistry is most elementary, involving only arithmetic and very simple algebra. Nevertheless, if you don't understand it, you can expect troubles before long. So, before we can start any chemistry, you must master the mathematical operations in the first six chapters.

As you study the solved problems in your text, in the lectures, and in this book, work through the mathematical operations. Then, if you run into anything you don't understand, clear it up before going ahead.

Exponents, Logarithms, and the Slide Rule

EXPONENTS AND DECIMAL POINTS

An exponent is a number that shows how many times another number (called the base) appears as a factor; exponents are written as superscripts. Thus 10^2 means $10 \times 10 = 100$: the number 2 is the *exponent*; the number 10 is the *base*, and is said to be raised to the second *power*. Likewise, 2^5 means $2 \times 2 \times 2 \times 2 \times 2 = 32$: here 5 is the exponent; 2 is the base, and is raised to the fifth power.

In the decimal system we use, it is simple to represent a number in exponential form as the product of some other number and a power of 10. For integral (whole) powers of 10, we have $10 = 1 \times 10^1$, $100 = 1 \times 10^2$, $1,000 = 1 \times 10^3$, etc. If a number is not an integral power of 10, we can express it as the product of two numbers, one of the two being an integral power of 10, and we can then write the integral power of 10 in exponential form. For example, 2,000 can be written as $2 \times 1,000$, and then put into the exponential form 2×10^3.

Notice that what we have done here is transform the expression

$$2,000.0 \times 10^0$$

(all numbers are understood to have a decimal point, and 10^0 is defined as equal to 1) into the expression

$$2.0 \qquad \times 10^3$$

Notice that we have shifted the decimal point *three* places to the *left*, and have also *increased* the exponent on the 10 by 3, the *same* number. If we now move the decimal point one place to the *right*, we must *decrease* the exponent by 1, to get

$$20.0 \quad \times 10^2$$

By following this rule — *left, increase; right, decrease* — you cannot go wrong. For example, suppose we have

$$3.0 \quad \times 10^{-3}$$

and we want to get rid of the negative exponent. Since we must *increase* the exponent by 3 to get 10^0, we must move the decimal point three places to the *left*, which leads to

$$.003 \quad \times 10^0$$

PROBLEM:
Write 3,500,000 in exponential form.

SOLUTION:
The decimal point can be set at any convenient place. Suppose we select the position shown below by the small x, between the digits 3 and 5. This gives as the first step

$$3_x500,000$$

Since we are moving the decimal point six places to the *left*, the exponent on 10^0 must be *increased* by 6, and so the answer is

$$3.5 \times 10^6$$

This last number might equally well be written as 35×10^5, 350×10^4, or 0.35×10^7, and so on. All of these are equivalent, and for each calculation we could arbitrarily set the decimal point at the most convenient place. However, the convention is to leave the lefthand number in the range between 1 and 10, that is, with a single digit before the decimal point; this is known as *standard* exponential form.

PROBLEM:
Write the number 0.0005 in standard exponential form.

SOLUTION:
We want to set the decimal point after the 5, as indicated by the small x:

$$0.0005_x$$

Since we are thus moving the decimal point four places to the *right*, we must *decrease* the exponent on the 10^0 by 4. Therefore the exponential form is 5×10^{-4}.

To use the exponential forms of numbers in mathematical operations, we need remember only the laws of exponents.

1. *Multiplication.* $X^a \cdot X^b = X^{a+b}$

2. *Division.* $\dfrac{X^a}{X^b} = X^{a-b}$

3. *Powers.* $(X^a)^b = X^{ab}$

4. *Roots.* $\sqrt[b]{X^a} = X^{a/b}$

Notice that since b (or a) can be a negative number, the first two laws are actually the same, and that since b can be a fraction, the third and fourth are also actually the same; that is, $X^{-b} = \dfrac{1}{X^b}$, and $X^{1/b} = \sqrt[b]{X}$.

In practice, we perform mathematical operations on numbers in exponential form according to the following simple rules.

1. To multiply two numbers, put them both in standard exponential form. Then multiply the two lefthand numbers by ordinary multiplication, and the two righthand numbers (powers of 10) by the multiplication law for exponents, that is, by adding their exponents.

PROBLEM:
Multiply 3,000 by 400,000.

SOLUTION:
Write each number in exponential form. This gives:

$$3,000 = 3 \times 10^3$$
$$400,000 = 4 \times 10^5$$

Multiply:

$$3 \times 4 = 12$$
$$10^3 \times 10^5 = 10^{3+5} = 10^8$$

The answer is 12×10^8.

If some of the exponents are negative, there is no difference in the procedure; the algebraic sum of the exponents is still the exponent of the answer.

PROBLEM:
Multiply 3,000 by 0.00004.

SOLUTION:

$$3,000 = 3 \times 10^3$$
$$0.00004 = 4 \times 10^{-5}$$

Multiply:

$$3 \times 4 = 12$$
$$10^3 \times 10^{-5} = 10^{3-5} = 10^{-2}$$

The answer is 12×10^{-2} (or 0.12).

2. To divide one number by another, put them both in standard exponential form. Divide the first lefthand number by the second according to the rules of ordinary division. Divide the first righthand number by the second according to the division law for exponents, that is, by subtracting the exponent of the divisor from the exponent of the dividend to obtain the exponent of the quotient.

PROBLEM:
Divide 0.0008 by 0.016.

SOLUTION:
Write in standard exponential form. This gives:

$$0.0008 = 8 \times 10^{-4}$$
$$0.016 = 1.6 \times 10^{-2}$$

Divide:

$$\frac{8}{1.6} = 5$$

$$\frac{10^{-4}}{10^{-2}} = 10^{-4-(-2)} = 10^{-4+2} = 10^{-2}$$

The answer is 5×10^{-2} (or .05).

3. To add or subtract numbers in exponential form, adjust the numbers to make all the exponents on the righthand numbers the same. Then add or subtract the lefthand numbers by the ordinary rules, making no further change in the righthand numbers.

PROBLEM:
Add 2×10^3 to 3×10^2.

SOLUTION:
Change one of the numbers to give its exponent the same value as the exponent of the other; then add the lefthand numbers.

$$2 \times 10^3 = 20 \times 10^2$$
$$3 \times 10^2 = \underline{3 \times 10^2}$$
$$23 \times 10^2$$

4. Since $10^0 = 1$ (more generally, for any number n, $n^0 = 1$), if the exponents in a problem reduce to zero, then the righthand number drops out of the solution.

PROBLEM:
Multiply 0.003 by 3,000.

SOLUTION:

$$0.003 = 3 \times 10^{-3}$$
$$3,000 = 3 \times 10^3$$

Multiply:

$$3 \times 10^{-3} \times 3 \times 10^3 = 9 \times 10^{3-3} = 9 \times 10^0 = 9 \times 1 = 9$$

The use of these rules in problems requiring both multiplication and division is illustrated in the following example.

PROBLEM:
Use exponents to solve

$$\frac{2,000,000 \times 0.00004 \times 500}{0.008 \times 20} = ?$$

SOLUTION:
First, rewrite all numbers in standard exponential form:

$$2,000,000 = 2 \times 10^6$$
$$0.00004 = 4 \times 10^{-5}$$
$$500 = 5 \times 10^2$$
$$0.008 = 8 \times 10^{-3}$$
$$20 = 2 \times 10^1$$

This gives

$$\frac{2 \times 10^6 \times 4 \times 10^{-5} \times 5 \times 10^2}{8 \times 10^{-3} \times 2 \times 10^1}$$

Dealing first with the nonexponential numbers, we find

$$\frac{2 \times 4 \times 5}{8 \times 2} = \frac{5}{2} = 2.5$$

The exponent of the answer is

$$\frac{10^6 \times 10^{-5} \times 10^2}{10^{-3} \times 10^1} = \frac{10^{6-5+2}}{10^{-3+1}} = \frac{10^3}{10^{-2}} = 10^{3-(-2)} = 10^{3+2} = 10^5$$

The complete answer is 2.5×10^5 or 250,000.

APPROXIMATE CALCULATIONS

A trained scientist often makes mental estimates of numerical answers to quite complicated calculations, with an ease that to the uninitiated appears to border on the miraculous. Actually, all he does is round off numbers and use exponents to reduce the calculation to a very simple form. It is quite useful for you to learn these methods. By using them you can save a great deal of time in homework problems and on tests.

PROBLEM:
We are told that the population of a city is 256,700 and that the assessed value of the property is $653,891,600. Find an approximate value of the assessed property per capita.

SOLUTION:
We need to evaluate the division

$$\frac{\$653,891,600}{256,700} = ?$$

First write the numbers in standard exponential form:

$$\frac{6.538916 \times 10^8}{2.56700 \times 10^5}$$

Round off to

$$\frac{6.5 \times 10^8}{2.6 \times 10^5}$$

Mental arithmetic gives

$$\frac{6.5}{2.6} = 2.5$$

$$\frac{10^8}{10^5} = 10^3$$

The answer is \$2.5 × 10³, or \$2,500. This happens to be a very close estimate; the value obtained with a calculator is \$2,547.29.

PROBLEM:
Find an approximate value for

$$\frac{2{,}783 \times 0.00894 \times 0.00532}{1{,}238 \times 6{,}342 \times 9.57}$$

SOLUTION:
First rewrite in exponential form, but instead of using standard form, set the decimal points to make each number as near to 1 as possible:

$$\frac{2.783 \times 10^3 \times 0.894 \times 10^{-2} \times 5.32 \times 10^{-3}}{1.238 \times 10^3 \times 6.342 \times 10^3 \times 0.957 \times 10^1}$$

Rewrite the numbers, rounding off to integers:

$$\frac{3 \times 10^3 \times 1 \times 10^{-2} \times 5 \times 10^{-3}}{1 \times 10^3 \times 6 \times 10^3 \times 1 \times 10^1}$$

Multiplication now gives

$$\frac{3 \times 1 \times 5 \times 10^{-2}}{1 \times 6 \times 10^7} = \frac{15}{6} \times 10^{-9} = 2.5 \times 10^{-9}$$

LOGARITHMS

The logarithm (abbreviated log) of a number N is the power to which 10 (called the base) must be raised to give N. The logarithm is therefore an exponent. Thus $\log 1 = 0$, $\log 10 = 1$, $\log 100 = 2$, and so on.

Since logarithms are exponents, we have the following *logarithmic laws* (compare these with the laws of exponents on p. 0):

$$\log AB = \log A + \log B$$

$$\log \frac{A}{B} = \log A - \log B$$

$$\log A^n = n \log A$$

$$\log \sqrt[n]{A} = \frac{\log A}{n}$$

The logarithm of a number N consists of two parts, called the characteristic and the mantissa. The characteristic is the number before the decimal point. It is determined by the position of the decimal point in N. The mantissa, read from a log table, is the number after the decimal point.

Finding the Log of a Number

To find the log of a number:
1. Write the number in standard exponential form.
2. Look up the mantissa in a log table. In finding the mantissa, pay no attention to where the decimal point of the number is.
3. The exponent is the characteristic of the log.

PROBLEM:
Find the log of 203.

SOLUTION:
1. Write the number as 2.03×10^2.
2. In the log table find 2.0 (or 20) in the left-hand column. Read across to the column under 3. This gives log 2.03 = 0.3075.
3. Since the exponent is 2, the characteristic is 2. Therefore, log 2.03×10^2 = 2.3075.

When a number is less than 1, the logarithm is negative, since log 1 = 0. We have a choice of two ways to express a negative log.
1. The mantissa is positive but the characteristic is negative.
2. Both characteristic and mantissa are negative.

PROBLEM:
Find the log of 0.000203.

SOLUTION:
Write the number as 2.03×10^{-4}. Log $2.03 \times 10^{-4} = 0.3075 - 4 = -3.6925$. If we wish to write the log as a positive mantissa, we indicate the negative characteristic by a bar over the number, that is $\bar{4}.3075$.

Interpolation

The log tables of this book show only three digits for N. If you want the log of a four-digit number, you must interpolate. An example will illustrate this process. To find the log of 2,032, proceed as follows:

$$\log 2{,}032 = \log 2.032 \times 10^3$$

mantissa 204 = 3096 (to save time, the decimal point before the mantissa is left out while making the interpolation)

mantissa 203 = 3075

difference = 21

$$\text{mantissa } 2{,}032 = 3075 + \left(\frac{2}{10} \times 21\right) = 3075 + 4 = 3079$$
$$\log 2{,}032 = 3.3079$$

Antilogs

The number that corresponds to a given logarithm is known as the anti-logarithm or antilog (also written $\log^{-1}$).

PROBLEM:
Find the antilog of 1.5502.

SOLUTION:
Locate 0.5502 (or the number nearest to it) in a log table, then find the value of N that has this log. We find N to be 3.55. Since the characteristic is 1, the number is $3.55 \times 10^1 = 35.5$.

Multiplication by Logs

To multiply numbers, find and add their logs, and find the antilog of the sum. For example, to multiply 167 by 0.00518:

$$\begin{aligned}
\log 167 &= \log 1.67 \times 10^2 = 2.2227 \\
\log 0.00518 &= \log 5.18 \times 10^{-3} = \overline{3}.7143 \\
\text{adding gives} & \qquad\qquad\qquad\quad \overline{1}.9370
\end{aligned}$$

$$\text{antilog } \overline{1}.9370 = 8.65 \times 10^{-1} \quad = 0.865$$

Division by Logs

To divide one number by another, subtract the log of the divisor from that of the dividend, and look up the antilog of the remainder to find the quotient. For example, to divide 167 by 0.00518:

$$\begin{aligned}
\log 167 &= 2.2227 \\
\log 0.00518 &= \overline{3}.7143 \\
& \quad\;\; 4.5084
\end{aligned}$$

$$\text{antilog } 4.5084 = 3.224 \times 10^4$$

Roots of a Number

To obtain the root of a number, find the log of the number, divide by the value of the root desired, and look up the antilog. For example, to find $\sqrt[5]{225}$:

$$\log 225 = 2.3522$$

$$\text{dividing by 5: } \frac{2.3522}{5} = 0.4704$$

$$\text{antilog } 0.4704 = 2.954 \times 10^0 = 2.954$$

Natural Logarithms

Although we have so far emphasized logarithms as the numbers to which the base 10 is raised, actually any number might be used as a base instead of 10. The most common base other than 10 that is used is e. For our purposes we need to know only that $e = 2.718$ . . ., an incommensurate number, like π, that can be written with as many numbers after the decimal point as one likes. It is a common base because e has mathematical significance, and many of the laws of chemistry and physics are derived mathematically from physical models and principles. It is traditional to write the logarithm of a number N to the base e as ln N, and to call it the *natural* logarithm. You can change from one base to the other by knowing that

$$\ln N = 2.303 \log N$$

or that

$$e^N = 10^{\frac{N}{2.303}}$$

SLIDE RULE

Every student of chemistry or physics should use a slide rule for arithmetical calculations. It is sheer waste of time and effort to carry out multiplications and divisions by longhand, when the slide rule will give satisfactory answers in a tenth of the time or less.

In general, a 10-inch rule is preferable to other sizes. A simple rule with A, B, C, D, K, and log scales is sufficient for the problems of the general chemistry course.

In learning to use a slide rule, it is helpful to realize that it is a graphic log table. The lengths of the scale divisions are proportional to the mantissas of the logarithms of the scale numbers. Therefore, to multiply two numbers, we add the scale lengths for multiplier and multiplicand; to divide two numbers, we subtract the scale length of the divisor from that of the dividend.

Construction of the Slide Rule

When operating a slide rule we add or subtract scale distances. For these operations the rule is constructed of three essential parts. There is a frame-like part that contains scales D, A, and K. A movable part known as the slide contains scale C and others with which we are not at present concerned. To mark scale readings, there is a glass or plastic cursor that carries an index line.

FIGURE 2-1
Portion of a slide rule, showing C and D scales.

Reading Scales

Multiplications and divisions are carried out on the C and D scales. The two are identical, with D fixed and C on the slide. Before you start to use a rule, you must learn how to read the scales accurately. Take your own rule and locate the C and D scales. Set the slide so that 1 on C coincides with 1 on D. The two scales should now coincide throughout.

Marked readings for a portion of the rule are shown in Figure 2-1. Use your own rule to locate the same readings. Set the index at 12 on the C and D scales, verify by checking with Figure 2-1, then move to the next reading. If you can locate each of the readings of Figure 2-1, you will have no further difficulty in reading the C and D scales.

The A and K scales are used for squares and cubes of numbers. Note that the A scale is a series of two scales, each half as long as the D scale and the K scale is a series of three, each one-third the length of the D scale.

Since the scale lengths of a slide rule are proportional to the mantissas of the logarithms of the scale numbers, the reading for a number is not influenced by the location of the decimal point. Thus the reading for 2, 20, 200, 0.02, and so on is the same. The user must supply his decimal point for the answer; the slide rule gives only the digits of the answer.

Multiplication

Take your own rule, and perform each step as directed. First, we will multiply 2 by 3.

1. Set 1 on C at 2 on D.

2. Move the hair-line index to 3 on C.

3. Read the answer 6 on D (beneath the hair line).

You will note that we have added the two scale distances; since these are proportional to the logs of the numbers, we are multiplying the numbers.

Without moving the slide, which is still set with 1 on C at 2 on D, move the index to 4 on C. Now read the index on D. The answer is 8. We have multiplied 2 by 4.

Move the index to 45 on C. It now reads 9 on D, which is the product of 2×45. Note that, although the answer is 90, the slide-rule reading is 9.

Suppose now that we wish to multiply 2 by 6. You find that 6 on C is beyond the end of the D scale. Whenever this occurs, reverse the slide by moving it to the left until the right-hand 1 on C lies at 2 on D. In this position move the index to 6 on C. The answer 12 can now be read on D.

Successive Multiplications

Suppose we wish to multiply $2 \times 3 \times 4$. Carry out the following operations.
1. Set 1 on C at 2 on D.
2. Move the index to 3 on C. Do not read on D.
3. Move the slide to bring the right-hand 1 on C to the position of the index.
4. Move the index to 4 on C.
5. Read the answer 24 on D.

Division

To illustrate division, we will divide 24 by 6. Go through the following steps.
1. Set the index at 24 on D.
2. Move the slide to bring 6 on C to the index.
3. Move the index to 1 on C.
4. Read the answer 4 on D.

Successive Operations

In a series of multiplications and divisions, we save moves by alternating the processes. To illustrate, we will solve the following problem:

$$475 \times (720/760) \times (273/298)$$

We could first multiply $475 \times 720 \times 273$, then divide the answer by 760 and divide that answer by 298. However, it is faster to carry out the operations in the following order.
1. Set the index at 475 on D.
2. Move the slide to bring 760 on C to the index.

3. Move the index to 720 on C.
4. Move the slide to bring 298 on C to the index.
5. Move the index to 273 on C.
6. Read the answer 412 on D.

Decimal Point

Slide-rule manuals give simple rules for determining the position of the decimal point, but generally it is easier to place a decimal point by an approximate calculation. In the last problem, for example, it is obvious that the answer is 412 and not 41.2 or 4,120 since we start with 475 and multiply it by two numbers, $\frac{720}{760}$ and $\frac{273}{298}$. Each of these is a fraction slightly smaller than 1.

In more complicated problems we set up the numbers in exponential form, then get an approximate answer to place the decimal point.

PROBLEM:
Use the slide rule to solve

$$\frac{0.00365 \times 5,470 \times 40}{23 \times 0.0083} = ?$$

SOLUTION:
Write in exponential form:

$$\frac{3.65 \times 10^{-3} \times 5.47 \times 10^3 \times 4 \times 10^1}{2.3 \times 10^1 \times 0.83 \times 10^{-2}}$$

The approximate answer is

$$\frac{4 \times 5 \times 4}{2 \times 1} \times \frac{10^{-3+3+1}}{10^{1-2}} = 40 \times 10^2 = 4,000$$

With the slide rule we find

$$\frac{365 \times 547 \times 4}{23 \times 83} = 418$$

Since the approximate solution gives an answer of 4,000, the slide-rule value is 4,180.

Square

Set the index at the number on the D scale. Read the answer on the A scale. To illustrate, set the index at 4 on D, and read the square, 16, on the second A scale. Determine the decimal point by inspection after writing the number in exponential form.

PROBLEM:
Find the square of 2,500.

SOLUTION:
Write the number as 2.5×10^3. Set the hairline at 2.5 on the D scale. Read the answer, 6.25, on the A scale. Square the exponential portion. This gives $(10^3)^2 = 10^6$. The answer is 6.25×10^6.

Square Root

Rewrite the given number in exponential form as an *even* power of 10 multiplied by a number between 1 and 100. If the number has one digit, set the index at the number on the first section of the A scale, and read the answer on D. If the number has two digits, use the second portion of the A scale. To obtain the exponent of the answer, divide the exponent by 2.

PROBLEM:
Find $\sqrt{0.025}$.

SOLUTION:
Write the number as 2.5×10^{-2}. Set the index at 2.5 on the first section of the A scale. The answer on D is 1.58. The exponent is $-2 \div 2 = -1$. Therefore the answer is $1.58 \times 10^{-1} = 0.158$.

PROBLEM:
Find $\sqrt{0.0025}$.

SOLUTION:
Write the number as 25×10^{-4}. Set the index at 25 on the second section of the A scale. Read the answer, 5, on the D scale. The exponent is -2. This gives $5 \times 10^{-2} = 0.05$ as the answer.

Cube

For a one-digit number (between 1 and 10) set the index at the given number on the D scale. Read the answer on the K scale. To illustrate, set the index at 3 on the D scale. Read the answer 27 on the second section of the K scale. The answer has one digit if it is on the first section, two if on the second, and three if on the third.

To find the cube of numbers not lying between 1 and 10, write the number in standard exponential form, use the slide rule to cube the lefthand number, and multiply the exponent by 3 to find the exponent of the answer.

PROBLEM:
Find the cube of 0.002.

SOLUTION:
Write the number as 2×10^{-3}. Find the cube of 2 on the slide rule. The answer is 8. The exponent of the answer is $3(-3) = -9$. The complete answer is 8×10^{-9}.

Cube Root

Rewrite the given number in exponential form as a number between 1 and 1,000, with the exponent divisible by 3. Set the index at the number on the proper section of the K scale. Read the answer on the D scale. For numbers between 1 and 10 use the first K scale, for numbers between 10 and 100 the second K scale, and for numbers between 100 and 1,000 the third K scale. Divide the exponent by 3 to get the exponent of the answer.

PROBLEM:
Find $\sqrt[3]{25,000}$.

SOLUTION:
Write the number as 25×10^3. Set the index at 25 on the second section of the K scale. The answer is 2.92 on the D scale. The exponent is $3 \div 3 = 1$. This gives $2.92 \times 10^1 = 29.2$. Note that, when the number is written as directed, the numerical value read on the D scale has only one integer before the decimal point.

Logarithms

Slide rules have a log scale, denoted by L. Often it is placed just beneath the D scale. To find the log of a number set the hairline at the number on the D scale and read the mantissa of the logarithm on the L scale; to find an antilog, set the hairline at the log value on the L scale and read the number on the D scale.

PROBLEMS A   Answers on page 331

1. Express each of the following as a power of 10:

   (a) $10^3 \times 10^2$ (e) $10^5 \times 10^4 \times 10^{-12}$

   (b) $10^3 \div 10^2$ (f) $\dfrac{1}{10^2} \times \dfrac{1}{10^{-3}}$

   (c) $10^2 \div 10^3$

   (d) $10^2 \times 10^{-2}$

2. Express each of the following in standard exponential form (for example, $5,280 = 5.28 \times 10^3$):

(a) 6,225,000   (d) 0.12

(b) 721,300,000   (e) 0.00257

(c) 324   (f) 0.00000000086

3. Find an approximate answer to each of the following, using mental arithmetic:

(a) $\dfrac{7,860}{0.0065}$ (b) $(3,785)^3$ (c) $\sqrt{4.674}$

(d) $\dfrac{475 \times 730 \times 273}{760 \times 298}$

(e) $\dfrac{(2.54)^3 \times 8.2}{6.02 \times 10^{23}}$

(f) $\dfrac{0.00852 \times 73,940 \times 23.16 \times 0.046}{1,637 \times 0.000045 \times 1.25 \times 0.8954}$

(g) $(3.6 \times 10^{-5})^2 \times 0.04856 \times (8.8)^3$

(h) $\sqrt[3]{\dfrac{9.5 \times 10^{-12} \times 765.6 \times 0.0003224}{6.02 \times 10^{23} \times 505}}$

(i) $(2.562 \times 10^{-8})^3 \times 751.2 \times 0.0822 \times 6.02 \times 10^{23} \times 0.00001813$

4. Use a slide rule to solve the following problems, and confirm your answers by mental arithmetic and approximate solution:

(a) 2×3 (l) $(30)^2$

(b) $2 \times 3 \times 4$ (m) $(4.7)^2$

(c) $2 \div 4$ (n) $(895)^2$

(d) $\dfrac{2 \times 6}{4}$ (o) $(0.015)^2$

(e) 2.5×3.4 (p) $\sqrt{4.9}$

(f) 1.2×4.7 (q) $\sqrt{49}$

(g) $\dfrac{1.7 \times 89}{43}$ (r) $\sqrt{490}$

(h) $\dfrac{1.89 \times 76.5}{5.75}$ (s) $\sqrt{16.8}$

(i) $\dfrac{19.75 \times 142.5}{1.025}$ (t) $\sqrt{0.0054}$

(j) $\dfrac{7.62 \times 0.926 \times 32.6}{175.5 \times 0.0105}$ (u) 3^3

(k) 3^2 (v) $(30)^3$

                                     (w) $(4.7)^3$

                                     (x) $(895)^3$

                                     (y) $(0.0105)^3$

                                     (z) $\sqrt[3]{2.7}$

5. Solve the problems of No. 3 using a slide rule.

6. Find the logs of the following numbers:

   (a) 2.54   (d) 9.5×10^{-12}

   (b) 852   (e) 3,785

   (c) 0.046   (f) 1,623,000

7. Find the antilogs of the following logarithms:

   (a) 3.9939   (d) 0.5141

   (b) $\overline{2}.0334$ (e) 1.8455

   (c) $\overline{5}.4771$ (f) $\overline{11}.6989$

8. Use logs without interpolation to obtain three-digit answers to the problems of No. 3.

9. Use logs with interpolation to obtain four-digit answers to the problems of No. 3.

10. Solve for the unknown quantity in each of the following equations. Consult an algebra book for the methods if necessary.

   (a) $\dfrac{7}{x} = \dfrac{4}{3}$ (d) $\frac{5}{9}(F - 32) = 45$

   (b) $\dfrac{31 \times 12}{20v} = \dfrac{3}{5}$ (e) $\dfrac{4\pi R^3}{3} = 720$

   (c) $\dfrac{8x}{3} = 24$ (f) $5P^{\frac{1}{2}} - 2 = 6$

   (g) $(0.4)(550)(T_2 - 20) + 35(T_2 - 20) = (0.1)(120)(100 - T_2)$

   (h) $(M - 5)^2 = 49$ (j) $3x + 5 = 4x^2 + 3$

   (i) $\dfrac{4x + 5}{32} = \dfrac{4 - 3x}{12}$ (k) $(x + 2)^2 = 3x + 6$

11. Express each of the following as a single number raised to some power:

   (a) $2^4 \times 2^5$

   (b) $4^2 \div 2^7$ (c) $3^5 \div \left(\dfrac{3^2}{81}\right)^3$

PROBLEMS B   No answers given

12. Express each of the following as a power of 10:

   (a) $10^4 \div 10^2$ (b) $10^6 \times 10^3$

(c) $10^{-6} \times 10^{-3}$

(d) $10^{-4} \times 10^2 \times 10^{-10}$

(e) $10^4 \times 10^{-4}$

(f) $\dfrac{1}{10^{10}} \times \dfrac{1}{10^{-4}} \times \dfrac{1}{10^{-7}}$.

13. Express each of the following in standard exponential form:

(a) 0.0572

(b) 0.0000001003

(c) 80,210

(d) 67,230,000,000

(e) 62.5

(f) 0.101

14. Find an approximate answer to each of the following, using mental arithmetic:

(a) 820×72.9

(b) $\sqrt{0.0118}$

(c) $\dfrac{3,721 \times 5.03 \times (210)^2}{47.6 \times 0.00326}$

(d) $(6.72)^3 (37.6)^{\frac{1}{3}}$

(e) $\dfrac{0.093 \times (76)^3 \times 4.96}{(52)^2 \times \sqrt{0.0038}}$

(f) $\left(\dfrac{5.73 \times 10^{-4} \times 3.8 \times 10^{10} \times 0.0067 \times 5.42 \times 10^6}{1.987 \times 0.082 \times 1.38 \times 10^{16}}\right)^{\frac{1}{2}}$

(g) $\dfrac{(3.78)^2 \times 5.8 \times 10^{-2}}{6.6 \times 10^{-27} \times (4.2)^4}$

(h) $\dfrac{8.035 \times 10^{-4} \times 0.000579 \times 45.45}{(3.51)^3 \times (4.2 \times 10^{-3})^2}$

(i) $\dfrac{(45.1)^{-2}(3.21)^3}{(8.1)^{-3}(5.07)^2}$

15. Use a slide rule to solve each of the following problems, and confirm by using mental arithmetic to get an approximate answer:

(a) 4×3

(b) $5 \times 2 \times 6$

(c) $3 \div 5$

(d) $\dfrac{3 \times 5}{6}$

(e) 36×4.2

(f) 0.027×63.6

(g) $\dfrac{94.2 \times 16.3}{275}$

(h) $\dfrac{0.0543 \times 89.7 \times 2.02}{0.00592 \times 6.08}$

(i) $\dfrac{4.02 \times 10^3 \times 1.07 \times 10^{-2} \times 3.43 \times 10^{-4}}{8.09 \times 10^{-3} \times 1.02 \times 10^6}$

(j) $\dfrac{0.00988 \times 5,310 \times 50.72}{71.3 \times 0.082}$

(k) $(4)^2$

(l) $(40)^2$

(m) $(5.5)^2$

(n) $(472)^2$

(o) $(0.0178)^2$

(p) $\sqrt{6.4}$

(q) $\sqrt{64}$

(r) $\sqrt{640}$

(s) $\sqrt{42.7}$ (w) $(5.5)^3$

(t) $\sqrt{0.000178}$ (x) $(472)^3$

(u) $(4)^3$ (y) $(0.0178)^3$

(v) $(40)^3$ (z) $\sqrt[3]{2.54}$

16. Solve the problems of No. 14 with a slide rule.

17. Look up the logs of the following numbers:

    (a) 47.4   (d) 8.73×10^{-7}

    (b) 367   (e) 1,572

    (c) 0.0052   (f) 3,627,000,000

18. Look up the antilogs of the following logarithms:

    (a) 7.7451   (d) $\bar{8}.9890$

    (b) 0.3003   (e) 2.0072

    (c) $\bar{6}.6368$ (f) $\bar{4}.5020$

19. By using logs, without interpolation, give three-digit answers to the problems of No. 14.

20. By using logs, with interpolation, give four-digit answers to the problems of No. 14.

21. Solve for the unknown quantity in each of the following equations:

    (a) $\dfrac{3}{x} = \dfrac{2.7}{5}$ (c) $\dfrac{9x}{5} = 36$

    (b) $\dfrac{45 \times 32}{17M} = \dfrac{5}{9}$ (d) $\frac{5}{9}(F - 32) = 90$

    (e) $(0.5)(4.78)(T_2 - 23) + 40(T_2 - 23) = (0.08)(200)(100 - T_2)$

    (f) $\dfrac{\pi D^3}{6} = 35$ (i) $2x^2 + 3x = 5$

                               (j) $3n^2 - 7n = 6$

    (g) $\frac{4}{5}(2x - 10) = 3(1 - 5x)$ (k) $(m - 3)^2 = 5 - 2m$

    (h) $(Y + 4)^2 - 3 = 22$

22. Express each of the following as a single number raised to some power:

    (a) $x^2 \div x^4$ (c) $\dfrac{216}{6^4} \div 6^{-6}$

    (b) $(5^{10} \times 5^{-9}) \div 125$

Use of Dimensions

When numbers are used to express the results of measurements, the units of measurement should always be given along with the numbers. Too frequently these units, or dimensions, are tacitly assumed but not shown.

In using dimensions in calculations we follow a few simple rules.

1. Every number that represents a measurement is given with its dimension, for example, 12 men, 16 feet, 5 miles.

2. Numbers that do not involve a measurement are written without a dimension. Examples are π (the ratio of the circumference of a circle to its diameter) and logarithms.

3. In addition and subtraction all numbers must have the same dimensions. We can add 2 apples to 3 apples, but cannot add 2 apples and 3 miles.

4. In multiplication and division the dimensions of the numbers are multiplied and divided just as the numbers are, and the product or quotient of the dimensions appears in the final result. Thus, the product of 6 men times 2 days is given as 12 man-days.

A term frequently used with numbers is *per,* which shows how many units of one measurement correspond to *one* unit of another. A common method of calculation that leads to this usage of per is to divide the total number of units of one property by the total number of units of another property to which it corresponds. Thus, if we are told that 4.2 gallons weigh 10.5 pounds, we express the relation by the ratio

$$\frac{10.5 \text{ lb}}{4.2 \text{ gal}} = 2.5 \frac{\text{lb}}{\text{gal}}$$

which is read as "2.5 pounds per gallon."

A few examples will help in understanding the use of units.

PROBLEM:
If apples cost 30 cents per dozen, how many can be bought for 50 cents?

SOLUTION:
Set up the equation that will give the number of apples.

$$\text{number of apples} = \frac{50 \text{ cents}}{30 \frac{\text{cents}}{\text{doz}}} \times 12 \frac{\text{apples}}{\text{doz}}$$

We start with a complex fraction. Simplifying this gives

$$\text{number of apples} = 50 \cancel{\text{cents}} \times \frac{\cancel{\text{doz}}}{30 \cancel{\text{cents}}} \times 12 \frac{\text{apples}}{\cancel{\text{doz}}}$$

Canceling the terms indicated gives

$$\text{number of apples} = (50/30) \times 12 \text{ apples} = 20 \text{ apples}$$

A somewhat more sophisticated method for setting up the problem is to use the negative exponent -1 for numbers that appear in the denominator. Thus the term "per dozen" may be written doz^{-1}.

The preceding problem can then be set up as

$$\text{number of apples} = \frac{50 \text{ cents}}{30 \text{ cents doz}^{-1}} \times 12 \frac{\text{apples}}{\text{doz}}$$

or as

$$\text{number of apples} = \frac{50 \text{ cents doz}}{30 \text{ cents}} \times 12 \text{ apples doz}^{-1}$$

PROBLEM:
Find the number of feet in 1.5 mi (one mile is 5,280 ft).

SOLUTION:
To find the number of feet, we need the product shown in the equation

$$\text{feet} = 1.5 \text{ mi} \times 5,280 \frac{\text{ft}}{\text{mi}}$$

$$= 7,920 \text{ ft}$$

Alternatively, we may use the solution

$$\text{feet} = 1.5 \text{ mi} \times 5,280 \text{ ft mi}^{-1}$$

$$= 7,920 \text{ ft}$$

PROBLEM:
Find the number of gallons in 5 cu yd (also written as yd³), using the conversion factors

$$231 \text{ in.}^3 = 1 \text{ gal}$$
$$3 \text{ ft} = 1 \text{ yd}$$
$$12 \text{ in.} = 1 \text{ ft}$$

SOLUTION:

$$\text{gallons} = 5 \text{ yd}^3 \times \left(3 \frac{\text{ft}}{\text{yd}}\right)^3 \times \left(12 \frac{\text{in.}}{\text{ft}}\right)^3 \times \frac{1 \text{ gal}}{231 \text{ in.}^3}$$

$$= 5 \text{ yd}^3 \times 27 \frac{\text{ft}^3}{\text{yd}^3} \times 1{,}728 \frac{\text{in.}^3}{\text{ft}^3} \times \frac{\text{gal}}{231 \text{ in.}^3}$$

$$= 1{,}010 \text{ gal (using a slide rule)}$$

PROBLEM:
Find a conversion factor, F, by which you can convert yd³ to gallons.

SOLUTION:
A conversion factor can be developed from other known factors by calculating a value of F that will satisfy the equation

$$\text{gal} = (\text{yd})^3[F]$$

F must have such units that, when they are substituted in this equation, they will cancel yd³ and yield only gal as the net result, as follows:

$$\text{gal} = (\text{yd})^3 \left[\left(\frac{3 \cancel{\text{ft}}}{\text{yd}}\right)^3 \left(\frac{12 \cancel{\text{in.}}}{\cancel{\text{ft}}}\right)^3 \left(\frac{1 \text{ gal}}{231 \cancel{\text{in.}^3}}\right)\right]$$

$$F = 202 \frac{\text{gal}}{\text{yd}^3}$$

Thus, if you want to convert 20 yd³ to gallons, you merely multiply 20 by 202, as follows:

$$\text{gal} = (20 \cancel{\text{yd}^3}) \left(202 \frac{\text{gal}}{\cancel{\text{yd}^3}}\right) = 4040 \text{ gal}$$

PROBLEM:
If a runner does the 100-yd dash in 10 seconds, what is his speed in miles per hour?

SOLUTION:
The first thing you should observe is that the units of the desired answer must be $\frac{\text{mi}}{\text{hr}}$; so the original information must be used in the same ratio of $\frac{\text{length}}{\text{time}}$, or as $\frac{100 \text{ yd}}{10 \text{ sec}}$. Once you have made the proper decision about how to use the units

of the original data, follow the same procedure as in finding a conversion factor, F:

$$\frac{mi}{hr} = \left(\frac{100 \ yd}{10 \ sec}\right)\left(\frac{1 \ mi}{1760 \ yd}\right)\left(\frac{60 \ sec}{min}\right)\left(\frac{60 \ min}{hr}\right) = 20.4 \ \frac{mi}{hr}$$

PROBLEMS A   Answers on page 332

1. Compute the number of seconds in the month of July.

2. Develop a factor to convert days to seconds.

3. A satellite is orbiting at a speed of 18,000 miles per hour. How many seconds does it take to travel 100 miles?

4. A traveler on a jet plane notes that in 30 seconds the plane passes 6 section-line roads (1 mile apart). What is the ground speed, in miles per hour?

5. A cubic foot of water weighs 62.4 lb. What is the weight of a gallon of water (231 cu in.)?

6. For each of the following pairs of units, work out a conversion factor, F, that will convert a measurement given in one unit to a measurement given in the other, and show the simple steps used in your work:
   (a) ounces to tons
   (b) cubic inches to cubic yards
   (c) feet per second to miles per hour
   (d) tons per square yard to pounds per square inch
   (e) cents per pound to dollars per ton
   (f) seconds to weeks
   (g) cubic feet per second to quarts per minute
   (h) miles to fathoms (1 fathom = 6 ft)
   (i) yards to mils (1 mil = 1/1,000 in.)

PROBLEMS B   No answers given

7. For each of the following pairs of units, follow the same procedure as for Problem 6:
   (a) cubic feet to gallons
   (b) ounces per square foot to pounds per square yards
   (c) gallons per second to cubic yards per minute
   (d) tons per cubic foot to pounds per cubic inch
   (e) yards per second to inches per hour
   (f) dollars per pound to nickels per ounce
   (g) miles to mils (1 mil = 1/1,000 inch)
   (h) knots to miles per hour (1 knot = 101.5 feet per minute)
   (i) degrees of arc per second to revolutions per minute

8. An acre-foot of water will cover an acre of land with a layer of water 1 foot deep. How many gallons are in an acre-foot? Use the factors: 1 acre = 4,840 yd^2; and 1 gal = 231 in.^{3}

9. Municipal water is sold at 21 cents per 100 cu ft. What is the price per acre-foot?

10. In the Bohr model of the hydrogen atom, an electron travels in a circular orbit about the nucleus at approximately 5×10^6 miles per hour. How many revolutions per second does the electron make if the radius of the orbit is 2×10^{-9} inches?

11. A light-year is the distance that light travels in one year at a velocity of 186,000 miles per second. How many miles is it to the galaxy in Andromeda, which is said to be 650,000 light-years away?

Units of Scientific Measurements

Scientific measurements range from fantastically large to incredibly small numbers, and units that are appropriate for one measurement may be entirely inappropriate for another. To avoid the creation of many different sets of units, it is common practice to vary the size of a fundamental unit by attaching a suitable prefix to it. Common metric prefixes and the values they indicate for any given unit of measurement are shown in Table 4-1. Thus, a kilometer is 1,000 meters, a microgram is 10^{-6} gram, and a nanosecond is 10^{-9} second.

TABLE 4-1
Metric Prefixes

Prefix	Factor	Symbol	Prefix	Factor	Symbol
tera	10^{12}	T	deci	10^{-1}	d
giga	10^{9}	G	centi	10^{-2}	c
mega	10^{6}	M	milli	10^{-3}	m
kilo	10^{3}	k	micro	10^{-6}	μ
hecto	10^{2}	h	nano	10^{-9}	n
deka	10^{1}	da	pico	10^{-12}	p
			femto	10^{-15}	f
			atto	10^{-18}	a

Except for temperature and time, scientific measurements are based on the metric system. Table 4-2 shows how some of the more familiar metric units are related to units commonly used in English-speaking countries * for nonscientific measurements. Four widely used metric units with special names are the Ångstrom (10^{-8} cm), the fermi (10^{-13} cm), the micron (10^{-4} cm) denoted by symbol μ, and the gamma or microgram (10^{-6}g) denoted by the symbol γ. The customary units of days, hours, minutes, and seconds are used in scientific measurements of time.

TABLE 4-2
Metric and English Units

Dimension measured	Metric unit	English unit	Conversion factor, F
Length	centimeter (cm)	inch (in.)	$2.540\ \dfrac{cm}{in.}$
Volume	liter	quart (qt)	$0.9463\ \dfrac{liter}{qt}$
Mass	gram (g)	pound (lb)	$453.6\ \dfrac{g}{lb}$
Temperature	degree Celsius (°C)	degree Fahrenheit (°F)	$°C = \dfrac{5}{9}(°F - 32)$
Time	second (sec)	second (sec)	—

At the time the metric system was devised, it was intended to simplify computations by defining the unit of mass in terms of the unit of length, that is, by defining the gram as the mass of 1 cubic centimeter of water at 3.98°C, its temperature of greatest density. Later, it was discovered that an error had been made, and that the old liter, instead of being exactly 1,000 cm³, was actually 1,000.027 cm³. In recent times, however, the liter has been redefined as exactly equal to 10^{-3} m³, and, as a result, 1 ml does in fact exactly equal 1 cm³.

Conversions of units are illustrated in the following problems. Note that in all such computations it is important to include the dimensions of the numbers, and that use of these dimensions helps avoid errors.

---

* Great Britain and Canada have decided to abolish the inches-pounds-Fahrenheit system and are converting to the metric system, mainly to facilitate international trade.

PROBLEM:
Convert $\frac{5}{16}$ inch to millimeters.

SOLUTION:

$$\frac{5 \text{ in.}}{16} \times \frac{2.54 \text{ cm}}{\text{in.}} \times \frac{10 \text{ mm}}{\text{cm}} = 8 \text{ mm}$$

PROBLEM:
The diameter of an atom is 3 Å. Express this in feet.

SOLUTION:

$$\text{diameter} = 3 \text{ Å} \times \frac{10^{-8} \text{ cm}}{\text{Å}} \times \frac{1 \text{ in.}}{2.54 \text{ cm}} \times \frac{1 \text{ ft}}{12 \text{ in.}}$$

$$= \frac{3 \times 10^{-8}}{2.54 \times 12} \text{ ft} = 0.98 \times 10^{-9} \text{ ft}$$

PROBLEM:
Find the number of cubic Ångstroms in a gallon (231 in.³).

SOLUTION:

$$\text{Å}^3 = 1 \text{ gal} \times 231 \frac{\text{in.}^3}{\text{gal}} \times \left(\frac{2.54 \text{ cm}}{\text{in.}}\right)^3 \times \left(\frac{10^8 \text{ Å}}{\text{cm}}\right)^3$$

$$= 231 \times (2.54)^3 \times (10^8)^3 = 3.78 \times 10^{27}$$

PROBLEM:
Find the weight, in pounds, of 1 gal of water.

SOLUTION:

$$\text{mass} = 1 \text{ gal} \times 231 \frac{\text{in.}^3}{\text{gal}} \times \left(\frac{2.54 \text{ cm}}{\text{in.}}\right)^3 \times \frac{1 \text{ g}}{\text{cm}^3} \times \frac{1 \text{ lb}}{454 \text{ g}}$$

$$= \frac{231 \times (2.54)^3}{454} = 8.35 \text{ lb}$$

Temperature

For measurements of temperature we must be familiar with three different scales: Celsius (C), absolute—also called Kelvin—(K), and Fahrenheit (F). In engineering work and in daily life, the Fahrenheit scale is used exclusively, but all scientific temperature measurements are expressed in degrees Celsius or degrees Kelvin. The three scales are compared in Table 4-3, which shows that 180°F cover the same range as 100°C or 100°K. Thus, 1°C and 1°K = 1.8°F, and 1°F = $\frac{100}{180} = \frac{5}{9}$°C or $\frac{5}{9}$°K.

TABLE 4-3
The Three Temperature Scales

Reference point	Scale		
	F	C	K
Boiling point of water	212°	100°	373°
Freezing point of water	32°	0°	273°
Difference (FP to BP)	180°	100°	100°

To convert Celsius temperature to the corresponding Fahrenheit temperature, we first multiply the Celsius temperature by the conversion factor, 1.8. This gives the number of degrees Fahrenheit above the freezing point of water. Then, since the freezing point is 32°F, we add 32° to obtain the Fahrenheit reading.

PROBLEM:
Convert 30°C to Fahrenheit.

SOLUTION:
Since 1°C = 1.8°F, the degrees Fahrenheit corresponding to 30°C are given by

$$F \text{ degrees} = 30°C \times \frac{1.8°F}{\text{degree C}} = 54°F$$

That is, a temperature 30°C above the freezing point of water is 54°F above the freezing point of water. The Fahrenheit temperature is obtained by adding 32°F:

$$32°F + 54°F = 86°F$$

PROBLEM:
Convert 115°F to Celsius and absolute temperatures.

SOLUTION:
First, find how many degrees Fahrenheit the given temperature is above the freezing point of water:

$$115°F - 32°F = 83°F \text{ above freezing point of water}$$

Convert to degrees Celsius:

$$83°F \times \frac{5°C}{9°F} = 46°C$$

Since the freezing point of water is at 0°C, the Celsius temperature is 46°. On the absolute scale the freezing point of water is 273°. Adding 46° to this gives

$$46°K + 273°K = 319°K$$

Note that the Celsius and absolute temperatures are in the same units, $1°C = 1°K$. The only difference is in the reference point taken as zero. Absolute zero or $0°K = -273°C$.

Convenient Formulas

Frequently we need to compute areas and volumes for solids of various geometric forms. The following formulas are useful for reference.

Area of circle of radius r:	$A = \pi r^2$
Area of cylinder of length L:	$A = 2\pi rL + 2\pi r^2$
Area of sphere of radius r:	$A = 4\pi r^2$
Area of triangle of base B and height H:	$A = HB/2$
Volume of cylinder of radius r and length L:	$V = \pi r^2 L$
Volume of sphere of radius r:	$V = \frac{4}{3}\pi r^3$

PROBLEMS A   Answers on page 333

When you work these problems, show the units in each step of the calculation, and show the units of the answers.

1. A brass bar is $2 \times 3 \times 6$ cm. Find its area and volume.

2. A cylindrical rod is 2 cm in diameter and 12 in. long. Find its area and volume.

3. First-class postage is 6 cents for each ounce or fraction thereof. How much postage is required for a letter weighing 98 g?

4. What is the weight, in pounds, of 20 kg of iron?

5. The distance from Paris to Rouen is 123 km. How many miles is this?

6. A regulation basketball may have a maximum circumference of $29\frac{1}{4}$ in. What is its diameter in centimeters?

7. The longest and shortest visible waves of the spectrum have wavelengths of 0.000067 cm and 0.000037 cm. Convert these values to (a) Ångstroms and (b) microns.

8. The wavelengths of X-rays characteristic of certain metallic targets are: (a) copper, 1.537395 Å; (b) chromium, 2.28503 Å; (c) molybdenum, 0.70783 Å; (d) tungsten, 0.20862 Å. Express these wavelengths in centimeters.

9. If the laboratory temperature is 21°C, what is the Fahrenheit temperature?

10. When the temperature gets to $-50°F$ in Siberia, what would the temperatures be on the centigrade and absolute scales?

11. Mercury freezes at $-38.87°C$. What are the freezing points on the Fahrenheit and absolute scales?

12. What is the Fahrenheit temperature at absolute zero ($0°K$)?

13. At what temperature do the Fahrenheit and Celsius scales have the same reading?

14. (a) A cube measures 1 cm on an edge. What is the surface area? (b) The cube is crushed into smaller cubes measuring 1 mm on an edge. What is the surface area after crushing? (c) Further crushing gives cubes measuring 100 Å on an edge. What is the surface area now? How many football fields (160 ft × 100 yd) would this make?

15. If 1 ml of water is spread out as a film 3 Å thick, what area in square meters will it cover?

16. The area of a powdered material is 100 m^2/g. What volume of water is required to form a film 10 Å thick over the surface?

17. An agate marble is placed in a graduated cylinder containing 35.0 ml of water. After the marble is added, the surface of the water stands at 37.5 ml. Find the diameter and surface area of the marble.

18. (a) If there are 6.02×10^{23} molecules in 18 ml of water, what is the volume occupied by 1 molecule? (b) If the molecules were little spheres, what would be the radius of a water molecule? (Give the answer in angstroms.)

19. The neck of a volumetric flask has an internal diameter of 12 mm. The usual practice is to fill a volumetric flask until the liquid level (meniscus) comes just to the mark on the neck. If by error 1 drop (0.05 ml) too much is added, at how many millimeters above the mark on the neck will the meniscus stand?

20. It has been found that the percentage of gold in sea water is 2.5×10^{-10}. How many tons of sea water would have to be processed in order to obtain 1 g of gold?

21. A solution contains 5 g of sodium hydroxide per liter. How many grams will be contained in 50 ml?

22. A solution contains 40 g of potassium nitrate per liter. How many milliliters of this solution will be needed in order to get 8 g of potassium nitrate?

23. The unit of viscosity (η) is called a poise. Viscosity is determined experimentally by measuring the length of time (t) for a certain volume (V) of liquid to run through a capillary tube of radius R and length L under a pressure P, according to the Poiseuille equation

$$\eta = \frac{P\pi t R^4}{8LV}$$

What are the units of a poise in the centimeter-gram-second system?

24. The energy of a quantum of light is proportional to the frequency (v) of the light. What must be the units of the proportionality constant (h) if $E = hv$? E is expressed in ergs, and v has the units of sec^{-1}.

25. Show that the product of the volume of a gas and its pressure has the units of energy.

26. If you should decide to establish a new temperature scale based on the assumptions that the melting point of mercury ($-38.9°C$) is $0°M$ and the boiling point of mercury ($356.9°C$) is $100°M$, what would be (a) the boiling point of water in degrees M and (b) the temperature of absolute zero in degrees M?

27. Light travels at a speed of 3×10^{10} cm/sec. A light-year is the *distance* that light can travel in a year's time. If the sun is 93,000,000 miles away, how many light-years is it from the earth?

28. A soap bubble is 3 inches in diameter and is made of a film that is 0.01 mm thick. How thick will the film be if the bubble is expanded to 15 inches in diameter?

PROBLEMS B   No answers given

29. What are the area and the volume of a bar measuring $2 \times 4 \times 20$ cm?

30. What are the area and the volume of a cylindrical rod with hemispherical ends if the rod is 1 inch in diameter and has an over-all length of 55.54 cm?

31. The driving distance between Los Angeles and San Francisco by one route is 420 miles. Express this in kilometers.

32. A common type of ultraviolet lamp uses excited mercury vapor, which emits radiation at 2,537 Å. Express this wavelength in (a) centimeters and (b) microns.

33. A certain spectral line of cadmium is often used as a standard in wavelength measurements. The wavelength is 0.000064384696 cm. Express this wavelength in (a) Ångstroms and (b) microns.

34. What must be the velocity in miles per hour of a jet plane if it goes at twice the speed of sound? (The speed of sound is 1,000 ft/sec under the prevailing conditions.)

35. The domestic airmail rate is 10 cents/oz. How much postage will be required for a letter weighing 76 g?

36. If the price of platinum is $102/oz (ounce avoirdupois), what is the cost of a crucible and cover weighing 12.356 g?

37. Liquid nitrogen boils at $-195.82°C$. What are the boiling points on the Fahrenheit and absolute scales?

38. Gallium is unusual in that it boils at 1,700°C and melts at 29.8°C. What are these temperatures on the Fahrenheit and absolute scales?

39. If 50 g of a substance S contains 6.02×10^{23} molecules, and if the cross-sectional area of each molecule is 20.0 Å², what is the surface area of a solid that needs 1.5 g of S to cover it with a layer 1 molecule thick?

40. (a) If there are 6.02×10^{23} molecules in 58.3 ml of ethyl alcohol, what is the volume occupied by 1 molecule? (b) If the molecules were little spheres, what would be the radius of an ethyl alcohol molecule? (Give the answer in Ångstroms.)

41. A regulation baseball is 9 inches in circumference. What is its diameter in centimeters?

42. A 10-ml graduated pipet has an 18-inch scale graduated in tenths of milliliters. What is the internal diameter of the pipet?

43. A manufacturer of a glass-fiber insulation material impresses his potential customers with the "fineness" of his product (and presumably with its insulating qualities too) by handing them glass marbles ½ inch in diameter and stating that there is enough glass in a single marble to make 96 miles of glass fiber of the type used in his product. If this is true, what is the diameter of the insulating glass fibers?

44. The volume of a red blood cell is about 90 μ^3. What is its diameter in millimeters? (Assume that the cell is spherical.)

45. A ½-inch-diameter marble is placed in a graduated cylinder containing 10.0 ml of water. To what level will the liquid rise in the cylinder?

46. A solution contains 0.05 g of salt per milliliter. How many milliliters of this solution will be needed if we are to get 100 g of salt?

47. A solution contains 1.0 g of sulfuric acid per 100 ml. How many grams of acid will be contained in 350 ml?

48. How many grams of sulfuric acid must be added to 500 g of water in order that the resulting solution be, by weight, 20 per cent sulfuric acid?

49. The average velocity of a hydrogen molecule at 0°C is 1.84×10^5 cm/sec. (a) How many miles per hour is this? Molecular gas velocities are proportional to the square root of the absolute temperature. (b) At what temperature will the velocity of a hydrogen molecule be 100,000 mph?

50. What will be the diameter, in inches, of the bottom portion (spherical) of a 500-ml round-bottom pyrex flask if the neck of the flask is 30 mm in diameter and 5 cm long? (Neglect the thickness of the glass walls.)

51. To what temperature must a bath be heated so that a Fahrenheit thermometer will have a reading that is three times as large as that on a Celsius thermometer?

52. According to Newton's law, the force exerted by an object is equal to its mass times its acceleration. The unit of force needed to accelerate a mass of 1 g by 1 cm/sec^2 is called a dyne. What are the units of a dyne in the cgs system?

53. A budding young chemist decided to throw tradition overboard and include time in the metric system. To do this he kept the unit "day" to refer to the usual 24-hr time interval we know. He then subdivided the day into decidays, centidays, millidays, and microdays. Solve the following problems. (a) A 100-yd dash done in 9.7 sec took how many microdays? (b) A 50-min class period lasts for how many centidays? (c) A car going 60 mph goes how many miles per deciday? (d) What is the velocity of light in miles per milliday if it is 186,000 mi/sec? (e) What is the acceleration of gravity in centimeters per microday^2 if it is 980 cm/sec^2?

Reliability of Measurements

In pure mathematics every number has an exact meaning—the figure 2, for example, means *precisely* two units, *not approximately* two units. This is not true when we use numbers to express the result of measurements, because no measurement is perfectly accurate. When we say that an object has a length of 2 m, we mean approximately 2 m. If carefully measured, the length would probably be between 1.9999 m and 2.0001 m, but it is unlikely that it would be exactly 2.000000 m.

Since measurements are not exact, we should use numbers in a way that indicates the reliability of the result. This can be illustrated by a simple experiment: Two persons are asked to measure the diameter of a dime with a centimeter scale. One might find it to be 1.75 cm, the other 1.76 cm. The reason for the disagreement is that the nearest 0.1 mm or 0.01 cm must be estimated. The two different results show that the estimation is uncertain and that the answer is *about* 1.75 cm. It would be correct to give either 1.75 cm or 1.76 cm, but it would not be correct to give the average value of 1.755 cm. This last figure implies that the measurements are known to the third decimal place.

If the measurements are repeated with a caliper having a vernier scale that reads to 0.001 cm, the two results might be 1.752 cm and 1.756 cm, giving an average value of 1.754 cm. It would now be proper to give the answer as

1.754 cm, since the third decimal place is reasonably well known. Or the answer might be stated as 1.754 ± 0.002 cm. The symbol $\pm$ is read as "plus or minus." It shows that results vary by 0.002 cm from the reported value.

The *accuracy* of measurements depends on how closely the average of the results agrees with the true value of the quantity that is measured. The precision of measurements depends on how reproducibly they are made. Measurements with high precision may actually be quite inaccurate, as illustrated by a clipped-off ruler. It is not infrequent that measurements involve *systematic errors,* which are reproducibly *repeated* by improper use, construction, or calibration of the equipment. Yet the precision of repeated measurements may give the illusion of accuracy. One way to check for the presence of systematic errors is to make the measurement by several entirely different methods; if the results agree well with each other, it is unlikely that systematic errors exist. An easier way to check the accuracy of a method is to use it with a "standard sample" whose value is certified by some reliable institution, such as the National Bureau of Standards. *Random errors* are the errors that are unpredictable and unreproducible, and they are usually associated with the limited sensitivity of the instruments, the quality of the scales that are read, the control of the environment (temperature, vibration, humidity, etc.), or human frailties (eyesight, hearing, etc.). We will have much more to say about random error later in this chapter.

Significant Figures

All digits of a number that are reasonably reliable are known as *significant figures*. The number 1.75 has three significant figures: 1, 7, and 5. The number 1.754 has four significant figures.

The position of the decimal point in a measured value has nothing to do with the number of significant figures. The diameter of a dime may be given as 1.754 cm or as 17.54 mm. In either case, four significant figures are used.

PROBLEM:
A student weighs a beaker on trip scales, finding values of 50.32 g, 50.31 g, and 50.31 g in successive weighings. Express the average weight to the proper number of significant figures.

SOLUTION:
To obtain the average, add the weights and divide by 3.

$$
\begin{array}{ll}
 & 50.32 \text{ g} \\
 & 50.31 \\
 & \underline{50.31} \\
\text{total} & = 150.94 \\
\text{average} = & 50.3133 \text{ g}
\end{array}
$$

Since the weighings disagree in the second decimal place, it is not proper to give the weight to more than the second decimal place. The average weight then, to the proper number of significant figures, is 50.31 g.

Final Zero as Significant Figure

Final zeros after decimal points are significant figures and are used to indicate the decimal place to which the measurements are reliable. Thus 1.0 cm indicates a length reliably known to tenths of a centimeter but not to hundredths of a centimeter, while 1.000 cm indicates a length reliably known to thousandths of a centimeter. A very common mistake is leaving out these zeros when the measured quantity has an integral value.

PROBLEM:
Give the value of a 10-g weight to the proper number of significant figures. The balance on which the weight is used will respond to weight differences of 0.0001 g.

SOLUTION:
The weight is given as 10.0000 g, to show that it is reliable to 0.0001 g.

When a number has no zeros after the decimal point, final zeros before the decimal point may or may not be significant, depending upon the usage. If we say that there are 1,000 students enrolled in a school, all the zeros are significant. But if the population of a city is given as 360,000 the last two or three zeros are not significant, for changes make the population uncertain by perhaps several hundred persons. The final zeros in this case are used to indicate the position of the decimal point. A convenient way to indicate reliability of a number that has final zeros before the decimal point is to express the number/in exponential form, using the correct number of significant figures in the nonexponential factor, for example, 3.60×10^5 for the city whose population is known only to the nearest 1,000 persons.

Zeros Before a Number

If a number is less than one, the zeros following the decimal point and preceding other digits are *not* significant. The number 0.0032 has two significant figures, 3 and 2. If "0.0032 m" is written as "3.2 mm," we have the same two significant figures. Remember, however, that if a zero is added after the last digit, it is significant.

PROBLEM:
State the number of significant figures in each of the following numbers: 275; 2.75; 0.0275; 0.027500; 27,500.

SOLUTION:

275 has 3 significant figures: 2, 7, 5.

2.75 has 3 significant figures: 2, 7, 5.

0.0275 has 3 significant figures: 2, 7, 5.

0.027500 has 5 significant figures: 2, 7, 5, 0, 0.

27.500, if we mean *exactly* 27,500, has 5 significant figures: 2, 7, 5, 0, 0.

Rounding off Numbers

In the beaker-weighing problem on p. 37 the number 50.3133 was changed to 50.31 by dropping the last two digits because they were not significant. The process of dropping figures that are not significant is known as "rounding off a number." The following conventions are usually observed:

1. When a number greater than 5 is dropped, increase the last remaining digit by 1; thus, 7.238 is rounded off to 7.24.

2. When a number less than 5 is dropped, the last remaining digit is left unchanged; thus, 7.243 is rounded off to 7.24.

3. If a 5 is dropped or if a 5 followed by zeros is dropped, the last remaining figure is left unchanged if it is even, or increased by 1 if it is odd; thus, 7.2450 is rounded off to 7.24, and 7.2550 is rounded off to 7.26.

Many times it is necessary to round off numbers to show the proper number of significant figures before making calculations. There are two rules for this.

1. In addition or subtraction round off all numbers to the same number of digits after the decimal point. If the numbers, all with the same units, are expressed in exponential form, the power of 10 must be the same for all.

Original number	Rounded number
13.8426 g	13.8 g
764.08 g	764.1 g
7.5 g	7.5 g
785.4226 g	785.4 g

2. In multiplication and division round off all numbers to the number of significant figures possessed by the least accurate number. A little care must be exercised in applying this rule. For example, to multiply

$$0.00296 \times 5845 \times 93$$

you might be inclined to round off each number to two significant figures. But if you did so, you would be using poor judgment. The number 93 *almost* has three significant figures; on a percentage basis there is very little more error with 1 in 93 than there is with 1 in 102, which has three significant figures. Further, since there is no indication whether the uncertainty in the last place of each number is 1, or 2, or 3, or just what, it makes more sense to round off all the numbers to three significant figures; i.e., you would multiply

$$(2.96 \times 10^{-3})(5.84 \times 10^3)(9.3 \times 10^1) = 1.61 \times 10^3$$

and express your answer with three significant figures. When a slide rule is used for multiplication and division, numbers are automatically rounded off to three significant figures because it is not possible to read the scale of a 10-in. rule more accurately than this; the readings at the far left end barely qualify as having four significant figures.

Pure numbers such as 3 and 4 have an unlimited number of significant figures (4.0000000 . . .), as do defined quantities such as π (3.14159 . . .). As a consequence, do not fall into the trap of excessively rounding off measurements that are to be used in equations containing pure or derived numbers. For example, if you want to find the volume of a sphere whose radius has been measured as 15.13 cm, you should not round off the radius to 2×10^1 cm just because you are going to use the formula $V = \frac{4}{3}\pi r^3$, in which 4, 3, and π each appear to have only one significant figure. Remember that it is the *measured* values which determine the number of significant figures.

Distribution of Errors

We have talked about significant figures and the general unreliability of the "last figure" of a measurement. Now we shall talk about just *how* unreliable these last figures are. Your past experience has shown that really gross errors rarely occur in a series of measurements. If you were able to make an infinite number of measurements on the same quantity, say X, you would not be surprised if, on plotting each observed value of X against the frequency with which it occurred, you obtained a symmetrical curve similar to that shown in Figure 5-1. One of the advantages of making an infinite number of measurements is that the *average* ($\bar{X}$) of the values will be equal to the "true value" (μ), represented by the dotted vertical line drawn from the peak of the curve. As expected, the more a value of X deviates from μ the less frequently it occurs. This curve is symmetrical because there is equal probability for + and − errors; it is called a *normal distribution curve*. If you had made your measurements in a more careless manner or with a less sensitive measuring device, you would have obtained a distribution curve more like that in Figure 5-2, shorter and broader but with the same general shape. The breadth of a distribution curve, the spread of results, is a measure of the reliability of the results. Two major ways of describing this spread are *average deviation* and *standard deviation*.

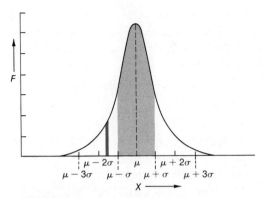

FIGURE 5-1
Normal distribution curve with a small standard deviation.

Average Deviation

The deviation of each individual measurement (X) from the average $(\bar{X})$ of all the measurements is found by simple subtraction; the deviation of the ith measurement is $X_i - \bar{X}$. We are interested in the *sizes* of the deviations, without regard to whether they are $+$ or $-$; that is, we are interested only in the absolute values of the deviation, $|X_i - \bar{X}|$. The average deviation is simply the average of these absolute values:

$$\text{average deviation} = \frac{|X_1 - \bar{X}| + |X_2 - \bar{X}| + |X_3 - \bar{X}| + \ldots + |X_n - \bar{X}|}{n}$$

Simple expressions for $\bar{X}$ and average deviation can be made by using the symbol Σ, which means "the sum of" whatever follows it:

$$\bar{X} = \frac{\Sigma X_i}{n}$$

and

$$\text{average deviation} = \frac{\Sigma |X_i - \bar{X}|}{n} \tag{5-1}$$

In each case, the sum of n values is understood. One important point to remember: $\bar{X}$ may be calculated for any number of measurements, but only for an infinite number of measurements will $\bar{X} = \mu$, the "true value."

PROBLEM:
Five persons measure the length of a room, getting values of 10.325 m, 10.320 m, 10.315 m, 10.313 m and 10.327 m. Find the average value and the average deviation.

SOLUTION:

Add the separate values and divide by 5 to get the arithmetical mean. Set opposite each value its deviation from the average, without regard to sign. Take the average of these deviations.

measurement X_i	*deviation* $|X_i - \bar{X}|$
10.325 m	0.005 m
10.320	0.000
10.315	0.005
10.313	0.007
10.327	0.007
$\Sigma X_i = 51.600$ m	0.024 m $= \Sigma|X_i - \bar{X}|$

$$\bar{X} = \frac{\Sigma X_i}{5} = 10.320 \text{ m}$$

$$\frac{\Sigma|X_i - \bar{X}|}{5} = 0.005 \text{ m} = \text{average deviation}$$

The average is 10.320 m, with an average deviation of 0.005 m. It is proper to write the average as 10.320 m, since the deviation affects digits in only the third decimal place. It is not correct to give the length as 10.32 m; this implies that the measurement is uncertain in the second decimal place. Average deviation is one of the simplest measures of reliability of measurements (the spread of experimental values), but a better estimate of reliability can be made with standard deviation.

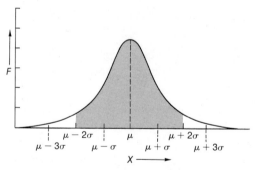

FIGURE 5-2
Normal distribution curve with a large standard deviation.

Standard Deviation

The standard deviation is the square root of the *variance* (S^2). The variance is *almost* the same as the average of the squares of the deviations of the measurements from the average ($\bar{X}$); it is defined as

$$\text{variance} = S^2 = \frac{(X_1 - \bar{X})^2 + (X_2 - \bar{X})^2 + (X_3 - \bar{X})^2 + \ldots + (X_n - \bar{X})^2}{n - 1}$$

$$= \frac{\Sigma(X_i - \bar{X})^2}{n - 1}$$

For reasons that we need not discuss here, $n - 1$ is used in the denominator instead of n. Of course, for very large values of n (say 1,000), there is no appreciable difference between n and $n - 1$, and thus for very large numbers of measurements you really can say that the variance is the average of the squares of the deviations. Since the standard deviation is the square root of the variance, we have

$$\text{standard deviation} = S = \sqrt{\frac{\Sigma(X_i - \bar{X})^2}{n - 1}} \qquad (5\text{-}2)$$

The standard deviation will have the same units as the original measurements, and the same units (but not the same value) as the average deviation.

PROBLEM:
Calculate the standard deviation (S) of the measurements made in the preceding problem.

SOLUTION:
As in the last problem, first find the average of the measurements, and then subtract it from each of the individual measurements to get the deviation. The sum of the squares of these deviations divided by $n - 1$ is the variance.

measurement X_i	deviation $(X_i - \bar{X})$	$(deviation)^2$ $(X_i - \bar{X})^2$
10.325 m	0.005 m	0.000025 m²
10.320	0.000	0.000000
10.315	−0.005	0.000025
10.313	−0.007	0.000049
10.327	0.007	0.000049

$\Sigma X_i = 51.600$ m $0.000148 \text{ m}^2 = \Sigma(X_i - \bar{X})^2$

$\bar{X} = 10.320$ m

$$\text{standard deviation} = S = \sqrt{\frac{\Sigma(X_i - \bar{X})^2}{n - 1}} = \sqrt{\frac{0.000148}{4}} = 0.0061 \text{ m}$$

In order to make the value of S useful for other calculations, it is customary to write it to one more decimal place than the last significant figure of the measurements; we shall do the same with $\bar{X}$.

Probability Distribution for Large Numbers of Measurements

In order to appreciate the usefulness of the standard deviation as a measure of reliability, we must take a closer look at the curves in Figures 5-1 and 5-2. The mathematical equation for these curves, when they represent a very large number of measurements, is called the Gaussian distribution equation, and in one of its forms the frequency of occurrence, F, is given by

$$F = \left[\frac{1}{\sigma(2\pi)^{\frac{1}{2}}}\right] e^{-\frac{(X_i - \mu)^2}{2\sigma^2}} \qquad (5\text{-}3)$$

where $\sigma = S$ (the standard deviation) when n is very large, just as $\mu = \bar{X}$ when n is very large. We use these special symbols for S and $\bar{X}$ to emphasize that this Gaussian equation does *not* apply to the curves we would find if we made only a small number of measurements. Note that Equation (5-3) is written with the base e instead of with the base 10 (see p. 12 for a discussion of natural logarithms). Some important characteristics of this distribution curve are discussed in the following paragraphs.

1. The size of σ and the shape of the curve. At the peak of the curve, $X = \mu$ and

$$F = \frac{1}{\sigma(2\pi)^{\frac{1}{2}}}$$

In other words, the maximum height of the Gaussian curve is determined solely by the value of σ and the constant 2π. If σ is small because the errors are relatively small, then F is large and the curve is tall (Figure 5-1). If σ is large because of relatively large errors, then F is small and the curve is short (Figure 5-2). For any other value of X, the exponent $\frac{(X_i - \mu)^2}{2\sigma^2}$ is large for a small value of σ and the sides of the curve fall off faster (as in Figure 5-1) than when σ is large (as in Figure 5-2).

2. The significance of the area under the curve. The very small blackened area in Figure 5-1 is bounded on one side by the infinitesimal distance dx and on the other side by the value of F for whatever value of X we have chosen. This area, $F dx$, represents the *number* of measurements of X that lie between X and $X + dx$. If we take the consecutive sum of all such small areas from one end of the curve to the other, we have the total area under the curve and we have included all our measurements. The total area under the curve, regardless of the size of σ, is always 1.000. This property, that the area under the curve is equal to unity, is characteristic of all *probability distribution curves*. A smaller area represents just part of the measurements; 20 per cent of them if the area is 0.20.

3. The relationship between σ and area under the curve. Regardless of whether deviations are relatively small (small σ) or relatively large (large σ),

the areas under the normal distribution curve that lie on either side of the true value by 1, 2, or 3 σ are *always* the same, as follows:

values of X between	area under curve
$\mu - \sigma$ and $\mu + \sigma$	0.683 (shaded area, Figure 5-1)
$\mu - 2\sigma$ and $\mu + 2\sigma$	0.954 (shaded area, Figure 5-2)
$\mu - 3\sigma$ and $\mu + 3\sigma$	0.997

This means that whenever you have a very large number of measurements, the probability is that 68.3 per cent of them (about two-thirds) will lie within ± 1 standard deviation of μ, that 95.4 per cent of them (about 19 out of 20) will lie within ± 2 standard deviations of μ, and that only 0.3 per cent of them (3 in 1,000) will lie outside ± 3 standard deviations.

4. The confidence interval and confidence level. If you wished to have the confidence that 80 per cent of your results would lie between certain limits, you would have to find the values of X on either side of μ that would give an area of 0.80 under the distribution curve. From what we said in part 3, you can see that these limits would be somewhere between 1σ and 2σ on either side of μ. In the general situation the limits will be $\mu - t\sigma$ and $\mu + t\sigma$, where the value of t will be determined by the per cent of confidence that is desired. In the last row (representing an infinite number of measurements) of the t-table, Table 5-1, you will find values of t that apply for a selected number of convenient confidence levels. The range of values that lies between $\mu - t\sigma$ and $\mu + t\sigma$ is called the *confidence interval*, and the probability that any measurement picked at random will lie within this interval is called the *confidence level*.

PROBLEM:
What is the confidence interval for a set of 1,000 measurements for which there is an 80 per cent probability that a measurement taken at random will lie within that interval? You are given that the true value is 2,756 and that the standard deviation is 13.0.

SOLUTION:
From the t-table we select a value of t from the column headed by 80 per cent and from the row corresponding to $n = \infty$. The value is 1.282 (a set of 1,000 measurements is so large that the value of t differs negligibly from that for an infinite number of measurements). The confidence interval is

$$2,756 \pm (1.282)(13.0) = 2,756 \pm 17$$

Either of the following statements could be made:
(a) The probability is 80 per cent that a value taken at random from the 1,000 values will lie within the interval $2,756 \pm 17$.
(b) Of the 1,000 measurements, 80 per cent lie within the interval $2,756 \pm 17$.

TABLE 5-1
The *t*-values for Various Sample Sizes and Confidence Levels

Sample size (*n*)	Per cent confidence level						
	50	60	70	80	90	95	99
2	1.000	1.376	1.963	3.078	6.314	12.706	63.657
3	.816	1.061	1.386	1.886	2.920	4.303	9.925
4	.765	.978	1.250	1.638	2.353	3.182	5.841
5	.741	.941	1.190	1.533	2.132	2.776	4.604
6	.727	.920	1.156	1.476	2.015	2.571	4.032
7	.718	.906	1.134	1.440	1.943	2.447	3.707
8	.711	.896	1.119	1.415	1.895	2.365	3.499
9	.706	.889	1.108	1.397	1.860	2.306	3.355
10	.703	.883	1.100	1.383	1.833	2.262	3.250
20	.688	.861	1.066	1.328	1.729	2.093	2.861
30	.683	.854	1.055	1.311	1.699	2.045	2.756
40	.681	.851	1.050	1.303	1.684	2.021	2.704
50	.680	.849	1.048	1.299	1.676	2.008	2.678
60	.679	.848	1.046	1.296	1.671	2.000	2.660
120	.677	.845	1.041	1.289	1.658	1.980	2.617
∞	.674	.842	1.036	1.282	1.645	1.968	2,576

Probability Distribution for Small Numbers of Measurements

It is often not practical to make a very large number of measurements; sometimes there may not be enough material to make more than a very few. The main objective of measurements is usually to find a true value, but with only a few measurements it is not possible to find a true value by taking an average. Everyone has had the experience of making a series of measurements and finding that the averages of selected groups of values differ from one another and depend on just which values were selected for the groups. Moreover, almost everyone feels that the more measurements one makes, the more likely it is that the average will be "correct." It is possible to make some statements about small numbers of measurements with some confidence, but with much less confidence than for those statements made in our discussion of large numbers of measurements.

For this discussion we use a distribution curve as before, but this time we do not have one represented by an equation as simple as the Gaussian equation. In fact, there is not just one curve; there are many, one for each size of sample (different numbers of measurements). It isn't practical to draw a different curve for each size of sample, but we can describe the changing nature of the distribution curves: as the size of the sample gets smaller, the corresponding distribution curve becomes shorter and broader than the one shown in Figure 5-1 for the same value of *S*. The same statement is true for Figure

5-2, where a larger value of S applies. This changing nature of the distribution curve is taken into account in the t-table, Table 5-1, which can be used in place of the curves.

The Precision of a Single Measurement

The calculation of standard deviation (Equation 5-2) is the same whether you have many measurements or only a few; sample size affects only the selection of the t-value. For a single measurement taken at random from a small number, n, of measurements, the confidence interval for the desired confidence level is

$$\bar{X} \pm tS \qquad (5\text{-}4)$$

This is a statement of the precision of a single measurement.

Note that since you don't know the true value (you didn't take many measurements), you must now use the average ($\bar{X}$) of your measurements as the best measure available as a substitute for the true value. Note also that the value of t you choose will depend on the sample size, n, as well as on the confidence level you desire. The confidence interval will get larger as the number of determinations gets smaller or as the confidence level increases.

The Precision of the Mean

Usually the main objective of making a series of measurements is to find the true value, and we would like to indicate the confidence we may have in the average of our values. Since the average of many measurements is more likely to be correct than the average of a few, our confidence interval should get smaller as we increase the number of measurements. This is commonly expressed quantitatively by using the *standard deviation of the mean*, defined as

$$\frac{S}{\sqrt{n}} = \text{standard deviation of the mean}$$

Note that the standard deviation of the mean decreases with the square root of n, not the first power. As a result, you cannot improve the quality of your average by a factor of 25 by making 100 measurements instead of 4; it will be improved only by a factor of $\sqrt{25}$, or 5. The useful statement we can make with the standard deviation of the mean is the following: for a series of n measurements and a specified confidence level, the *true value* of X will lie in the interval

$$\bar{X} \pm t\left(\frac{S}{\sqrt{n}}\right) \qquad (5\text{-}5)$$

This is a statement of the precision of the mean.

PROBLEM:

The density of a liquid is measured by filling a 50-ml flask as close as possible to the index mark and weighing. In successive trials the weight of the liquid is found to be 45.736 g, 45.740 g, 45.705 g, and 45.720 g. For these weights calculate the average deviation, the standard deviation, the 95 per cent confidence interval for a single value, and the 95 per cent confidence interval for the mean.

SOLUTION:

Since the weighings are all for the same measured volume, we first average the weights. Let X refer to the weight measurement.

weight X_i	deviation $|X_i - \bar{X}|$	(deviation)^2 $(X_i - \bar{X})^2$
45.736 g	0.011 g	0.000121 g^2
45.740	0.015	0.000225
45.705	0.020	0.000400
45.720	0.005	0.000025
$\Sigma X_i = 182.901$ g	$\Sigma|X_i - \bar{X}| = 0.051$ g	$\Sigma(X_i - \bar{X})^2 = 0.000771$ g^2

$$\bar{X} = \frac{\Sigma X_i}{4} \qquad \text{average deviation} = \frac{\Sigma|X_i - \bar{X}|}{4} \qquad S = \sqrt{\frac{\Sigma(X_i - \bar{X})^2}{3}}$$

$$= 45.725 \text{ g} \qquad\qquad = 0.0128 \text{ g} \qquad\qquad = 0.0161 \text{ g}$$

From the t-table the t-value of 3.182 is found in the row for sample size of 4 and in the column for 95 per cent confidence level. The precision of a single value is therefore

$$45.725 \pm (3.182)(0.0161) = 45.725 \pm 0.0513 \text{ g}$$

There is a 95 per cent probability that any weight value picked at random has a value that lies within 0.0513 g of the average. The precision of the mean is given by

$$45.725 \pm (3.182)\frac{(0.0161)}{\sqrt{4}} = 45.725 \pm 0.0256 \text{ g}$$

That is, there is a 95 per cent probability that the true value of the weight of 50 ml of this liquid lies within the interval 45.725 ± 0.0256 g.

Relative Error

It is frequently convenient to express the degree of error on a relative basis, rather than on an absolute basis as was done above. A relative basis has the advantage of making a statement independent of the size of the measurements that were made. For example, the statement that a solid contains 10 per cent silver is a *relative* statement; it says that one-tenth of the solid is silver, and it is understood that you would naturally get more silver from a large sample of the solid than a small one. Percentage is "parts per hundred" and it is found for this example by multiplying the fraction of the sample that is silver (in this case, 0.1) by 100. It would be as correct to multiply the fraction by 1,000 and call it "100 parts per thousand," or to multiply the fraction by 1,000,000 and call it "100,000 parts per million." The choice of parts per hundred (percentage) or parts per million, and so on is determined by convenience. If the fraction were very small, say 0.00005, it would be more convenient to say 50 parts per million than 0.005 parts per hundred or 0.005 per cent.

PROBLEM:
Express the 95 per cent confidence level of the standard deviation of the mean obtained in the previous problem as percentage, as parts per thousand, and as parts per million.

SOLUTION:
The 95 per cent confidence level of the standard deviation of the mean was found to be ± 0.0256 g where the weight itself was (on the average) 45.725 g. The *fraction* of the total weight that might be error is

$$\frac{0.0256 \text{ g}}{45.725 \text{ g}} = 0.00056$$

The relative error will then be given by:

$(0.00056)\ (100) = 0.056$ parts per hundred $= 0.056$ per cent
$(0.00056)\ (10^3)\ = 0.56$ parts per thousand (p.p.t.)
$(0.00056)\ (10^6) = 560$ parts per million (p.p.m.)

Computation Assistance

Persons who must handle very large quantities of data find it useful to re-arrange Equation (2), which defines the standard deviation, and write it in the form:

$$S = \sqrt{\frac{\Sigma X_i^2 - \dfrac{(\Sigma X_i)^2}{n}}{n-1}} \tag{5-6}$$

The advantage of this form is that one need not find the average before proceeding to find the deviations. If you use a calculator or computer, you will normally accumulate ΣX_i and $(\Sigma X_i)^2$ during the course of the computation; at the end of the computation you can divide ΣX_i by n, or the second term in the numerator of Equation (5-6) by ΣX_i, and obtain $\bar{X}$ even though you did not have to find it before calculating the deviations. When a calculator or computer is used for computational assistance, students often have an urge to carry far more figures in their reported results than their data justify, simply because of the effortless way in which they were obtained. Computing machines cannot improve experimental reliability, and *it is the student's responsibility* to round off the final answer to the proper number of significant figures.

PROBLEMS A   Answers on page 334

1. State the number of significant figures in each of the following numbers:
(a) 374; (b) 0.0374; (c) 3,074; (d) 0.0030740; (e) 3,740; (f) 3.74×10^5;
(g) 75 million; (h) 21 thousand; (i) 6 thousandths; (j) 2 hundredths.

2. Express the answer in each of the following calculations to the proper number of significant figures:

(a) $3.196 + 0.0825 + 12.32 + 0.0013$

(b) $721.56 - 0.394$

(c) $525.3 + 326.0 + 127.12 + 330.0$

(d) $5.23 \times 10^{-2} + 6.01 \times 10^{-3} + 8 \times 10^{-3} + 3.273 \times 10^{-2}$

(e) $\dfrac{3.21 \times 432 \times 650}{563}$

(f) $\dfrac{8.57 \times 10^{-2} \times 6.02 \times 10^{23} \times 2.543}{361 \times 907}$

(g) $\dfrac{4.265 \times (3{,}081)^2 \times 8.275 \times 10^{-8}}{0.9820 \times 1.0035}$

(h) $\dfrac{6.327 \times 10^{-5} \times 7.056 \times 10^{-7} \times 9.0038 \times 10^{-9}}{6.022 \times 10^{23} \times 27.00 \times 10^{-2}}$

3. Water analysts often report trace impurities in water as "parts per million" — that is, parts by weight of impurity per million parts by weight of water. When 2.5 liters of a water sample are evaporated to a very small volume in a platinum dish and the residue is treated with a sensitive reagent that develops a red color whose intensity is a measure of the amount of nickel present, it is found that there is 0.41 mg of nickel. How many parts per million of nickel were present in the original sample of water? (Assume that the density of water is 1 g/ml.)

4. State the precision, in both parts per thousand and percentage, with which each of the following measurements is made:
(a) 378 with a standard deviation of 2
(b) 0.0378 with a standard deviation of 0.0002
(c) 3,078 with a standard deviation of 2
(d) 0.003078 with a standard deviation of 0.00003
(e) 0.0030780 with a standard deviation of 0.0000001

(f) 3,780 with a standard deviation of 50

(g) 3.78×10^5 with a standard deviation of 2×10^3

5. A radioactive sample showed the following counts for 1-minute intervals: 2,642; 2,650; 2,649; 2,641; 2,641; 2,637; 2,651; 2,636. Find the average deviation, the standard deviation, and the 90 per cent confidence interval for a single value and for the mean.

6. A student wished to calibrate a pipet by weighing the water it delivered. A succession of such measurements gave the following weights: 5.013 g; 5.023 g; 5.017 g; 5.019 g; 5.010 g; 5.018 g; 5.021 g. Calculate the average deviation and the 95 per cent confidence interval for the mean.

7. In determining the viscosity of a liquid by measuring the time required for 5.00 ml of the liquid to pass through a capillary, a student recorded the following periods: 3 min 35.2 sec; 3 min 34.8 sec; 3 min 35.5 sec; 3 min 35.6 sec; 3 min 34.9 sec; 3 min 35.3 sec; 3 min 35.2 sec. Find the average deviation and the 70 per cent confidence interval for a single value, expressing both as a percentage.

8. A student wished to determine the mole weight of a gas by measuring the time required for a given amount of the gas to escape through a pinhole. He observed the following time intervals: 97.2 sec; 96.6 sec; 96.5 sec; 97.4 sec; 97.6 sec; 97.1 sec; 96.9 sec; 96.4 sec; 97.3 sec; 97.0 sec. Find the average deviation and the 99 per cent confidence interval for the mean, expressing both in parts per thousand.

9. The height (h) to which a liquid will rise in a capillary tube is determined by the force of gravity (g), the radius (r) of the tube, and the surface tension (γ) and density (d) of the liquid at the temperature of the experiment. The relationship is $\gamma = \frac{1}{2}hdgr$. A student decided to determine the radius of a capillary by measuring the height to which water would rise at 25.0°C. In several attempts he observed: 75.7 mm; 75.6 mm; 75.3 mm; 75.8 mm; 75.2 mm. Find the average deviation of these heights and the 80 per cent confidence interval for the mean, expressing both in parts per thousand.

10. At 25.00°C the density and surface tension of water are 0.997044 g/ml and 71.97 dynes/cm, respectively. What actual values for these properties should be used with the data of Problem 9 to determine the radius of the capillary?

11. A sample of a copper alloy is to be analyzed for copper by first dissolving the sample in acid and then plating out the copper electrolytically. The weight of copper plated is to be measured on a balance that is sensitive to 0.1 mg. The alloy is approximately 5 per cent copper. What size of sample should be taken for analysis so that the error in determining the weight of copper plated out does not exceed 1 part per thousand? (Remember that two weighings are needed in order to find the weight of copper.)

PROBLEMS B   No answers given

12. State the number of significant figures in each of the following numbers.
    (a) 6,822; (b) 6.822×10^{-3}; (c) 6.82; (d) 682; (e) 0.006820; (f) 6.82×10^6;
    (g) 0.0682; (h) 34 thousandths; (i) 167 million; (j) 62 hundredths.

13. Express the answer in each of the following calculations to the proper
    number of significant figures:

    (a) $0.0657 + 23.77 + 5.369 + 0.0052$

    (b) $365.72 - 0.583$

    (c) $365.2 + 27.3 + 968.45 + 5.62$

    (d) $4.27 \times 10^{-5} + 1.05 \times 10^{-6} + 5 \times 10^{-6} + 1.234 \times 10^{-5}$

    (e) $\dfrac{65.4 \times 1.23 \times 464}{231}$

    (f) $\dfrac{6.55 \times 10^{-27} \times 2.045 \times 7.34 \times 10^5}{565 \times 432}$

    (g) $\dfrac{5,280 \times (2,885)^3 \times 6.570 \times 10^{-12}}{4.6295 \times 0.8888}$

    (h) $\dfrac{96.08 \times 4.712 \times 10^{-5} \times 7.308 \times 10^{-3}}{6.547 \times 10^{-27} \times 6.022 \times 10^{23}}$

14. Water analysts often report trace impurities in water as "parts per million"
    —that is, parts by weight of impurity per million parts by weight of
    water. A swimming pool whose dimensions are $20 \times 50 \times 9$ m has 14 lb of
    chlorine added as a disinfectant. How many parts per million of chlorine
    are present in this swimming pool? (Assume that the density of water is
    1 g/ml.)

15. State the precision, in both percentage and parts per thousand, with which
    each of the following measurements is made:
    (a) 6,822 with a standard deviation of 4
    (b) 6.822×10^{-3} with a standard deviation of 0.0004
    (c) 6.82 with a standard deviation of 0.4
    (d) 682 with a standard deviation of 4
    (e) 0.00682 with a standard deviation of 0.000004
    (f) 6.82×10^6 with a standard deviation of 4,000
    (g) 0.0682 with a standard deviation of 0.004

16. A radioactive sample showed the following counts for 1 minute intervals:
    3,262; 3,257; 3,255; 3,265; 3,257; 3,264; 3,259. Calculate the average
    deviation and the 60 per cent confidence interval for the mean.

17. A student wished to calibrate a pipet by weighing the water it delivered. A
    succession of such determinations gave the following weights in grams:
    4.993; 4.999; 4.991; 4.994; 4.995; 4.995. Find the average deviation, the
    standard deviation, and the 70 per cent confidence interval for a single
    value and for the mean.

18. If the viscosity of a given liquid is known, the viscosity of another may be determined by comparing the time required for equal volumes of the two liquids to pass through a capillary. To do this, a student made the following observations of time intervals: 4 min 9.6 sec; 4 min 8.8 sec; 4 min 10.2 sec; 4 min 9.8 sec; 4 min 9.0 sec. Find the average deviation and the 90 per cent confidence interval for the mean, expressing both in parts per thousand.

19. The time required for a given amount of gas to effuse through a pinhole under prescribed conditions is a measure of its molecular weight. A student making this determination observed the following effusion times: 1 min 37.3 sec; 1 min 38.5 sec; 1 min 36.9 sec; 1 min 37.2 sec; 1 min 36.5 sec; 1 min 38.7 sec; 1 min 37.0 sec. Find the average deviation and the 99 per cent confidence interval for a single value, expressing both as a percentage.

20. The surface tension of a liquid may be determined by measuring the height to which it will rise in a capillary of known radius. A student made the following observations of capillary rise with an unusual liquid that he had just prepared in the laboratory: 63.2 mm; 63.5 mm; 62.9 mm; 62.8 mm; 63.7 mm; 63.4 mm. Find the average deviation of these heights and the 90 per cent confidence interval for the mean, expressing both in parts per thousand.

21. How accurately should the values of liquid density and capillary radius be known if all of the figures in the measurements in Problem 20 are to be considered significant?

22. A graduated tube arranged to deliver variable volumes of a liquid is called a buret. If a buret can be read to the nearest 0.01 ml, what total volume should be withdrawn so that the volume will be known to a precision of 3 parts per thousand? (Remember that two readings of the buret must be made for every volume of liquid withdrawn.)

23. At some time or another nearly everyone must decide whether to reject a suspicious-looking result or to include it in the average of all the other results. There is no agreement on what criteria should be used, but, lacking information about errors made in the experimental procedure, a rejection may be made on the following *statistical* basis: If d and r be, respectively, the differences between the questionable result and the values closest to it and farthest from it, then there is a 90 per cent probability that the questionable result is grossly in error and should be rejected if

for 5 values, $d/r > 0.64$
for 4 values, $d/r > 0.76$
for 3 values, $d/r > 0.94$

Which, if any, result should be rejected from the following series of measurements?

(a) 9.35, 9.30, 9.48, 9.40, 9.28   (c) 2534, 2429, 2486
(b) 9.35, 9.30, 9.48, 9.32, 9.28   (d) 2534, 2429, 2520

Graphical Representation

Measurements made on many chemical and physical systems are often presented in graphical form, because the graph may give one a much better feeling for how different variables are related to each other, or how the change in one variable affects the change in another. Another reason for graphical representation is that an important *derived* property can be obtained from, say, the slope or intercept of the graph if it is a straight line. In this chapter we will discuss only those kinds of graphs and graph papers that are normally used at an elementary level in chemistry.

Constructing a Graph

If the following guidelines are followed, the resulting graph will generally have maximum usefulness.

1. Use a large enough scale to cover as much as possible of the full page of graph paper. Do not cramp a miserable little graph into the corner of a large piece of graph paper.

2. Use a convenient scale, to simplify the plotting of the data and the reading of the graph.

3. Do not place the "zero" or origin of the coordinate system at one corner of the graph paper if doing so would make a very small, cramped, and inconvenient scale.

4. Label the axes (the vertical axis is the ordinate, the horizontal axis the abscissa) with both units and dimensions.

5. When possible and desirable, simplify the scale units in order to use simple figures. For example, if you wanted to plot as scale units 1,000 min, 2,000 min, 3,000 min, 4,000 min, etc., it would be simpler to use the figures 1, 2, 3, 4, etc., and then label the axis as $\min \times 10^{-3}$. Such labeling states that the *actual* figures, in minutes, have been multiplied by 10^{-3} in order to give the simple figures shown along the axis.

6. Draw a smooth curve that best represents all the points; such a curve may not necessarily pass through any of the points. Straight-line segments should never be drawn to connect consecutive points, unless there is reason to believe that discontinuities (breaks) in the curve really should occur at the experimental points; such reasons almost never exist.

Properties of a Straight-line Graph

Whenever possible, cast data into such a form that, when they are plotted, a straight-line graph results. A straight line is much easier to draw accurately than a curved one, and often one can obtain important information from the slope or intercept of the line. If the two variables under discussion are x and y (the convention is to plot x as the abscissa and y as the ordinate), and if they are *linearly* related (i.e., if the graph is a straight line), the form of the mathematical equation that represents this line is

$$y = mx + b$$

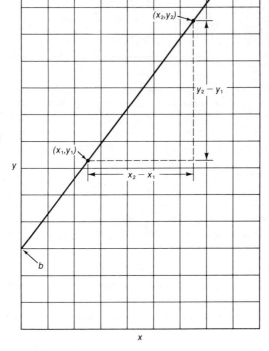

FIGURE 6-1

No matter what the value of m, when $x = 0$, then $y = b$. It is for this reason that b is called the "y-intercept," the point at which the line intersects the y axis (see Figure 6-1).

If two arbitrary points, (x_1, y_1) and (x_2, y_2), are selected from this line, both sets of points must satisfy the general equation for the line, so consequently we have two specific equations:

$$y_2 = mx_2 + b$$

and

$$y_1 = mx_1 + b$$

If we subtract the second equation from the first, we obtain

$$y_2 - y_1 = mx_2 - mx_1 = m(x_2 - x_1)$$

which, when rearranged, gives

$$m = \frac{y_2 - y_1}{x_2 - x_1}$$

By looking at Figure 6-1, you can see that $(y_2 - y_1)$ and $(x_2 - x_1)$ are the two sides of the right triangle made by connecting the two arbitrary points. The *ratio* of the sides, $\frac{y_2 - y_1}{x_2 - x_1}$, is the slope of the curve; this ratio has the same value whether the two arbitrary points are taken close together or far apart. It is for this reason that m is said to be the "slope" of the line.

PROBLEM:

The junction of two wires, each made of a different metal, constitutes a *thermocouple*. Some pairs of metals can generate a significant electrical voltage that varies substantially with the temperature of the junction. The following values of voltage (expressed in millivolts) were observed at the temperatures (°C) shown for a junction of two alloys, chromel and alumel.

T, temperature, in °C	25	50	75	100	125	150	175	200
V, voltage, in mv	0.23	1.20	2.24	3.32	4.34	5.35	6.33	7.31

Plot these values of millivolts and temperature, and from the resulting graph determine the mathematical equation that describes V as a function of T.

SOLUTION:

The graph of these values of V and T is shown in Figure 6-2. Note that the best line passes through only one of the experimental points. In order to include the y-intercept, one must take some negative values along the V-axis. The value of the y-intercept can then be taken directly from the graph; it is −0.80. The slope

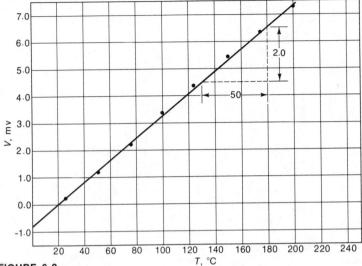

FIGURE 6-2

of the line is calculated by selecting two convenient points *on* this line, as shown, then taking the ratio of the two sides of the triangle formed from these points; in the figure it is the ratio $\frac{2.0}{50} = 0.040$. Knowing the slope and the *y*-intercept, we can directly write the mathematical equation for the relationship between *V* and *T* as

$$V = 0.040\ T - 0.80$$

Students commonly tend to select two of the *experimental* points in order to calculate the slope of a straight-line function; doing so is usually very bad practice, because there are experimental errors inherent in the individual data points. The straight line that best represents *all* the points minimizes the experimental errors, and the slope calculated from this best straight line will thus usually be more reliable than one calculated from two randomly selected experimental points.

Frequently, in order to use a convenient scale and to construct a graph of a reasonable size, one must *not* place the origin of the *x*-axis at the lower left corner of the graph paper. When you have such a graph, you cannot obtain the value of *b*, the *y*-intercept, directly from the graph. This situation need cause no difficulty, however, because *m*, the slope, can still be obtained from the graph, and, once obtained, it can be used, along with any arbitrarily chosen point on the line, to *solve* for the value of *b*. The following problem illustrates this.

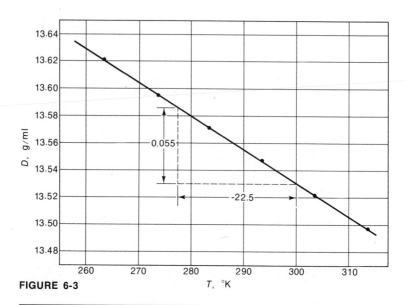

FIGURE 6-3

PROBLEM:
The following data show how the density (D) of mercury varies with the absolute temperature (T). Plot these data, and from the resulting graph determine the mathematical equation that shows how D varies with T.

T, in °K	263	273	283	293	303	313
D, in g/ml	13.6201	13.5955	13.5708	13.5462	13.5217	13.4971

SOLUTION:
The graph of these values of D and T is shown in Figure 6-3. Note that a greatly expanded scale is needed along the y-axis in order to show in a reasonable way the small changes in D with temperature, and note how far from absolute zero the scale along the x-axis starts. The slope, m, is calculated by taking two convenient points *on* the line, as shown, then taking the ratio of the two sides of the triangle formed from these points; in the figure it is the ratio $\dfrac{0.055}{-22.5} =$ $-0.00244 = m$. Note also that the sign of the slope is *negative*, as it must always be when the line slopes downward from left to right, that is, when y decreases with increasing values of x. Since the mathematical equation for this line must be of the form

$$D = mT + b$$

we can take any point on the line (300 and 13.530 for convenience) and use it with the value of $m = -0.00244$ to solve for the value of b:

$$b = D - mT$$
$$b = 13.530 - (-0.00244)(300) = 13.530 + 0.732$$
$$b = 14.262$$

The equation that relates D to T for mercury over this range of temperatures is thus

$$D = -0.00244T + 14.262$$

It is often desirable to recast data into a different form before making a graph, in order to obtain a straight-line graph. The data in Table 11-1 can be used to illustrate this. With the data plotted just as they are given in the table (vapor pressure of water expressed in torr, and the corresponding temperatures in °C), the graph shown in Figure 6-4 is obtained.

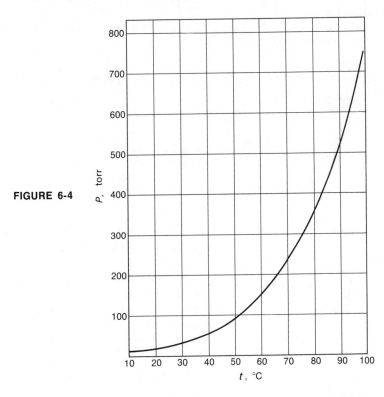

FIGURE 6-4

It is known from theoretical principles, however (and your text may explain this), that the vapor pressure of a liquid is related to its heat of vaporization (ΔH_V), which is a physical constant characteristic of the liquid, to the gas constant ($R = 1.987 \ \dfrac{\text{cal}}{\text{mole deg}}$), and to the absolute temperature (T), by the equation

$$\log P = -\frac{\Delta H_V}{2.303R} \times \frac{1}{T} + B$$

From this it follows that a straight-line graph should be obtained if you let $y = \log P$ and $x = \dfrac{1}{T}$, for then the equation will be of the form

$$y = mx + b$$

with the slope composed of a collection of constants, $m = -\dfrac{\Delta H_V}{2.303R}$, and the intercept, $b = B$.

PROBLEM:

Using the data of Table 11-1, plot a graph from which you can calculate the heat of vaporization of water, ΔH_V.

SOLUTION:

Table 6-1 shows the data given in Table 11-1, along with the values of $\log P$ and $\dfrac{1}{T}$ calculated from the data.

TABLE 6-1
Temperature and Pressure Data

t	P	$1/T$	$\log P$
20	17.5	0.003412	1.243
30	31.8	0.003301	1.502
40	55.3	0.003195	1.743
50	92.5	0.003095	1.966
60	149.4	0.003004	2.174
70	233.7	0.002915	2.369
80	355.1	0.002832	2.550
90	525.8	0.002755	2.721
100	760	0.002681	2.881

Figure 6-5 shows the plot of $\log P$ against $\dfrac{1}{T}$. The slope of this straight-line graph is

$$m = \frac{0.500}{-0.000225} = -2.22 \times 10^3 \text{ deg}$$

And since $m = -\dfrac{\Delta H_V}{2.303R}$,

$$\Delta H_V = -(m)\,(2.303)\,(R) = -(-2.22 \times 10^3 \text{ deg})\,(2.303)\,(1.987\,\frac{\text{cal}}{\text{mole deg}})$$

$$= +\,10{,}120 \text{ cal/mole}$$

FIGURE 6-5

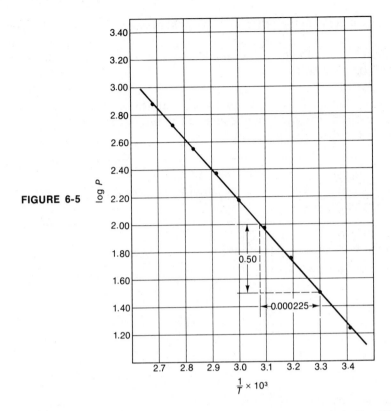

If you also want to find B, then choose a simple pressure, such as $P = 100$ torr (so that $\log P = 2.000$) and the corresponding value of $\frac{1}{T}$ (it turns out to be 0.003075), and substitute them into the basic equation along with the calculated value of the slope (-2.22×10^3), to give:

$$\log P = -\frac{\Delta H_V}{2.303R} \times \frac{1}{T} + B$$

$$2.000 = (-2.22 \times 10^3)(0.003075) + B$$

$$B = 2.000 + 6.826 = 8.826$$

The general equation that shows how the vapor pressure of water varies with temperature is thus

$$\log P = -\frac{2,220}{T} + 8.826$$

In a plot such as that in Figure 6-5, one can eliminate the extra labor of converting all the pressure values to the corresponding logarithms by using semilog graph paper, which has a logarithmic scale in one direction and the usual linear scale in the other. Semilog paper is available in what is known as one-cycle, two-cycle, three-cycle, etc., paper, each cycle being able to handle a 10-fold spread of data. For example, one-cycle paper can handle data from 0.1 to 1.0, or 1.0 to 10.0, or 10^4 to 10^5, etc., while two-cycle paper can handle data from 10^{-4} to 10^{-2}, or 0.1 to 10, or 10 to 1,000, etc. Always choose the paper with the smallest number of cycles that will do the job, in order to have the largest possible graph on the paper. The use of this kind of paper is illustrated in the following problem.

PROBLEM:
Calculate the heat of vaporization of water using the data of Table 11-1 in a semilog plot.

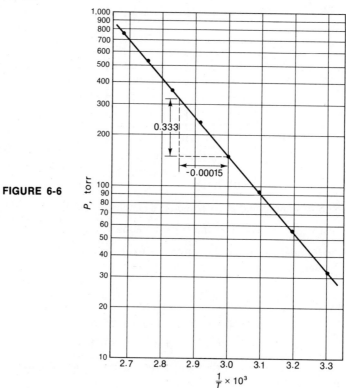

FIGURE 6-6

SOLUTION:
It is still necessary to convert the Celsius temperature values to reciprocal degrees absolute, as in the previous problem, but the pressure values can be used directly, as shown in Figure 6-6.

At this point, take great care in calculating the slope of the line, because, although a logarithmic *scale* was used, the *values* shown along the ordinate are *not* logarithmic values. In calculating the slope, therefore, select two convenient points on the line, but convert the two pressure values selected into their logarithms *before* making the calculations. For example, in Figure 6-6 when $\frac{1}{T} \times 10^3$ is 2.85, the value of P is 323 and log P is 2.509. Similarly, when $\frac{1}{T} \times 10^3$ is 3.00, P is 150 and log P is 2.176. Thus, the slope is given by

$$m = \frac{2.176 - 2.509}{0.00300 - 0.00285} = -2.22 \times 10^3$$

and, as before,

$$\Delta H_V = -(m)\,(2.303)\,(R) = -(-2.22 \times 10^3)\,(2.303)\,(1.987)$$
$$= +10,120 \text{ cal/mole}$$

PROBLEMS A   Answers on page 334

1. Construct graphs for each of the following functions, plotting q as ordinate and p as abscissa:

   (a) $q = 6p$
   (b) $q = 6p + 10$
   (c) $qp = 20$
   (d) $q = 10p^2 + 5$
   (e) $q = 5.0 \times 10^{-3p}$

2. Rearrange those equations in Problem 1 that do not give a straight-line relationship in such a way that, when plotted in a different fashion, they will yield a straight line. Plot each of these new equations, showing which function extends along each axis.

3. (a) Plot the solubility of $Pb(NO_3)_2$ in water as a function of temperature; solubility, S, is given in g per 100 g of H_2O. The experimental data are as follows:

t, in °C	20	40	60	80	100
S	56.7	74.8	94.0	113.1	132.0

   (b) From the graph made in (a), determine the mathematical equation for S as a function of t.

4. In colorimetric analysis, it is customary to use the fraction of light absorbed by a given dissolved substance as a measure of the concentration of the substance present in solution; monochromatic light must be used, and the length of the absorbing light path must be known or must always remain the same. The incident light intensity is I_o and the transmitted light intensity is I; the fraction of light transmitted is I/I_o.

   (a) The following data were obtained for the absorption of light by MnO_4^- ion at a wavelength of 5,250 Å in a cell with a 1.00 cm light path. Construct a graph of I/I_o against C (as abscissa).

C, concentration of MnO_4^-, in mg of Mn/100 ml	1.0	2.0	3.0	4.0
I/I_o	0.400	0.158	0.063	0.025

(b) Replot the given data on semilog paper, again plotting concentration along the abscissa, but plotting I_0/I along the ordinate.

(c) From the graph made in (b), find the mathematical equation for the relationship between I_0/I and C.

5. When radioactive isotopes disintegrate, they obey a rate law that may be expressed as $\log N = -kt + K$, where t is time and N is the number of radioactive atoms (or something proportional to it) present at time t. A common way to describe the number of radioactive atoms present in a given sample is in terms of the number of disintegrations observed per minute on a Geiger counter (this is referred to as "the number of counts per minute," or simply as cpm).

(a) The following cpm were obtained for a sample of an unknown isotope at 10-minute intervals, beginning with $t = 0$: 10,000; 8,166; 7,583; 6,464; 5,381; 5,023; 4,466; 3,622; 2,981; 2,690; 2,239; 2,141; 1,775; 1,603; 1,348; 1,114; 1,048. Plot these data, using semilog graph paper, and from this graph evaluate the constants k and K.

(b) What is the physical significance of K?

(c) Using the equation for the rate law, show the relationship between the "half-life" $(t_{1/2})$ of the isotope and the value of k. The half-life is defined as the time required for one-half the original isotope sample to disintegrate.

(d) Using the relationship that you found in (c), find the half-life of the isotope in this problem.

6. When the rates of chemical reaction are studied, it is common to determine the "rate constant," k, which is characteristic of a given reaction at a given temperature. Still further information can be obtained by determining the value of k at several different temperatures, because these values of k are related to the absolute temperature (T) at which they were measured according to the equation

$$\log k = -\frac{\Delta H_a}{2.3R} \times \frac{1}{T} + Q$$

In this equation ΔH_a is the so-called "energy of activation," and Q is a constant characteristic of the reaction.

(a) Data were obtained for the reaction

$$N_2O_5 \rightarrow 2\ NO_2 + \frac{1}{2} O_2$$

as follows:

t, °C	25	35	45	55	65
k	3.46×10^5	1.35×10^6	4.98×10^6	1.50×10^7	4.87×10^7

Plot these data on semilog graph paper, and from the resulting graph determine the mathematical equation for the relationship between k and T.

(b) Calculate the value of the activation energy for this reaction.

(c) What is the physical significance of the constant Q?

7. Regardless of how fast a chemical reaction takes place, it usually reaches an equilibrium position at which there appears to be no further change, because the reactants are being reformed from the products at the same rate at which they are reacting to form the products. This position of equilibrium is commonly characterized at a given temperature by a constant, K_e, called the equilibrium constant (see Chapter 14). The equilibrium constant is commonly measured at several different temperatures for a given reaction, because these values of K_e are related to the absolute temperature (T) at which they are measured; the relationship is

$$\log K_e = -\frac{\Delta H}{2.3R} \times \frac{1}{T} + Z$$

In this equation, ΔH is the so-called "energy (or enthalpy) of reaction," and Z is a constant characteristic of the reaction.

(a) Data were obtained for the reaction

$$H_2 + I_2 \rightarrow 2\ HI$$

as follows:

t, °C	340	360	380	400	420	440	460
K_e	70.8	66.0	61.9	57.7	53.7	50.5	46.8

Plot these data on semilog graph paper, and from the resulting graph determine the mathematical equation for the relationship between K_e and T.

(b) Calculate the value of the energy of the reaction for this reaction.

(c) What is the physical significance of the constant Z?

PROBLEMS B   No answers given

8. Construct graphs for each of the following functions, plotting m as ordinate and n as abscissa:

(a) $m = 2.5\pi n$

(b) $m = 20 - 4n$

(c) $mn = 1/4$

(d) $m = 4(n-1)^2$

(e) $m = 4e^{-9.2n}$

9. Rearrange those equations in Problem 8 that do not give a straight-line relationship in such a way that, when plotted in a different fashion, they will yield a straight line. Plot each of these new equations, showing which function extends along each axis.

10. (a) Plot the solubility of K_2SO_4 in water as a function of temperature; solubility, S, is given as g per 100 g of H_2O. The experimental data are:

t, °C	20	40	60	80	100
S	9.9	14.0	18.0	21.9	26.0

(b) From the graph made in (a), determine the mathematical equation for S as a function of t.

11. (a) The nature of colorimetric analysis is described in Problem 4. With this in mind, use the following data, obtained for the absorption of light by $CrO_4^=$ ions at a wavelength of 3660 Å in a cell with a 1.00 cm light path, to construct a graph of I/I_o against C (as abscissa).

C, concentration of $CrO_4^=$, in moles of $CrO_4^=$/liter	0.8×10^{-4}	1.2×10^{-4}	1.6×10^{-4}	2.0×10^{-4}
I/I_o	0.42	0.275	0.175	0.110

(b) Replot the given data on semilog paper, again plotting concentration along the abscissa, but plotting I_o/I along the ordinate.

(c) From the graph made in (b), find the mathematical equation for the relationship between I_o/I and C.

12. (a) The nature of the radioactive disintegration of unstable isotopes is described in Problem 5. Bearing this in mind, use the following data, obtained for a given sample of an unknown radioactive isotope, and taken at five-minute intervals, beginning with $t = 0$: 4,500; 3,703; 2,895; 2,304; 1,507; 1,198; 970; 752; 603; 496; 400; 309; 250; 199. Plot these data, using semilog graph paper, and from this graph evaluate the constants k and K.

(b) What is the physical significance of K?

(c) Using the equation for the rate law, show the relationship between the "half-life" $(t_{1/2})$ of the isotope and the value of k.

(d) Using the relationship that you found in (c), find the half-life of the isotope in this problem.

13. Pertinent statements about rate constants for chemical reactions are given in Problem 6. Keeping these statements in mind, use the data that were obtained for the reaction

$$CO + NO_2 \rightarrow CO_2 + NO$$

as follows:

t, °C	267	319	365	402	454
k	0.0016	0.021	0.12	0.63	2.70

(a) Plot these data on semilog graph paper, and from the resulting graph determine the mathematical equation for the relationship between k and T.

(b) Calculate the value of the activation energy for this reaction.

(c) What is the physical significance of the constant Q?

14. (a) Some fundamental statements are made about chemical equilibria in Problem 7. Keeping these statements in mind, use the data for the reaction

$$H_2 + CO_2 \rightarrow CO + H_2O$$

as follows:

t, °C	600	700	800	900	1000
K_e	0.39	0.64	0.95	1.30	1.76

Plot these data on semilog graph paper, and from the resulting graph determine the mathematical equation for the relationship between K_e and T.

(b) Calculate the value of the energy of the reaction for this reaction.

(c) What is the physical significance of the constant Z?

Density and Specific Gravity

Density is defined as the mass per unit volume, or

$$\text{density} = \frac{\text{mass}}{\text{volume}}$$

In scientific work the densities of solids and liquids are customarily expressed in grams per cubic centimeter or in grams per milliliter; densities of gases are usually given in grams per liter or in milligrams per milliliter. In engineering work densities of solids and liquids are usually stated in pounds per cubic foot.

When one is experimentally determining the densities of solids or liquids, it is frequently convenient to find the ratio between the weight of a given sample and the weight of an equal volume of some reference liquid, usually water. This ratio is known as the specific gravity (sp gr):

$$\text{sp gr} = \frac{\text{weight of sample}}{\text{weight of an equal volume of water}}$$

For gases, the specific gravity is often given with reference to air:

$$\text{sp gr} = \frac{\text{weight of gas sample}}{\text{weight of an equal volume of air}}$$

Density of a Solid

If a solid object has a regular geometric form, the density may be computed from its weight and volume.

PROBLEM:
A cylindrical rod weighing 45.0 g is 2.0 cm in diameter and 15.0 cm in length. Find the density.

SOLUTION:
Mass = 45.0 g. Volume = $\pi r^2 L$ = 3.14 × (1 cm)² × 15.0 cm = 47.2 cm³ (or cc).

$$\text{density} = \frac{\text{mass}}{\text{volume}} = \frac{45.0 \text{ g}}{47.2 \text{ cc}} = 0.955 \text{ g/cc}$$

When a solid object is irregular in shape, it is usually not possible to find its volume by measurements of its length, breadth, and thickness. A convenient procedure is to immerse the object in a liquid and determine its volume by measuring the volume of the liquid it displaces.

PROBLEM:
A 5.7-g sample of metal pellets is put into a graduated cylinder which holds 5.0 ml water. After the pellets are added, the water level stands at 7.7 ml. Find the density of the pellets.

SOLUTION:
The volume of water displaced by 5.7 g of solid is 7.7 − 5.0 = 2.7 ml.

$$\text{density} = \frac{\text{mass}}{\text{volume}} = \frac{5.7 \text{ g}}{2.7 \text{ ml}} = 2.1 \text{ g/ml}$$

Since the volume in a graduated cylinder cannot be read to better than 0.1 ml, this method is not highly accurate. The method of weighing a solid in air, then reweighing while it is suspended in a liquid, is much more precise, but what is obtained is the specific gravity, not the density.

PROBLEM:
A metal bar is suspended by a thread from one arm of a balance and weighed. A beaker of water is now placed so that the bar is totally immersed in water, and the bar is again weighed. From the following data, compute the specific gravity of the bar:

weight in air = 20.562 g

weight in water = 15.331 g

SOLUTION:
The difference between the weight in air and the weight in water is 5.231 g. This is the weight of water displaced by the bar.

$$\text{sp gr of bar} = \frac{\text{weight of metal}}{\text{weight of water displaced}} = \frac{20.562 \text{ g}}{5.231 \text{ g}} = 3.94$$

Note that specific gravity is a dimensionless number.

If a sample consists of several pieces that cannot conveniently be suspended from an arm of a balance, we may use the method of the following example.

PROBLEM:

A small volumetric flask is weighed empty, then filled with water to a designated mark and reweighed. The flask is emptied, and a known weight of the solid sample is put in the dry flask; water is then added to the graduation mark, and the flask is again weighed. From the following data compute the specific gravity of the sample:

(A) weight of empty flask $= 27.32$ g
(B) weight of flask filled to mark with water $= 52.42$ g
(C) weight of sample $= 21.22$ g
(D) weight of flask + sample + water to mark $= 63.45$ g

SOLUTION:

The weight of water to fill the empty flask is B − A or 52.42 g − 27.32 g = 25.10 g.

The weight of (flask + sample) is A + C or 27.32 g + 21.22 g = 48.54 g.

The weight of water to fill a flask containing 21.22 g of sample is D − (A + C) or 63.45 g − 48.54 g = 14.91 g.

Since the empty flask holds 25.10 g water and the flask with sample holds 14.91 g water, the sample displaces 25.10 g − 14.91 g = 10.19 g.

$$\text{sp gr} = \frac{\text{weight of sample}}{\text{weight of equal volume of water}} = \frac{21.22 \text{ g}}{10.19 \text{ g}} = 2.082$$

Specific Gravity of Liquids

The specific gravity of a liquid is measured by comparing the weight of a given volume of the liquid with that of an equal volume of water. To do this accurately we employ a container that can be filled to a reference mark, such as a volumetric flask or pycnometer.

PROBLEM:

A volumetric flask is weighed dry, filled with water and reweighed, then emptied, dried, filled with a liquid sample and again weighed. From the following data, find the specific gravity of the liquid sample:

(A) weight of dry empty flask $= 28.31$ g
(B) weight of flask filled with water $= 53.36$ g
(C) weight of flask filled with liquid sample $= 48.43$ g

SOLUTION:

$$\text{weight of water to fill flask} = B - A = 25.05 \text{ g}$$

$$\text{weight of liquid sample} = C - A = 20.12 \text{ g}$$

$$\text{sp gr of liquid} = \frac{20.12 \text{ g}}{25.05 \text{ g}} = 0.803$$

Conversion of Specific Gravity to Density

Since specific gravity is measured in terms of water at a specified temperature, it can be converted to density by using density tables for water at various temperatures. Table 7-1 is an abbreviated list with densities given to four decimal places. Density values can be measured to very high precision, and the more complete tables in chemistry handbooks give many more decimal places.

PROBLEM:
Experimentally it is found that the specific gravity of a metal-shot sample is 2.375 at 25°C. Find the density at 25°C.

SOLUTION:
The metal shot are 2.375 times as heavy as water at 25°C. At this temperature (see Table 7-1) 1.0000 ml water weighs 0.9970 g/ml.

$$\text{density} = 2.375 \times 0.9970 \text{ g/ml} = 2.368 \text{ g/ml}$$

Note that density has the dimensions of mass per unit volume, whereas specific gravity is a dimensionless number.

TABLE 7-1

**Density of Water at Various Temperatures
(weight of 1.0000 ml of water, in vacuo)**

$t°C$	Density (g/ml)	$t°C$	Density (g/ml)
15	0.9991	23	0.9975
16	.9989	24	.9973
17	.9988	25	.9970
18	.9986	26	.9968
19	.9984	27	.9965
20	.9982	28	.9962
21	.9980	29	.9959
22	.9978	30	.9956

Experimental weighings to measure specific gravity need not be made with the sample and water at the same temperature. One might, for example, calibrate a flask with water at one temperature, then use it at a different temperature for measuring the specific gravity of a liquid. It is customary, therefore, to list with a specific gravity value both the temperature of the sample and the temperature of the water. The expression

$$\text{sp gr } \frac{25°}{15°}$$

indicates that the sample is measured at 25°C against an equal volume of water at 15°C. To compute the density we use the density of water at 15°.

Density of Air and its Effect on Weighing

We have noted that if an object is weighed in air and then suspended in water and reweighed, the two values differ by the weight of the water displaced by the object (Archimedes' Principle).

Air, too, is a fluid that exerts a buoyant effect on any object it surrounds. At ordinary conditions the density is 0.0012 g/ml, and in some weighings it is necessary to take into account the buoyant effect of the displaced air, that is, to compute the weight in vacuo, W_v. This is done by computing the weight of air displaced by the object, the weight of air displaced by the weights, and adding the difference between these two values to the weight in air, W_a. That is,

$$W_v = W_a + \text{(weight of air displaced by object)} -$$
$$\text{(weight of air displaced by weights)}$$

A moment's reflection shows that the correction is the difference between the two weights of air, since the air exerts a lifting effect on the object at one end of the balance beam and a lifting effect on the weights at the other end of the beam. Consequently, if we pumped air from the balance case, we would have to add or subtract weights to restore a condition of balance.

To illustrate, assume that an object with a volume of 10 ml is just balanced in air by weights of volume 1.0 ml. We have (taking $D_{air} = 0.0012$ g/ml)

$$W_v = W_a + \left(10 \text{ ml} \times 0.0012 \frac{g}{ml}\right) - \left(1.0 \text{ ml} \times 0.0012 \frac{g}{ml}\right)$$

$$= W_a + 0.0108 \text{ g}$$

PROBLEM:
A 50.000-g sample of water is weighed in air with brass weights. What is the weight in vacuo? Assume $D_{H_2O} = 1.0$ g/ml, $D_{brass} = 8.0$ g/ml, $D_{air} = 0.0012$ g/ml.

SOLUTION:

$$\text{volume of water} = \frac{50.0 \text{ g}}{1.0 \frac{\text{g}}{\text{ml}}} = 50.0 \text{ ml}$$

$$\text{volume of weights} = \frac{50.0 \text{ g}}{8.0 \frac{\text{g}}{\text{ml}}} = 6.2 \text{ ml}$$

$$\text{weight of air displaced by water} = 50.0 \text{ ml} \times 0.0012 \frac{\text{g}}{\text{ml}}$$

$$= 0.060 \text{ g}$$

$$\text{weight of air displaced by weights} = 6.2 \text{ ml} \times 0.0012 \frac{\text{g}}{\text{ml}}$$

$$= 0.007 \text{ g}$$

$$W_v = 50.000 \text{ g} + 0.060 \text{ g} - 0.007 \text{ g}$$

$$= 50.053 \text{ g}$$

We see that, in this weighing, the correction for buoyancy amounts to 0.053 g in 50 g or about 1 part per thousand. A correction of this size is important in some measurements.

Calibration of Volumetric Apparatus

Glass apparatus for accurate measurements of volume is *calibrated* by weighing the amount of water or other liquid delivered or contained at fixed marks. Calibration of a volumetric flask "to contain" is illustrated in the following problem.

PROBLEM:
A 50-ml volumetric flask is weighed, then filled with water just to the graduation mark and reweighed. From the data compute the true volume of the flask:

$$\text{weight of empty flask} = 65.37 \text{ g}$$

$$\text{weight of flask} + \text{water} = 115.24 \text{ g}$$

$$\text{temperature of water} = 20°C$$

SOLUTION:

weight of water in air = 115.24 g − 65.37 g = 49.87 g

weight of water in vacuo = 49.92 g (computed as in the preceding problem)

In Table 7-1 we find that the density of water at 20°C is 0.9982 g/ml.

$$\text{volume of water} = \frac{49.92 \text{ g}}{0.9982 \frac{\text{g}}{\text{ml}}} = 50.01 \text{ ml}$$

PROBLEMS A Answers on page 339

Show units in all calculations.

1. The density of mercury is 13.54 g/ml. How many milliliters of mercury are needed to give 454 g?

2. The density of a sulfuric acid solution is 1.540 g/ml. How much does 1 liter of this solution weigh?

3. If 17.5 g of brass filings (density 8 g/ml) are put into a dry 10-ml graduated cylinder, what volume of water is needed to complete the filling of the cylinder to the 10-ml mark?

4. A ring weighing 7.3256 g in air weighs 6.9465 g when suspended in water at 24°C. Is the ring made of gold (density = 19.3 g/ml) or brass (density = 8 g/ml)?

5. The weight of a metal sample is measured by finding the increase in weight of a volumetric flask when the metal sample is placed in it. The volume of the metal sample is measured by finding how much less water the volumetric flask holds when it contains the metal sample. Compute the density of the metal sample from the following data, assuming that the density of water is 1 g/ml:

$$\text{weight of empty flask} = 26.735 \text{ g}$$
$$\text{weight of flask} + \text{sample} = 47.806 \text{ g}$$
$$\text{weight of flask} + \text{sample} + \text{water} = 65.408 \text{ g}$$
$$\text{weight of flask} + \text{water (no sample)} = 50.987 \text{ g}$$

6. A chemical is soluble in water but insoluble in benzene. As a consequence, benzene may be used to determine its density. The density of benzene is 0.879 g/ml at 20°C. From the following data (obtained as in Problem 5) compute the density of the sample:

$$\text{weight of empty flask} = 31.862 \text{ g}$$
$$\text{weight of flask} + \text{sample} = 56.986 \text{ g}$$
$$\text{weight of flask} + \text{sample} + \text{benzene to fill} = 75.086 \text{ g}$$
$$\text{weight of flask} + \text{benzene (without sample)} = 52.175 \text{ g}$$

7. A sample of benzene ($D = 0.88$ g/ml) weighs 25.3728 g in air with brass weights. What is the weight in vacuo?

8. A platinum crucible ($D = 21.5$ g/ml) weighs 56.3724 g in air with brass weights. What is the weight in vacuo?

9. A brass sample weighs 16.3428 g in air with brass weights. What is the weight in vacuo?

10. The density of water at 25°C is 0.9970 g/ml. What volume is occupied by 1.0000 g of water weighed in air with brass weights?

11. A chemical of density 2.50 g/ml is weighed in air with brass weights. The observed weight is 0.2547 g. Show by calculation that it is not necessary to make a correction of the weight to vacuo if the weighing is reliable only to the nearest 0.2 mg.

12. What weight of water at 25°C, weighed in air with brass weights, should be delivered by a 25.00 ml pipet if it is accurately graduated?

13. A liquid has a specific gravity of 0.8042 at 25°C. What is its density?

14. The density of a liquid is 1.6045 g/ml at 25°C. What is its specific gravity?

15. Find the weight of 1 cu ft of air at 21°C, assuming a density of 0.0012 g/ml at this temperature.

16. (a) The specific gravity at 24°/24° for gallium is 5.885; find sp gr 24°/4°.
    (b) Specific gravity 24°/4° for nickel bromide is 4.640; find sp gr 24°/28°.
    (c) Which values in *a* and *b* are the densities?

17. Mercury (density = 13.54 g/ml) is sold by the "flask," which holds 76 lb of mercury. If the cost is $280 per flask, how much does 1 ml of mercury cost?

18. A glass bulb with a stopcock weighs 54.9762 g when evacuated, and 54.9845 g when filled with a gas at 25°C. The bulb will hold 50.0 ml of water. What is the density of the gas at 25°C?

19. (a) A column of mercury (density = 13.54 g/ml) 730 mm high in a tube of 8-mm inside diameter is needed to balance a gas pressure. What weight of mercury is in the tube? (b) If the same gas pressure were balanced by mercury in a tube 16-mm inside diameter, what weight of mercury would be needed?

20. In normal whole blood there are about 5.4×10^9 red cells per milliliter. The volume of a red cell is about 90 μ^3, and the density of a red cell is 1.096 g/ml. How many pints of whole blood would we need in order to collect 8 oz (avoirdupois) of red cells?

21. How far (in centimeters) does a 1-cm cube of wood stick out of the water if its density is 0.85 g/ml?

22. A spherical balloon 100 ft in diameter is filled with a gas having a density one-fifth that of air. The density of air is 1.20 g/liter. How many pounds, including its own weight, can the balloon lift? (The lift is the difference between the weight of the gas and that of an equal volume of air.)

23. A piece of Invar (density = 8.00 g/ml) weighs 15.4726 g in air and 13.9213 g when suspended in liquid nitrogen at a temperature of −196°C. What is the density of liquid nitrogen at the temperature? (Invar has a very small coefficient of thermal expansion, and its change in density with temperature may be neglected in this problem.)

24. A student wishes to make a hydrometer (see Figure 7-1) that will determine specific gravities at 25°C. (In use, a hydrometer is placed in a liquid

and allowed to float. The scale reading at the meniscus of the liquid in which it floats is the specific gravity of the liquid.) The container is to be made of glass. The volume of section A is 30.00 ml. Section B is made of 3.0-mm (outside diameter) tubing and the scale is marked on paper which is glued to the inside of the B section. The total weight of the glass and the scale is 14.523 g. Determine (a) what volume of mercury (density = 13.54 g/ml) must be used at C and (b) how far apart the scale divisions must be in order for the hydrometer to read as shown.

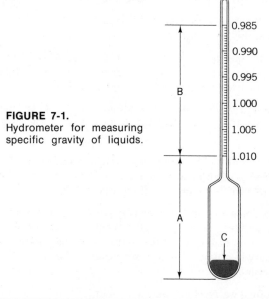

FIGURE 7-1.
Hydrometer for measuring specific gravity of liquids.

PROBLEMS B   No answers given

25. The density of sodium hydroxide solution is 1.1589 g/ml. How much does 1 liter of this solution weigh?

26. The density of carbon tetrachloride is 1.595 g/ml. How many milliliters of carbon tetrachloride are needed to give 500 g?

27. A liquid has a specific gravity of 0.9302 at 23°C. What is its density?

28. The density of a liquid is 0.7926 g/ml at 21°C. What is its specific gravity?

29. The density of benzene is 0.879 g /ml. How many grams of benzene will be needed to fill a 25-ml graduated cylinder?

30. A glass bulb with a stopcock weighs 66.3915 g evacuated, and 66.6539 g when filled with xenon gas at 25°C. The bulb holds 50.0 ml ofwater. What is the density of xenon at 25°C?

31. If 20 g of magnalium lathe turnings of density 2.50 g/ml are put into a 25-ml graduated cylinder, what volume of water will be needed to complete the filling of the cylinder to the 25-ml mark?

32. What is the weight in pounds, of 1 cu ft of aluminum (density $= 2.70$ g/ml)?

33. The density of Dowmetal, a magnesium alloy, is 1.78 g/ml. Find the weight, in grams, of a rod $\frac{1}{2}$ inch in diameter and 2 ft long.

34. Neptunium has a density of 17.7 g/ml. What would be the radius of a sphere of neptunium that weighed 500 g?

35. A metal earring weighs 2.6321 g when suspended in air. When immersed in water at 22°C, it weighs 2.3802 g. What is the density of the earring?

36. (a) Specific gravity 20°/20° for lead is 11.337; find sp gr 20°/4°. (b) Specific gravity 18°/4° for lanthanum chloride is 3.947; find sp gr 18°/18°. (c) Which values in *a* and *b* are the densities?

37. Gallium (density $= 3.01$ g/ml) may be purchased at the rate of $3.25 per gram. How much does 1 cu ft of gallium cost?

38. A thousandth of a milliliter is called a lambda (λ). A certain biochemical procedure calls for the addition of 5 λ of a 2 per cent solution of sodium chloride (density $= 1.012$ g/ml). How many milligrams of sodium chloride will be added from the 5 λ pipet?

39. The weight of a metal sample is measured by finding the increase in weight of a volumetric flask when the metal sample is placed in it. The volume of the metal sample is measured by finding how much less water the volumetric flask holds when it contains the metal sample. Compute the density of the metal sample from the following data, assuming that the density of the water is 1 g/ml:

weight of empty flask $= 23.482$ g

weight of flask + sample $= 40.375$ g

weight of flask + sample + water $= 63.395$ g

weight of flask + water (without sample) $= 48.008$ g

40. (a) Repeat the calculations of Problem 39, using an exact density value for water (refer to Table 7-1) and assuming a temperature of 68°F. (b) In which decimal place does it make a significant difference if the exact density for water is used instead of the approximate value of 1 g/ml?

41. A chemical is soluble in water but insoluble in kerosene. As a consequence, kerosene may be used to determine its density. The density of kerosene is 0.735 g/ml at 22°C. From the following data (obtained as in Problem 39) compute the density of the sample:

weight of empty flask $= 28.176$ g

weight of flask + sample $= 40.247$ g

weight of flask + sample + kerosene to fill = 51.805 g

weight of flask + kerosene (without sample) = 45.792 g

42. Calculate the specific gravity for (a) the metal in Problem 40, (b) the chemical in Problem 41.

43. A geologist often measures the density of a mineral by mixing two dense liquids, carbon tetrachloride and acetylene tetrabromide, in such proportions that the mineral grains will *just* float. He then determines the density of the liquid mixture, which is equal to the density of the solid. When a sample of the mixture in which calcite (calcium carbonate) just floats is put in a special density bottle, the weight is 6.2753 g. When empty, the bottle weighs 2.4631 g, and when filled with water it weighs 3.5441 g. What is the density of this calcite sample? (The temperature of these measurements is 25°C.)

44. When 235 g of uranium 235 disintegrates by nuclear fission, 0.205 g is converted into 4.4×10^{12} calories (4.4 million million calories!) of energy. What volume of uranium is converted to energy if its density is 18.9 g/ml?

45. How many pounds, including its own weight, can a hydrogen-filled balloon lift in air if it is 50 ft in diameter? (The density of air is 1.205 g/liter, and the density of hydrogen is 0.0833 g/liter. The lift is the difference between the weight of the gas and that of an equal volume of air.)

46. What percentage of your body (density approximately 1.03 g/ml) would be out of the water while you floated on your back in Great Salt Lake (density of water approximately 1.19 g/ml)?

47. (a) Find a general factor by which pounds per cubic foot could be multiplied to convert to density in grams per milliliter. (b) Find a factor to convert grams per milliliter to pounds per cubic foot.

48. How far into a 2-inch cube of wood (density = 0.90 g/ml) must a 14-g iron screw, $\frac{1}{4}$ inch in diameter, be driven in order that the block will just float? (Assume that the block does not change in size when the screw is driven into it. The density of iron is 7.60 g/ml. Assume that the density of water is 1 g/ml.)

49. A "Cartesian diver" is a hollow bulb made of thin glass such that the density is just a little less than 1 g/ml. As a consequence, it floats in water. If pressure is applied to the gas over the water in which the bulb floats, the bulb collapses a little and then sinks. When the pressure is released, it reexpands and floats again. A glass (density = 2.20 g/ml) sphere made on this principle weighs 3.25 g. (a) What is its radius? (b) What is the thickness of the glass?

50. A sample of liquid of density 0.85 g/ml weighs 32.3524 g in air with brass weights. What is the weight in vacuo?

51. A platinum dish ($D = 21.5$ g/ml) weighs 65.2364 g in air with brass weights. What is the weight in vacuo?

52. A crude brass weight of nominal value 100 g weighs 99.9986 g in air, with brass weights. What is its true weight in vacuo?

53. A chemist uses water to calibrate several pieces of equipment. To save time, he first finds a factor by which he can multiply the weight of water in air to obtain the weight in vacuo. What is the numerical value of this factor? Brass weights are used.

54. A buret is calibrated by filling with water to the zero mark, withdrawing to the 25-ml mark and weighing the water withdrawn. From the following data, compute the correction that must be applied at the 25-ml mark (similar corrections at other intervals may be used to construct a correction curve for the entire range of the buret):

$$\text{final reading} = 25.00 \text{ ml}$$
$$\text{initial reading} = 0.03 \text{ ml}$$
$$\text{weight of flask} + \text{water delivered} = 81.200 \text{ g}$$
$$\text{weight of empty flask} = 56.330 \text{ g}$$
$$\text{temperature of water} = 21.5°C$$

Formulas and Nomenclature*

At first thought it may seem difficult to learn the formulas and names of the hundreds of chemical compounds used in the first chemistry course. Actually the job is not so hard if you start in a systematic way, by learning the atomic building blocks that make up the compounds and the rules for naming the compounds.

One of the many ways to classify inorganic compounds is into electrolytes, nonelectrolytes, and weak electrolytes. When *electrolytes* are dissolved in water the resulting solution is a good conductor of electricity; for *nonelectrolytes* the water solutions do not conduct electricity; and for *weak electrolytes* the solutions are very poor conductors. Water itself is an extremely poor conductor of electricity. A flow of current is a movement of electrical charges caused by a difference in potential (voltage) between the two ends of the conductor.

In metals, electrons are the structural units that carry the electrical charge, a negative one. But since electrons cannot exist for any significant length of time as independent units in water, some other kind of charged structural unit must be present in solutions of electrolytes. The general term for this charged structural unit is *ion*. A negatively charged ion is an atom (or a group of atoms) carrying one or more electrons it has received from other atoms. And, naturally, those atoms or groups of atoms that gave up electrons are no longer electrically neutral; they are positively charged ions. At all times the solutions or crystals that contain ions are electrically neutral, because the

---

*The approved system for naming complex ions is given on page 277.

total number of negative charges gained by one group of atoms always exactly equals the total number of positive charges created in the groups from which the electrons came. Ions tend to stay in the vicinity of each other because of the attraction of opposite electrical charges. Ions that contain more than one atom in stable combinations are often referred to as *radicals*. In addition to explaining the electrical conductivity of water solutions, ions are important because a tremendous number of chemical reactions take place between them.

Inorganic compounds may also be classified as acids, bases, and salts. This classification is particularly useful as a basis for naming the chemicals with which we shall deal.

The names of the elements, and many of the symbols used to represent them, are traditional, rather than part of a logical system. The electrical charges that the ions usually carry can be reasoned out, but chemists do not work through such reasoning every time they want to use the charges or talk about them; they just know them as characteristic properties.

Once you have learned the symbols for the elements, you will easily recognize and understand formulas. A *formula* is the shorthand notation used to identify the composition of a molecule. It includes the symbol of each element in the molecule, with numerical subscripts to show how many atoms of each element are present if there are more than one. For example, the formula for sulfuric acid, H_2SO_4, shows that this molecule has 2 hydrogen atoms, 1 sulfur atom, and 4 oxygen atoms. Note that a molecular formula does not tell how the atoms are bound together, only the kinds and numbers of atoms.

Listed in the paragraphs that follow are the names, symbols, and common electrical charges for 30 common positive ions, and the names and formulas for 37 common acids that will be frequently referred to in this text. You should memorize these so as to have them at instant recall; the use of flash cards or other foreign language learning aids is recommended. Aside from their direct intrinsic value, you can *reason out* from them the names and formulas of almost 50 bases and over 1600 different salts, *none* of which should be memorized.

POSITIVE IONS WHOSE CHARGES DO NOT VARY*

The ions listed in Table 8-1 carry exactly the same names as the elements from which they are derived. For example, Na and Mg are sodium and magnesium atoms, while Na^+ and Mg^{++} are sodium and magnesium ions. These elements do not normally form ions that have charges other than those shown.

---

*Some of the elements listed here do exhibit other charges under very unusual conditions, but the infrequency with which they occur does not warrant our knowing them now. The principal exception, H^-, is listed on p. 87.

TABLE 8-1
Positive Ions Whose Charges Do Not Vary

Single charge		Double charge		Triple charge	
hydrogen	H^+	beryllium	Be^{++}	aluminum	Al^{+++}
lithium	Li^+	magnesium	Mg^{++}		
sodium	Na^+	calcium	Ca^{++}		
potassium	K^+	strontium	Sr^{++}		
rubidium	Rb^+	barium	Ba^{++}		
cesium	Cs^+	zinc	Zn^{++}		
silver	Ag^+	cadmium	Cd^{++}		
ammonium	NH_4^+				

TABLE 8-2
Positive Ions Whose Charges Vary

IUPAC name	Traditional name		
	Root	-ous ending	-ic ending
Copper(I) and (II)	cupr-	Cu^+	Cu^{++}
Gold(I) and (III)	aur-	Au^+	Au^{+++}
Mercury(I) and (II)	mercur-	$Hg^+(Hg_2^{++})$	Hg^{++}
Chromium(II) and (III)	chrom-	Cr^{++}	Cr^{+++}
Manganese(II) and (III)	mangan-	Mn^{++}	Mn^{+++}
Iron(II) and (III)	ferr-	Fe^{++}	Fe^{+++}
Cobalt(II) and (III)	cobalt-	Co^{++}	Co^{+++}
Nickel(II) and (III)	nickel-	Ni^{++}	Ni^{+++}
Tin(II) and (IV)	stann-	Sn^{++}	Sn^{++++}
Lead(II) and (IV)	plumb-	Pb^{++}	Pb^{++++}
Cerium(III) and (IV)	cer-	Ce^{+++}	Ce^{++++}
Arsenic(III) and (V)	arsen-	As^{+++}	As^{+++++}
Antimony(III) and (V)	antimon-	Sb^{+++}	Sb^{+++++}
Bismuth(III) and (V)	bismuth-	Bi^{+++}	Bi^{+++++}

POSITIVE IONS WHOSE CHARGES VARY

The atoms of some metals can lose different numbers of electrons under different conditions. For these atoms it has been traditional to add the suffix -*ous* to the atom's root name for the lower charge state, and the suffix -*ic* for the higher charge state. Thus the aurous ion is Au^+ and the auric ion is Au^{+++}. There is difficulty when an element has more than two charge states. To overcome this difficulty and to clarify some subtler points, the International Union of Pure and Applied Chemistry (IUPAC) has in recent years recommended the adoption of a system that, when applied to a positive ion, requires that the name of the element be used and that it be followed immediately by its charge in Roman numerals in parentheses. For example, Au^+ would be gold (I) and Au^{+++} would be gold (III); they would be read as "gold one" and "gold three." Until there is unanimous adoption of the IUPAC system, chemists will have to use both systems. In any case, neither system relieves the student of the responsibility of knowing the usual charge states of the common elements.

BASES

We shall define a *base* as any substance that can accept or react with a hydrogen ion, H^+. This definition includes a wide variety of compounds, but for the present it is convenient to limit our discussion to one special type of base called a *hydroxide*. A hydroxide is any compound that has one or more replaceable hydroxide ions (OH^-). Any of the positive ions in the two groups cited just above might combine with the hydroxide ion, the principle being that the resulting compound must be electrically neutral. Naturally there must be as many OH^- ions as there are positive charges on the other ion. In naming, the word "hydroxide" is preceded by the name of the positive ion; for example,

NaOH   sodium hydroxide
$Co(OH)_2$ cobaltous hydroxide or cobalt(II) hydroxide
$Sn(OH)_4$ stannic hydroxide or tin(IV) hydroxide

ACIDS

We shall define an *acid* as any substance that has one or more replaceable hydrogen ions (H^+); we often say that an acid can donate one or more hydrogen ions to a base. No matter how many H atoms a molecule might have, if none of them can be replaced by some positive ion, then the molecule does not qualify as an acid. On the other hand, if a molecule does have several H atoms, of which only one is replaceable, that alone is enough to qualify it as an acid. In writing the formula of an acid it is traditional to show the replaceable H ions at the beginning of the formula, separated from those that are not replaceable. For example, in acetic acid, $HC_2H_3O_2$, there is one replaceable H^+ and three H atoms that are not replaceable. Besides identifying the number of replaceable H atoms, this traditional method also provides a simple way of knowing the charge of the negative ions; for every H^+ that is removed from an acid there must be one negative charge left on the residue. For example, if HBr loses its H^+, the Br^- has a minus one charge; if H_2SO_4 loses its two H^+, the $SO_4^=$ has a minus two charge, etc. The names and formulas of acids that you should know in this course are listed in two separate groups below, because the traditional method of naming them depends on whether or not the molecule contains oxygen.

1. Acids That Do Not Contain Oxygen

These acids are named by putting the prefix *hydro-* before the rest of the name of the characteristic element (or elements) and adding the suffix *-ic.*

HF	hydrofluoric acid	HCN	hydrocyanic acid
HCl	hydrochloric acid	H_2S	hydrosulfuric acid
HBr	hydrobromic acid	HN_3	hydrazoic acid
HI	hydriodic acid		

2. Acids That Contain Oxygen

If the characteristic element forms only one oxygen-acid, the name is that of the characteristic element followed by the suffix *-ic.* Thus H_2CO_3 is carbonic acid. If the characteristic element forms two oxygen-acids, the name of the one with the larger number of oxygen atoms ends in *-ic,* and the name of the one with the smaller number of oxygen atoms ends in *-ous.* Thus HNO_3 is nitric acid and HNO_2 is nitrous acid.

If there are several oxygen-acids, there is a systematic terminology to indicate more or less oxygen atoms than the number assigned to the acid whose name ends in *-ic.* This can be illustrated by the oxygen-acids of chlorine:

$HClO_4$	perchloric acid	more oxygen than *-ic* acid
$HClO_3$	chloric acid	arbitrarily given *-ic* ending some time ago
$HClO_2$	chlorous acid	less oxygen than the *-ic* acid
$HClO$	hypochlorous acid	less oxygen than the *-ous* acid

Table 8-3 includes the common acids that contain oxygen.

SALTS

One general type of chemical reaction is that which occurs when a hydroxide reacts with an acid. This reaction, like all chemical reactions, can be represented by a *chemical equation* in which the reactants are separated by "+" signs to indicate that they are mixed together, the products are separated by "+" signs to indicate that they are produced as a mixture, and the products are separated from the reactants by an arrow to show that the reactants are producing the products. In order to write a chemical equation for the general

TABLE 8-3
Acids That Contain Oxygen

Formula	Name	Formula	Name
H_2CO_3	Carbonic acid	$HClO_4$	Perchloric acid
H_3BO_3	Boric acid	(HIO_4 is similar to $HClO_4$)	
H_4SiO_4	Silicic acid	$HClO_3$	Chloric acid
HNO_3	Nitric acid	(HIO_3 and $HBrO_3$ are	
HNO_2	Nitrous acid	similar to $HClO_3$)	
H_2SO_4	Sulfuric acid	$HClO_2$	Chlorous acid
H_2SO_3	Sulfurous acid	$HClO$	Hypochlorous acid
$H_2S_2O_3$	Thiosulfuric acid	(HIO and $HBrO$ similar to $HClO$)	
H_2CrO_4	Chromic acid	$HMnO_4$	Permanganic acid
$H_2Cr_2O_7$	Dichromic acid	$HOCN$	Cyanic acid
H_3PO_4	Phosphoric acid	$HSCN$	Thiocyanic acid
H_3PO_3	Phosphorous acid	$H_2C_2O_4$	Oxalic acid
H_3AsO_4	Arsenic acid	$H_2C_8H_4O_4$	Phthalic acid
H_3AsO_3	Arsenious acid	$HC_2H_3O_2$	Acetic acid
		$H(NH_2)SO_3$	Sulfamic acid

reaction between an acid and a hydroxide, we need to know first that in every case the acid donates an H^+ to each OH^- of the hydroxide to form water (H_2O), and second that the electrically neutral combination of the positive ions from the hydroxide and the negative ions from the acid is what constitutes a *salt*. If we express this in very general terms, this type of reaction can be written as

$$\text{hydroxide} + \text{acid} \rightarrow \text{salt} + \text{water}$$

Now that we know the names and formulas of some acids and bases we can also write equations for specific examples of this type of reaction, using our shorthand notation. For example, if potassium hydroxide reacts with hydrobromic acid, we write

$$KOH + HBr \rightarrow KBr + H_2O$$

If cupric hydroxide reacts with perchloric acid, we write

$$Cu(OH)_2 + 2\ HClO_4 \rightarrow Cu(ClO_4)_2 + 2\ H_2O$$

And if aluminum hydroxide reacts with sulfuric acid, we write

$$2\ Al(OH)_3 + 3\ H_2SO_4 \rightarrow Al_2(SO_4)_3 + 6\ H_2O$$

Note that the chemical equation shows not only which chemicals react and which are produced, it shows relative numbers of molecules needed in the reaction. In the above reactions we had to use the proper number of acid molecules and base molecules so that the salt that was formed would be electrically neutral. In the third reaction, for example, the two Al^{+++} ions require three $SO_4^=$ ions in order to have electroneutrality. Besides showing the correct chemical formulas of all the reactants and products, a *balanced* chemical equation always abides by a principle of conservation, which might be stated as: "atoms are never created or destroyed in a chemical reaction." In other words, no matter how drastically the atoms are rearranged in a chemical reaction, there must always be the same number of each kind of atom in the collection of products (the right-hand side of the equation) as there is in the collection of reactants (the left-hand side of the equation).

The name of a salt is determined by whether the acid from which it is derived contains oxygen.

1. If the acid does not contain oxygen, the salt is named by replacing the prefix *hydro-* by the name of the positive ion and changing the suffix *-ic* to *-ide*; for example,

KBr   potassium bromide
Sb_2S_3 antimonous sulfide or antimony(III) sulfide
$Hg(CN)_2$ mercuric cyanide or mercury(II) cyanide

2. If the acid does contain oxygen, the salt is named by giving the name of the positive ion followed by the name of the acid, but changing the suffix *-ic* to *-ate* or the suffix *-ous* to *-ite*; for example,

$SrCO_3$ strontium carbonate
$Cu(ClO_4)_2$ cupric perchlorate or copper(II) perchlorate
$(NH_4)_2SO_3$ ammonium sulfite
$Co(BrO)_2$ cobaltous hypobromite or cobalt(II) hypobromite

By changing the proportions in which some acids and bases are mixed, it is possible to make salts in which only some of the replaceable hydrogen atoms are actually replaced. For example, with H_3PO_4 and $NaOH$, we could get:

$$3\ NaOH + H_3PO_4 \rightarrow Na_3PO_4 + 3\ H_2O$$
$$2\ NaOH + H_3PO_4 \rightarrow Na_2HPO_4 + 2\ H_2O$$
$$NaOH + H_3PO_4 \rightarrow NaH_2PO_4 + H_2O$$

Each of these reactions is possible and each equation is balanced. Obviously we need a way to distinguish between the three different kinds of sodium phosphate salts. Of the several ways to accomplish this, the most highly recommended and unequivocal is to include as part of the name the number of H^+ ions that have not been replaced in salt formation. In the example above we would have

Na_3PO_4 sodium phosphate
Na_2HPO_4 sodium monohydrogen phosphate
NaH_2PO_4 sodium dihydrogen phosphate

If the acid contains only two H^+, then only two different salts are possible; they are most acceptably named by the method just described. However, another traditional method that will probably continue in use for many years is to use the prefix *bi-* for the salt of the acid with just half of the hydrogen atoms replaced, as illustrated for the sodium bisulfate that is produced by the reaction

$$NaOH + H_2SO_4 \rightarrow NaHSO_4 + H_2O$$

BINARY COMPOUNDS NOT DERIVED FROM ACIDS

The atoms listed in this section may combine with many metals to form binary compounds (compounds made up of two elements) that are salt-like in nature but not derived from acids. For purposes of naming, it is convenient to assign negative charges to these atoms. Except in the names of the oxides, the suffixes *-ous* and *-ic* are not used with metals forming compounds in this group. The names of all these compounds end in *-ide*. Only the metal oxides of this group are common.

H^-	LiH	lithium hydride
$O^=$	FeO	ferrous oxide, Fe_2O_3 ferric oxide
$N^{\equiv}$	Sn_3N_4	tin nitride
$P^{\equiv}$	Ba_3P_2	barium phosphide
$As^{\equiv}$	Na_3As	sodium arsenide
$C^{\equiv}$	Al_4C_3	aluminum carbide
$Si^{\equiv}$	Mg_2Si	magnesium silicide

BINARY COMPOUNDS COMPOSED OF TWO NONMETALS

Allowing for a bit of quibbling one way or the other, there are only twenty nonmetallic elements. Of these twenty elements, six are so unreactive that until recently it had been categorically stated that they never combine with other elements. If for practical considerations we eliminate these six elements and hydrogen (which we've already dealt with in various forms above), this leaves just thirteen nonmetallic elements (B, C, N, O, F, Si, P, S, Cl, Se, Br, Te, I) that can combine with each other. These nonmetallic binary compounds are designated by the names of the two elements followed by the ending -*ide*. Before the name of the second element there is a prefix to indicate how many atoms of it are in the molecule. Unfortunately, this is not usually done for the first element. The following are examples of some common compounds of this type:

CO	carbon monoxide	SO_3	sulfur trioxide
CO_2	carbon dioxide	CCl_4	carbon tetrachloride
Cl_2O	chlorine monoxide	PF_5	phosphorus pentafluoride
ClO_2	chlorine dioxide	SF_6	sulfur hexafluoride
Cl_2O_7	chlorine heptoxide	N_2O_4	nitrogen tetroxide

PROBLEMS A   Answers on page 340

Name the following compounds (proper spelling is required).

1. $Ca(OH)_2$
2. Ag_3PO_4
3. $AgSCN$
4. $MgC_8H_4O_4$
5. $(NH_4)_2SO_4$
6. ZnS
7. $Cd(CN)_2$
8. $Ba(IO_3)_2$
9. $CuSO_3$
10. CuI
11. $Fe(NO_3)_3$

12. FeC_2O_4
13. Hg_2Cl_2
14. $MnCO_3$
15. $Mn(OH)_3$
16. $Ni(ClO)_2$
17. $CrAsO_4$
18. $SnBr_4$
19. CrF_2
20. $Pb(MnO_4)_2$
21. Na_4SiO_4
22. Bi_2O_5

23. $Al(ClO_4)_3$
24. $Hg(C_2H_3O_2)_2$
25. $CsClO_3$
26. $Sr(IO)_2$
27. Rb_3AsO_3
28. Be_3N_2
29. $Ca(HCO_3)_2$
30. $Sb(NO_3)_3$
31. PCl_3
32. $Bi(OCN)_3$
33. $Al_2(S_2O_3)_3$

Without consulting a text, give the formulas for the following compounds.

34. aluminum bromate
35. mercurous phosphate
36. bismuth(III) oxide
37. strontium bicarbonate
38. aurous iodide
39. chromium(III) iodate
40. manganous hydroxide
41. lithium arsenide
42. arsenic(III) sulfate
43. stannic chloride
44. nickelous periodate
45. chlorine heptoxide
46. silver oxalate

47. chromium(II) borate
48. antimonous sulfide
49. aluminum acetate
50. calcium oxalate
51. sodium chlorite
52. tin(II) azide
53. mercury(II) cyanide
54. ammonium sulfite
55. cobalt(II) permanganate
56. plumbous carbonate
57. zinc phosphide
58. cupric silicate
59. barium hypoiodite

60. Given that selenium (Se) is similar in properties to sulfur (S) and francium (Fr) is similar to sodium (Na), write the formulas for the following compounds: (a) zinc selenide; (b) francium phosphate; (c) cobalt(II) selenite; (d) selenium dioxide; (e) francium selenate; (f) selenium hexafluoride; (g) francium hydride.

PROBLEMS B   No answers given

Name the following compounds (proper spelling is required).

61. Ag_3PO_4
62. $CoCl_3$
63. $Be(NO_2)_2$
64. $Fe(MnO_4)_3$
65. NH_4NO_2
66. Al_2S_3
67. $Zn(IO_4)_2$
68. $Pb_3(BO_3)_2$
69. $As(CN)_3$

70. $Ni_3(AsO_3)_2$
71. I_2O_7
72. Ba_2Si
73. $AuHSO_3$
74. $Ba(BrO)_2$
75. CaH_2
76. N_2O_5
77. Sb_2S_3
78. MgC_2O_4

79. ICl
80. Rb_4SiO_4
81. SF_6
82. Ca_3P_2
83. MnO_2
84. $Cu(C_2H_3O_2)_2$
85. $CrBr_2$
86. $BaCrO_4$
87. $Cd(SCN)_2$

Without consulting a text, give the formulas for the following compounds.

88. potassium oxalate
89. cupric arsenate
90. bismuthous carbonate
91. manganese(III) oxide
92. mercurous sulfate
93. nitrogen tri-iodide
94. cobalt(II) borate
95. cesium hypoiodite
96. boron nitride
97. cadmium dichromate
98. ammonium acetate
99. zinc cyanide
100. tin(II) phosphate

101. iron(III) bromate
102. arsenic(V) perchlorate
103. magnesium monohydrogen borate
104. boron trifluoride
105. strontium silicate
106. beryllium hydroxide
107. stannic oxide
108. gold(III) fluoride
109. ferric chromate
110. iodine pentoxide
111. lithium thiocyanate
112. silver thiosulfate
113. antimonic permanganate

114. Given that astatine (At) is similar in properties to chlorine (Cl) and gallium (Ga) is similar to aluminum (Al), write the formulas for the following compounds: (a) potassium astatate; (b) barium astatide; (c) gallium sulfate; (d) hydrastatic acid; (e) gallium thiocyanate; (f) gallium hypoastatite.

Sizes and Shapes of Molecules

Atoms commonly enter into chemical reaction in accord with the Octet Rule. There is undoubtedly a fairly detailed discussion of the Octet Rule in your text, but, in essence, it may be described as the tendency for atoms to lose, gain, or share electrons in order to achieve an s^2p^6 electron configuration in the outermost shell. The simplest atoms (H, Li, Be, etc.) tend to achieve a $1s^2$ configuration, according to what might be called the Duet Rule. In Chapter 8 we emphasized the loss and gain of electrons leading to the formation of electrically charged ions, such as Na^+ and Cl^-. When electrons are shared, a molecule is formed, and the atoms are connected by a covalent bond. In this chapter we emphasize the approximate shapes, interatomic distances, and bond energies of molecules and molecule ions that are held together by covalent bonds.

COVALENT BOND ENERGIES

The strengths of the bonds that hold the atoms together in a molecule can be determined in a variety of ways: by direct calorimetric measurement, by dissociation equilibrium measurements, or by spectral measurements, for example.

We define *bond energy* as the energy change for the chemical process in which one mole of a given bond is broken when both the reactants and the

TABLE 9-1
Covalent Bond Radii and Energies

Element	Radius, Å	Bond energy, kcal/mole	Electronegativity of element
H—H	0.30	104.2	2.1
B—B	0.88	62.7	2.0
C—C	0.772	83.1	2.5
C=C	0.667	147	
C≡C	0.603	194	
Si—Si	1.17	42.2	1.8
Si=Si	1.07		
Ge—Ge	1.22	37.6	1.8
Sn—Sn	1.40	34.2	1.8
N—N	0.70	38.4	3.0
N=N	0.60	100	
N≡N	0.55	226.2	
P—P	1.10	51.3	2.1
P=P	1.00	117	
As—As	1.21	32.1	2.0
Sb—Sb	1.41	30.2	1.9
Bi—Bi	1.52	25	1.9
O—O	0.66	33.2	3.5
O=O	0.55		
S—S	1.04	50.9	2.5
S=S	0.94		
Se—Se	1.17	44.0	2.4
Se=Se	1.07		
Te—Te	1.37	33.0	2.1
F—F	0.64	36.6	4.0
Cl—Cl	0.99	58.0	3.0
Br—Br	1.14	46.1	2.8
I—I	1.33	36.1	2.5

products are in the hypothetical ideal-gas state of 1 atm and 25°C. For a diatomic molecule the bond energy is identical to the energy required to dissociate the *gaseous* molecule into its respective *gaseous* atoms. For the dissociation of Cl_2 gas this corresponds to the reaction

$$Cl_{2(g)} \rightleftarrows 2\ Cl_{(g)}$$

for which the Cl—Cl bond energy is 58.0 kcal.

For a polyatomic molecule of the type AB_n, which possesses n A—B bonds, our definition of bond energy implies that each bond is the same and that it corresponds to $1/n$ of the total energy required to dissociate the gaseous AB_n molecule into $A + nB$ gaseous atoms. This is a useful definition except when studying the detailed steps of a chemical reaction. For example, the total binding energy in a CH_4 molecule is 396 kcal/mole, and by our definition of bond energy, the C—H bond energy $= 396/4 = 99$ kcal. Extensive, complicated, and detailed studies have shown, however, that each H atom is not equally easily removed from carbon in this molecule; it is estimated that the individual bond energies are 103 kcal for CH_3—H, 87 kcal for CH_2—H, 125 kcal for CH—H, and 81 kcal for C—H, with a total of 396 kcal. Normally such detailed information is not available; neither is it normally needed except in discussion of the individual steps involved in chemical reactions.

The atoms of some elements, such as C, N, and O, are able to share more than one pair of electrons between them, to form single, double, or triple bonds, depending on whether one, two, or three pairs of electrons are shared. In general, the bonding energy increases and the internuclear distance decreases as the number of bonds between a pair of atoms increases. The data for Table 9-1 were taken from compounds that include identical atoms and therefore pure covalent bonds. Thus, for the single bonds, H_2 might be used for H—H, N_2H_4 for N—N, H_2O_2 for O—O, etc. As in so many sharing processes, the pair of electrons in a covalent bond is often not equally shared by the two atoms. The atom with the greater electron affinity will hold the pair closer to its nucleus, with the result that its end of the bond (and its end of the molecule) will be somewhat more negative than the other end. When this happens we say that the bond is partially ionic, and because opposite charges attract each other, this partially ionic bond will be stronger than it would have been with equal sharing. Whenever a strictly covalent bond is formed between two atoms, the bond energy can be estimated impressively well by taking the average of the covalent single-bond energies for the separate atoms. Carbon and iodine form such a bond; its energy would be $\frac{1}{2}(83.1 + 36.1) = 59.6$ kcal.

ELECTRONEGATIVITY

Pauling made a careful study of a tremendous number of partially ionic covalent bonds, and came to the conclusion that, as a measure of its electron affinity, each element could be assigned an *electronegativity value* (ϵ) that would make it possible to estimate the energy of these bonds. To calculate the approximate bond energy $(\Delta H^\circ_{A—B})$ between the atoms A and B, his formula requires knowledge of the A—A bond energy $(\Delta H^\circ_{A—A})$, the B—B bond energy $(\Delta H^\circ_{B—B})$, and the electronegativities of A and B $(\epsilon_A$ and $\epsilon_B)$.

The covalent bond energies and the electronegativities are listed in Table 9-1. Pauling's formula (with units in kcal) is

$$\Delta H^\circ_{A-B} = \tfrac{1}{2}[\Delta H^\circ_{A-A} + \Delta H^\circ_{B-B}] + 23.06\,(\epsilon_A - \epsilon_B)^2$$

You can see that if A and B have equal electronegativities, then ΔH°_{A-B} is simply the average of the two covalent bond energies. The greater the difference in electronegativities, the greater the percent ionic character and the stronger the bond. If the difference in electronegativity becomes too great, the bond is essentially ionic and the atoms are held together by electrostatic forces as in an ionic crystal; the concept of a molecule disappears.

From this brief discussion you can see that more often than not most covalent bonds will be "partially ionic," and most ionic bonds will be "partially covalent." We shall describe a bond as being ionic or covalent according to its predominant characteristic. Some chemists like to say that a bond possesses a certain "percent ionic character." One way of calculating an approximate value for this is by the expression

$$\log\,(1 - F_i) = -0.11\,(\epsilon_A - \epsilon_B)^2$$

where F_i is the fraction of ionic character.

PROBLEM:
Calculate the bond energy and per cent ionic character for the C—Cl bond in CCl_4.

SOLUTION:
From Table 9-1 we find:

$$\Delta H^\circ_{C-C} = 83.1 \text{ kcal/mole}$$
$$\Delta H^\circ_{Cl-Cl} = 58.0 \text{ kcal/mole}$$

and
$$\epsilon_C = 2.5$$
$$\epsilon_{Cl} = 3.0$$

With the Pauling equation we obtain:

$$\Delta H^\circ_{C-Cl} = \tfrac{1}{2}[83.1 + 58.0] + (23.06)(3.0 - 2.5)^2$$
$$= 70.6 + 5.8 = 76.4 \text{ kcal/mole}$$

and
$$\log\,(1 - F_i) = -(0.11)(3.0 - 2.5)^2 = -0.027$$
$$1 - F_i = 10^{-0.027} = 10^{.973} \times 10^{-1} = 0.94$$
$$F_i = 0.06$$
$$\text{per cent ionic character} = 6\%$$

COVALENT BOND RADII

Before considering shape principles, we should take a look at the *sizes* of molecules, particularly their internuclear distances. Experimental measurements of these distances have been made by various methods. An interatomic distance is defined as the *average* internuclear distance between two atoms bonded together. Since these two atoms vibrate, the distance between them alternately lengthens and shortens in rapid succession, but if the distance is averaged for a period of time, the atoms appear to be separated by some fixed distance. Further, the atoms of a given element always appear to have the same radius regardless of the kinds of atoms to which they are bound. Some typical covalent bond radii are listed in Table 9-1. Because of the additive nature of covalent radii, we can use the values in this table to determine the interatomic distances in covalently bonded molecules and ions.

PROBLEM:
Calculate the C—Cl distance in the CCl_4 molecule.

SOLUTION:
The interatomic distance is the sum of the covalent radii. Therefore,

$$\text{interatomic distance} = R_C + R_{Cl}$$
$$= 0.77 + 0.99 = 1.76 \text{ Å}$$

We have already observed in a previous problem that the energy of the C—Cl bond is 76.4 kcal/mole and that it possesses ~6% ionic character.

SHAPES OF MOLECULES

For predicting the shape of a molecule, we can apply a set of simple principles outlined below, but before we begin you should realize that there is very little predictive value in them when you try to answer the question, "What molecule is formed when two or more elements react?" From the same elements, but under differing conditions, many different molecules may often be prepared, not just one. Instead, our principles will answer the question, "What is the shape of the molecule whose formula is_____?"

The concept of the "central atom" is convenient to use in discussing the shapes of molecules other than the simple ones. In a given simple molecule, one of the atoms is usually "central" to the whole molecule. For example, in CCl_4 the central atom is C, the one to which all the other atoms are attached. In general, excluding H, the central atom is the least electronegative atom of the group. We shall use the term *ligand* in its broadest sense to refer to those atoms that are attached to the central atom. What we mean by the shape of a molecule or radical is the geometric arrangement of the ligands about the central atom.

For the purpose of our discussion we shall assume that each central atom's valence electrons are spin-paired (the few atoms having an odd number of valence electrons are exceptions), and that these pairs of electrons will repel each other in such a way that they occupy positions of minimum repulsion, that is, positions of minimum potential energy. The electron pairs will try to get as far away from each other as they can and still stay in the molecule. Unlike oppositely charged ions with spherical charge distribution, which are held together by electrostatic attraction, the atoms in a molecule are held together by pairs of valence electrons, at least one pair of electrons per bond. Each pair will occupy a *molecular orbital*.

In order to apply the principles of electron-pair repulsion, we must first find the number of electron pairs that are associated with the valence shell of the central atom. One simple way to do this is to draw classical "electron-dot" structures, trying to adhere as far as possible to the Octet Rule (or Duet Rule for H) for all atoms in the molecule. Typical examples are shown in Figure 9-1.

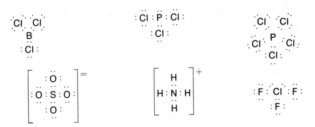

FIGURE 9-1
Electron-dot structures.

The central atom frequently violates the Octet Rule; the ligands *rarely* do.

We can formalize these pictorial representations in a convenient way, using the following symbols:

P = pairs of valence electrons associated with the central atom in the molecule.

V_M = number of valence electrons contributed to the molecule by the central atom (M). When numbering from left to right (but omitting the d block), this corresponds to the column of the periodic table in which the element is located.

V_L = number of electrons contributed to the central atom by the ligands (L) for the purpose of bonding.

Since it takes two electrons to make a pair,

$$P = \tfrac{1}{2}(V_M + V_L)$$

Those pairs of electrons not used for bonding are called lone pairs (LP), and those used for bonding are bond pairs (BP). It follows that

$$P = \text{LP} + \text{BP}$$

We must first discuss the "electron-pair geometry" of the central atom and then subsequently show how this determines the "molecular geometry," or shape, of the molecule. The P electron pairs will repel each other symmetrically around the central atom so as to occupy equilibrium positions of minimum potential energy. These "equilibrium positions" are, of course, molecular orbitals with specific directional characteristics pointing out from the central atom. There is one characteristic set of equilibrium positions, that is, one characteristic "electron-pair geometry," for each value of P.

As we describe the electron-pair geometries for the molecular systems with two or more pairs of electrons, imagine each pair of electrons to be attached to the central atom by a weightless string that permits free movement within the confines of the tether. Under these conditions it would be natural to expect that with only two such pairs of electrons, the pairs would lie diametrically opposite each other with M in the center, for, within the confines of the string, any other position would bring them closer together in a more repulsive position. Thus, when $P = 2$, the electron-pair geometry is always linear, not angular (that is, the angle, p-M-p, is always 180°). The letter p refers to a single pair of electrons.

When $P = 3$, the two most likely electron-pair geometries are a triangle pyramid with M at the apex, and a triangle-coplanar structure, in which M lies at the center of an equilateral triangle and in the same plane as the electron pairs, which lie at the corners of the triangle. A little reflection quickly leads you to the triangle-coplanar structure as the one with less electron-pair repulsion, for the electron pairs are farther apart in this configuration. The p-M-p angle is 120°.

When $P = 4$, the two most likely electron-pair geometries are: a square-coplanar structure, in which M lies at the center of the square and in the same plane as the pairs of electrons, which lie at each corner of the square; and a tetrahedral structure with M at the center of the tetrahedron outlined by the four pairs of electrons, one pair at each apex of the tetrahedron. Again, for a given length of string, the pairs of electrons will be further from each other in the tetrahedron, in which the p-M-p angle is 109°28′, than in the square-coplanar structure, where the p-M-p angle is 90°; the tetrahedral electron-pair geometry will thus always be expected with four pairs of electrons.

When $P = 5$, there immediately come to mind the possibilities of (1) a pentagonal-coplanar structure or (2) a triangular bipyramid formed by placing two triangle-based pyramids base-to-base, with M centered in the plane where the two bases come together. Again, for a given length of string, the repulsion between the electrons will be greater in the coplanar pentagon, where the p-M-p angle is 72°, than in the triangular bipyramid, where the p-M-p angle is either 90°, 120°, or 180°, depending on which two pairs of electrons are under consideration. The triangular bipyramid will always be the expected electron-pair geometry.

At least three reasonable electron-pair geometries might occur to you for the $P = 6$ case. They are (1) coplanar hexagon (with M centered in the plane), (2) triangular prism (two triangular pyramids, apex-to-apex, with base edges parallel to each other and M located at the apex-to-apex contact), and (3) octahedron (a square bipyramid formed by placing two square-based pyramids base-to-base with M centered in this base plane). The same arguments used above lead to the conclusion that, when $P = 6$, the expected electron-pair geometry will always be the octahedron. Note that all the p-M-p angles between adjacent p positions are the same, 90°, which means that all the electron-pair positions are equivalent in the octahedron, in contrast to the nonequivalent positions in the triangular bipyramid.

Having established the electron-pair geometry for the common numbers of electron pairs, let us now say that for the more complicated case when $P = 7$, the expected electron-pair geometry is a pentagonal bipyramid with M centered in the common pentagonal base plane. And for $P = 8$, three electron-pair geometries are possible: a cube with M at the center; a square antiprism (two square-based pyramids, apex-to-apex, with the base of one rotated 45° relative to the other, and M located at the apex-to-apex contact); and a dodecahedral arrangement, which can be regarded as a distorted cubic arrangement. The square antiprism and the distorted cubic are about equally probable and involve less electron-pair repulsion than the simple cubic. The number of compounds in which M has $P = 7$ or 8 is actually rather limited.

Now, let us look at molecular geometries or shapes. These are determined by the electron pairs that are used as *bond* pairs (BP), for these bond pairs will lie in molecular orbitals between M and L in positions determined by P, as we have just discussed. The total number of pairs will often (usually) be equal to the number of ligands attached to M, and when this is true, that is, when $P = $ BP, the molecular geometry will be identical to the electron-pair geometry. In the sketches, the electron-pair geometry is shown by shaded planes, and bond pairs of electrons by solid lines.

When $P = 1$ we have a trivial case (such as HCl, in which H is considered to be the central atom), in which the only pair is of necessity a bond pair or there would be no molecule at all, and the shape is typical of all diatomic molecules regardless of the number of electron pairs—it is linear.

When $P = 2$, as in $HgCl_2$, the molecular geometry is *linear,* since both pairs are bond pairs repelling each other at 180°. In this case, $P = 2$ because $V_M = 2$ and $V_L = 2$.

Linear

FIGURE 9-2
Linear molecule.

The compound $SnCl_2$, which appears superficially to be the same as $HgCl_2$, is actually different because we have $V_M = 4$ and $V_L = 2$, which leads to a triangular coplanar electron-pair geometry with $P = 3$. Here, however, BP $= 2$ and LP $= 1$, and the net result is that $SnCl_2$ is an *angular* molecule, not a linear one.

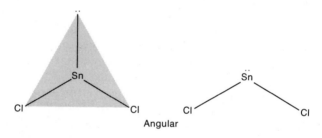
Angular

FIGURE 9-3.
Angular molecule.

A triatomic molecule is never called triangular or planar, even though it is always both; it should always be called either linear or angular. In either case, the three atoms will lie in the same plane, because any three noncollinear points always determine a plane.

Boron trichloride also has three electron pairs because $V_M = 3$ and $V_L = 3$, but all three are bond pairs and the molecule is triangular-coplanar.

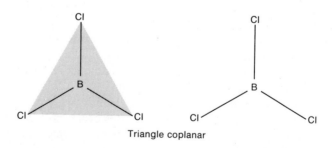
Triangle coplanar

FIGURE 9-4.
Triangular-coplanar molecule.

For CCl_4, $P = 4$ because $V_M = 4$ and $V_L = 4$, and the molecular structure is tetrahedral with C at the center, since BP is also 4.

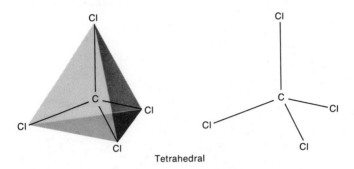

FIGURE 9-5.
Tetrahedral molecule.

The ammonia molecule is another in which $P = 4$ because $V_M = 5$ and $V_L = 3$, but the molecule is *pyramidal*, not tetrahedral, because there are only three bond pairs. At the apex of the pyramid is the N atom, and the lone pair of electrons occupies a tetrahedral position in the electron-pair geometry. The term "pyramid" is meant to be applied to *irregular* tetrahedra, "tetrahedron" to regular (equal-edged) tetrahedra.

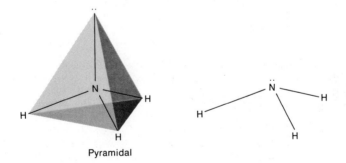

FIGURE 9-6.
Pyramidal molecule.

Still another type of molecule exists for $P = 4$; water is an example. Here, $V_M = 6$, $V_L = 2$, BP $= 2$, and LP $= 2$. The net result is an angular molecule with the two lone pairs occupying tetrahedral positions in the electron-pair geometry.

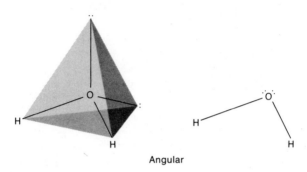

Angular

FIGURE 9-7.
Angular molecule.

Note the change in *molecular* geometry that occurs when a proton (H⁺ with no electrons) is bonded to an NH_3 or an H_2O molecule through a coordinate covalent bond to form NH_4^+ or H_3O^+; in each, P still equals 4, but the number of *bond* pairs has increased.

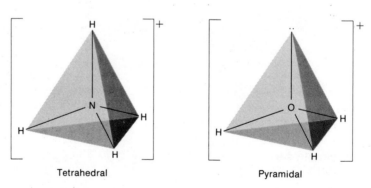

Tetrahedral   Pyramidal

FIGURE 9-8.
Tetrahedral molecule (left) and pyramidal molecule (right).

A common type of molecule is PCl_5, in which $P = 5$, because $V_M = 5$ and $V_L = 5$. Since all of the pairs are bond pairs, it follows that the molecular geometry will be the same as the electron-pair geometry, a triangular bipyramid.

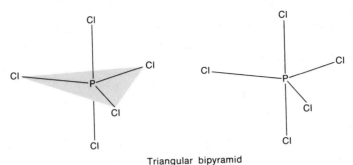

Triangular bipyramid

FIGURE 9-9.
Triangular-bipyramidal molecule.

Note that all of the P—Cl bond distances are the same, but that the Cl—Cl distances (*not* bonds) are greater between any two Cl atoms in the plane than between an apical Cl and a Cl atom in the plane. Up to this point, it has made no difference to molecular shape which pair of electrons in a given set of electron-pairs was used as a lone pair. But when the electron-pair positions are not equivalent, then the lone-pair positions are all-important in determining the molecular shape.

A bond pair of electrons has a less effective negative charge than a lone pair, because the former's charge is more reduced by its lying between two positive nuclei than the latter's charge is by its being attached to only one nucleus. As a result, we would expect a lone-pair lone-pair repulsion to be greater than a lone-pair bond-pair repulsion, and this in turn to be greater than a bond-pair bond-pair repulsion; that is,

$$LP\text{–}LP > LP\text{–}BP > BP\text{–}BP.$$

To simplify the application of these differences in repulsion to the determination of molecular shapes, we can, for practical purposes, ignore the repulsions between pairs of electrons that lie at angles greater than 90° to each other in comparison with those that lie at angles of less than 90°.

Let us apply these principles to the molecule $TeCl_4$, in which $V_M = 6$, $V_L = 4$, and $P = 5$; BP = 4 and LP = 1. Two molecular shapes are possible. The one that actually exists is the one with the least repulsion between the electron pairs.

> Model (a) has 3 LP–BP repulsions at 90° and
> 3 BP–BP repulsions at 90°.
> Model (b) has 2 LP–BP repulsions at 90° and
> 4 BP–BP repulsions at 90°.

In both models all other repulsions are at angles greater than 90° and can be ignored. Since LP–BP repulsion is greater than BP–BP repulsion, it follows that (b) has less electron-pair repulsion than (a), and $TeCl_4$ has the shape of a seesaw, as observed.

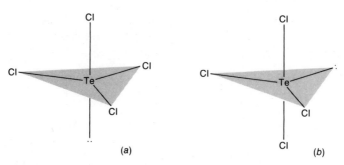

FIGURE 9-10.
Two possible shapes of $TeCl_4$.

In ClF_3, where $V_M = 7$, $V_L = 3$, and $P = 5$, we have to make a more complicated decision involving three possible structures. Application of the same principles as above shows that:

> Model (a) has 6 LP–BP repulsions at 90°;
> Model (b) has 1 LP–LP repulsion at 90°,
> 3 LP–BP repulsions at 90°, and
> 2 BP–BP repulsions at 90°.
> Model (c) has 4 LP–BP repulsions at 90° and
> 2 BP–BP repulsions at 90°.

Models (a) and (b) both have more electron-pair repulsion than model (c); hence we would choose the structure of ClF_3 to be a T-shaped molecule as in (c), which in fact it is.

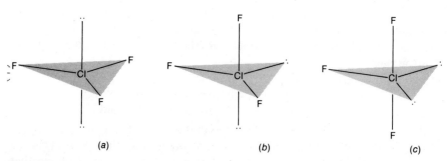

FIGURE 9-11.
Three possible shapes of ClF_3.

When $P = 6$, as in SF_6, where $V_M = 6$ and $V_L = 6$, the structure is simple, because all the pairs are bond pairs and the molecular geometry is the same as the electron-pair geometry; it is octahedral.

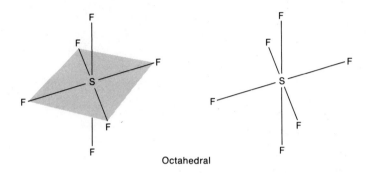

Octahedral

FIGURE 9-12.
Octahedral molecule.

The problem is not significantly more difficult for a molecule such as IF_5, where P also equals 6 because $V_M = 7$ and $V_L = 5$. With an electron-pair geometry that is octahedral, all positions are equivalent and it does not matter which position is occupied by the one lone pair. No matter how you look at it, IF_5 is a square-based pyramid with the I atom centered in the base.

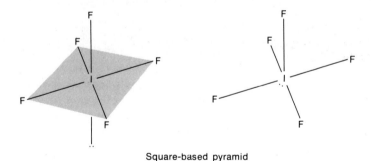

Square-based pyramid

FIGURE 9-13.
Square-based pyramidal molecule

Again, because all octahedral positions are equivalent, the shape of the ion ICl_4^- can easily be determined. Here, $P = 6$ because $V_M = 7$ and $V_L = 5$ (3 from 3 Cl atoms and 2 from a Cl^-, the latter giving the ICl_4^- ion a charge of -1). The two possible structures are shown in Figure 9-14.

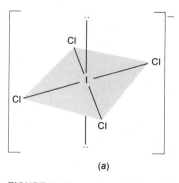 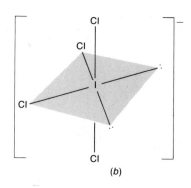

(a)   (b)

FIGURE 9-14.
Two possible shapes of ICl_4^-.

In evaluating the minimum repulsion, we find that:

Model (a) has 8 LP–BP repulsions at 90° and
4 BP–BP repulsions at 90°;
Model (b) has 1 LP–LP repulsion   at 90°,
6 LP–BP repulsions at 90°, and
5 BP–BP repulsions at 90°.

Offhand the minimum repulsion in the two structures might seem similar, but the decrease in repulsion by having the very strong LP–LP repulsion go from 90° as in (b) to 180° as in (a) is so great that it more than compensates for the smaller simultaneous increase in repulsion caused by having one BP–BP repulsion at 90° as in (b) go to a LP–BP repulsion also at 90° as in (a). As a consequence ICl_4^- is a square coplanar molecule as shown in (a).

A good example of a pentagonal-bipyramidal molecule is IF_7, because $P = 7$, $V_M = 7$, $V_L = 7$. All the pairs are bond pairs.

FIGURE 9-15.
Pentagonal-bipyramidal molecule.

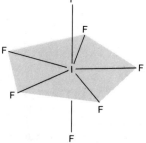

Pentagonal bipyramid

SbBr$_6^=$ has an especially irregular shape, which can be deduced from the fact that $P = 7$, caused by $V_M = 5$ and $V_L = 9$. The value nine arises because three of the Br ligands contribute three electrons as atoms and the other three contribute six electrons as bromide ions. Of the two possible shapes in Figure 9-16, the irregular octahedron (b) is more likely, as may be deduced from the following considerations.

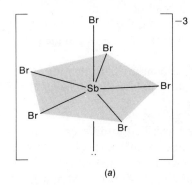 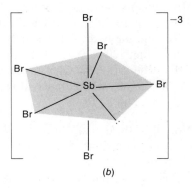

FIGURE 9-16.
Two possible shapes of SbBr$_6^=$.

> Model (a) has 5 LP–BP repulsions at 90°,
> 5 BP–BP repulsions at 72°, and
> 5 BP–BP repulsions at 90°.
> Model (b) has 2 LP–BP repulsions at 72°,
> 2 LP–BP repulsions at 90°,
> 3 BP–BP repulsions at 72°, and
> 8 BP–BP repulsions at 90°.

The difference between the two models comes down to this:

> Model (a) has 3 LP–BP repulsions at 90° and
> 2 BP–BP repulsions at 72°;
> Model (b) has 2 LP–BP repulsions at 72° and
> 3 BP–BP repulsions at 90°.

Qualitatively, the difference is that model (a) has a partial LP–BP repulsion at 90°, and model (b) has a partial BP–BP repulsion at 90°. Since the LP–BP repulsion is greater, we choose model (b).

Another consequence of the unequal repulsion between lone pairs and bond pairs of electrons is the distortion of the molecule from the perfectly regular geometric shapes we have discussed. In CCl$_4$, the Cl—C—Cl angle is the tetrahedral value 109°28', as expected. In NH_3, however, the H—N—H angle is only 106°45', caused in part by the stronger LP–BP repulsion forcing the H atoms closer together against a weaker BP–BP repulsion. The same effect can be seen in:

$SnCl_2$, where the Cl—Sn—Cl angle is less than 120°;

H_2O, where the angle is only 104°27' (instead of 109°28');

$TeCl_4$, where the two apical Cl atoms are not quite linear with respect to Te, and the two Cl atoms in the plane have a Cl—Te—Cl angle of a little less than 120°;

ClF_3, where the T-shaped molecule has a bent cross to the T;

IF_5 molecule, where the I atom is slightly below the plane of the 4 F atoms; and so on.

Not all of these distortion effects can be ascribed to differences in LP–BP and BP–BP repulsions. Some distortion is due to differences in electronegativity between M and the ligands. If the two have equal electronegativity, then, of course, the bond pairs are shared equally between M and L. If M is more electronegative than L, however, the bond pairs will be held more closely to M. But as they are drawn closer to M, the bond pairs are also drawn closer to each other, and, in an effort to reduce this added repulsion, they tend to widen the angle between them. The net result is a distortion. Compare the following sets of bond angles, which reflect both the differences in LP–BP and BP–BP repulsions and the differences in electronegativity between M and L; H—M—H angles are cited:

NH_3, 106°45'; PH_3, 93°50'; AsH_3, 91°35'; SbH_3, 91°30';

H_2O, 104°27'; H_2S, 92°20'.

If the ligands are more electronegative than M, then the bond pairs are drawn farther from M and away from each other, a situation that assists the lone pair in making the L—M—L angle smaller as it operates against this weaker BP–BP repulsion. For example, compare NH_3 (106°45') with NF_3 (102°9'), and H_2O (104°27') with OF_2 (101°30').

RESONANCE

A surprising thing is noted when we examine interatomic distances for certain ions and molecules, such as $CO_3^=$, NO_2^-, or C_6H_6. If we draw the usual electron-dot formulas for these, we get the patterns in Figure 9-17.

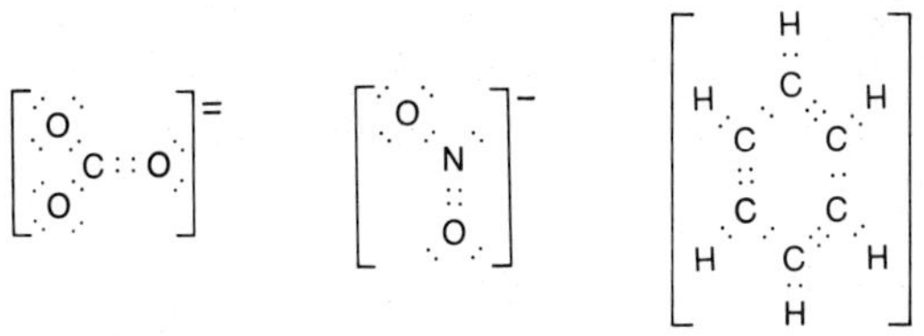

FIGURE 9-17.
Electron-dot formulas.

Using Table 9-1, we would say that: in $CO_3^=$ there is one C=O bond of length 1.22 Å and two C—O bonds of length 1.43 Å; in NO_2^- one N—O bond of length 1.36 Å and one N=O bond of length 1.15 Å; in C_6H_6, three C—C bonds of length 1.54 Å and three C=C bonds of length 1.34 Å. *In actual fact,* the bond lengths in any one of these three molecules are the same: for $CO_3^=$ it is 1.30 Å; for NO_2^- it is 1.24 Å; and for C_6H_6 it is 1.40 Å. They are the weighted averages of single and double bond lengths. It appears that in such molecules or ions, there is no real preference for single or double bonds to be located between any specific atoms. How should this situation be described? Pauling suggested that the actual state of the molecule is a *resonance hybrid* of all the separate forms that can be written in the classical way. The principal resonance forms for our three examples would be as shown in Figure 9-18.

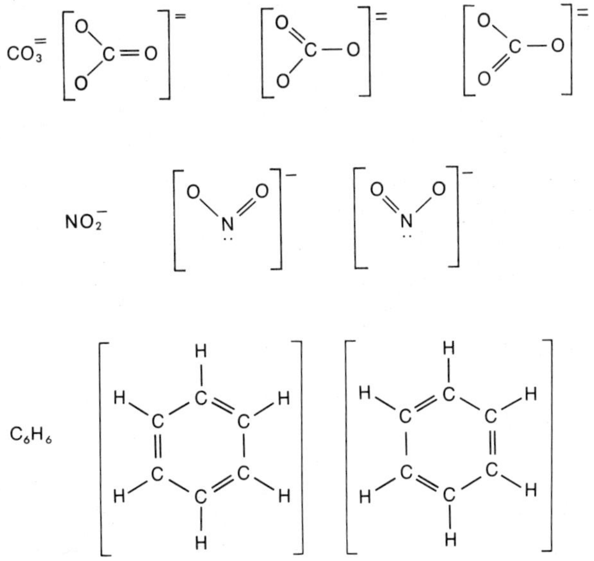

FIGURE 9-18.
Resonance forms.

It must be emphasized that none of these resonance forms actually exists, but the superposition of all of them for a given molecule could serve as a better representation, and show that each bond possesses both single-bond and double-bond characteristics.

A MOLECULAR ORBITAL DESCRIPTION

So far we have looked at molecular orbitals in a simplified way as electron-pairs that try to seek locations of minimum potential energy. When double bonds are also involved, it is profitable to add an additional concept that carries over from our discussion of atomic orbitals. A molecular orbital might be considered as being formed from the "end-overlap" of two half-filled atomic orbitals (one from each atom), or of the "end-overlap" of one empty atomic orbital on one atom with a filled (lone-pair) orbital on the other. In HCl the molecular orbital could be looked at as the end-overlap of the half-filled s orbital of H with the one half-filled p orbital of Cl.

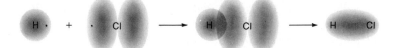

FIGURE 9-19.
HCl molecular orbital.

In ICl, the molecular orbital is the end-overlap of two half-filled p orbitals.

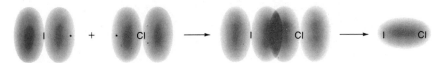

FIGURE 9-20.
ICl molecular orbital.

In making an NH_4^+ ion from H^+ and NH_3, the new bond is the molecular orbital resulting from the end-overlap of the empty s orbital of H and the filled (lone-pair) molecular orbital of NH_3. These end-overlap molecular orbitals are called σ (sigma) bonds.

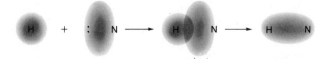

FIGURE 9-21.
NH_4^+ molecular orbital.

   Double and triple bonds are usually formed only if both atoms have one or two half-filled p atomic orbitals. If it is imagined that under proper reaction conditions one (or both) of these half-filled atomic p orbitals retains its *atomic* integrity (and electron), leaving the other valence electrons to form the usual molecular orbitals, then one of the following possibilities may occur.

1. With *one* half-filled p orbital on each atom, a double bond may form that consists of:
   (a) one ordinary filled σ molecular orbital between the two atoms formed by "end-overlap";
   (b) one π molecular orbital between the two atoms formed by "side overlap" of the two half-filled p orbitals on each atom. The electrons in a π bond are delocalized. They do not lie between the atoms being bonded, and they are loosely associated with both M and L in an orbital that places one electron in a cloud on one side of the M—L σ bond and the other diametrically opposite, as shown in Figure 9-22.

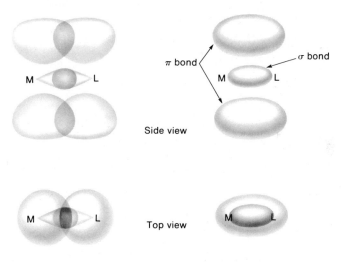

FIGURE 9-22.
Double bond.

2. With *two* half-filled p-orbitals on each atom, a triple bond may form that consists of:
   (a) one σ bond;
   (b) two π bonds at right angles to each other, because the atomic p orbitals of each atom are at right angles to each other.

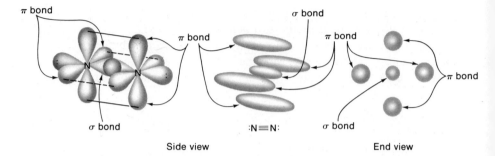

FIGURE 9-23.
Triple bond.

There are several molecular phenomena that can be interpreted in terms of σ and π bonds, but we shall limit ourselves now to the topic at hand — molecular shape. Since π bonds are always oriented parallel to the σ bonds with which they are associated, it follows that the molecular shape will be determined solely by the number of σ bonds that the central atom has. This, in turn, means that *to calculate P, the number of pairs of valence electrons associated with the central atom, all double and triple bonds must be counted as only one pair of electrons corresponding to the σ bond pair.*

Applying this new principle to $CO_3^=$, NO_2^-, and C_6H_6, we see that $CO_3^=$ is triangular coplanar, because C has $P = 3$; NO_2^- is angular, because N has $P = 3$; and C_6H_6 is entirely coplanar, because each C (acting as a central atom) has $P = 3$.

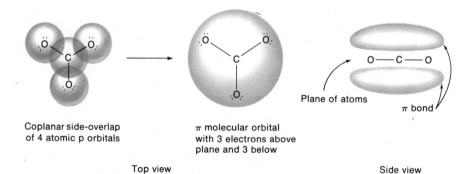

Coplanar side-overlap
of 4 atomic p orbitals

π molecular orbital
with 3 electrons above
plane and 3 below

Top view
(bottom view the same)

Side view

FIGURE 9-24.
π molecular orbital of $CO_3^=$.

Looking a little more closely, we can imagine for the $CO_3^=$ ion that the half-filled p orbital of carbon extends perpendicularly above and below the plane of the paper at C. Also extending perpendicularly above and below the plane of the paper is a p orbital at each oxygen atom, two of them filled and one half-filled — but who is to say which? It makes just as much sense to say that there is side-overlap of *four* p-orbitals (one from each O and one from C) with three electrons in the delocalized π electron cloud above the plane of the paper, and three electrons in the delocalized π electron cloud below the paper.

For the NO_2^- ion we can imagine that each O is bonded to the N with a σ bond and, as a result of side-overlap between the p orbitals that rise perpendicularly out of the plane of the paper at each atom, a double banana-shaped π molecular orbital is formed that contains two electrons in the delocalized cloud above the paper and two in the cloud below.

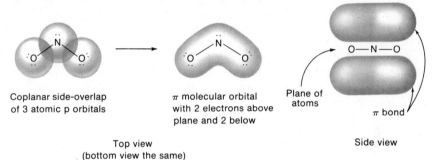

Coplanar side-overlap π molecular orbital   Plane of
of 3 atomic p orbitals   with 2 electrons above   atoms
                               plane and 2 below π bond

Top view   Side view
(bottom view the same)

FIGURE 9-25.
π molecular orbital of NO_2^-.

Concerning C_6H_6 it is again impossible to say between which three pairs of C atoms a π bond is formed. Instead, it is simpler to think that there is side-overlap between *six* p orbitals (one from each C) with three electrons in the delocalized π electron cloud above the plane of the molecule and three electrons in the delocalized π electron cloud below the plane.

As the radii of atoms get larger, the possibility of side-overlap becomes less, and the occurrence of double and triple bonds decreases, and may not even exist for the really large atoms.

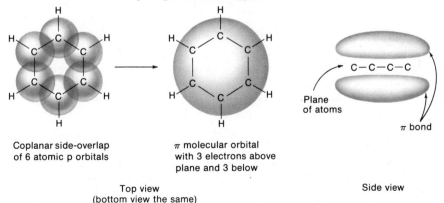

Coplanar side-overlap π molecular orbital
of 6 atomic p orbitals   with 3 electrons above
                               plane and 3 below

Top view   Side view
(bottom view the same)

FIGURE 9-26.
π molecular orbital of C_6H_6.

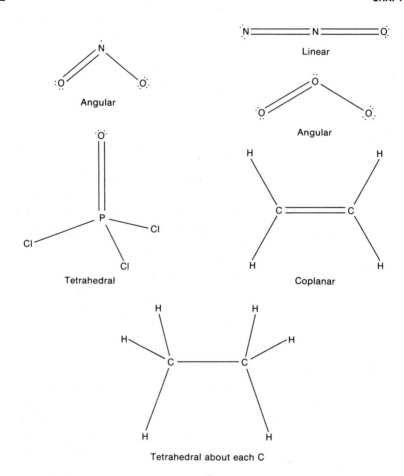

FIGURE 9-27.
Examples of molecules with various bond characteristics.

In connection with the retention by an atom of one or two p electrons in atomic orbitals for use in π bonding, we should mention that one of the standard ways to describe an atom's molecular orbitals is in terms of the atomic orbitals from which they can be considered to be "made" mathematically. With this approach, molecular orbitals are said to be mathematical *hybrids* of the atomic orbitals. On this basis, four tetrahedral molecular orbitals are the hybrids of one s and three p atomic orbitals, and they are often referred to as sp³ orbitals. When a double bond is formed which uses one of the p orbitals in π bonding, the remaining s and two p orbitals can be hybridized to give three molecular orbitals that, as we have seen, always have triangular-coplanar geometry. These are often called sp² or trigonal molecular orbitals. Finally, when two p atomic orbitals are used in forming a triple bond, the remaining s and p orbitals can be hybridized to give two molecular orbitals that result in linear geometry; they are also called sp or digonal molecular orbitals. Many other types of molecular orbitals may be "made" by hybridizing s, p, and d atomic orbitals.

Some additional examples of double-bonded molecules are given in Figure 9-27.

It appears that for shape-prediction purposes the single unpaired electron occasionally associated with a central atom should be placed in a lone-pair molecular orbital (as in NO_2 in Figure 9-27), where it will exert a repulsive effect but much less so than a whole lone pair. You would expect the O—N—O angle (134° by experiment) in NO_2 to be greater than the O—N—O angle (115° by experiment) in NO_2^-, for example. You can also see why two NO_2 molecules will so readily react with each other to form N_2O_4; the two half-filled lone-pair orbitals will form a σ bond between the two N atoms.

You should also realize that the NO_2 and O_3 molecules involve resonance, for there is no reason to believe that one bond in each should be single and the other double; both bond lengths in each molecule will be equal.

Finally, we might comment on one other interesting property that is associated with a double or triple bond. The π bond keeps the two halves of a double bond from freely rotating about the σ bond axis that joins them. This property is of the utmost importance in explaining the shape and properties of many organic compounds. In C_2H_4, the two $=CH_2$ groups are unable to rotate freely with respect to each other. In C_2H_6, the two —CH_3 groups tend to arrange themselves to give an end-on view as in Figure 9-28, with each H atom at maximum possible distance from the others. The activation energy for rotation about the C—C bond is only 3 kcal, so that rotation occurs relatively easily.

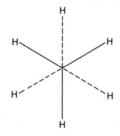

FIGURE 9-28.
End-on view of C_2H_6.

COMPLEX IONS

The complex ions that are studied in Chapter 22 consist of metal ions acting as the "central atom" to which several ligands are attached. The metal ion is very frequently a transition metal ion. In these complex ions the pairs of electrons by which the ligands are attached are *all* furnished by the ligands, and there are *no lone pairs*. This leads to a great simplification in predicting their shapes. The number of electron pairs (P) around the transition metal ion is equal to the number of ligands attached; if this is 4 the ionic shape is tetrahedral, or if it is 6 it is octahedral, and so on. In spite of the fact that these statements represent great oversimplification (especially of the effect of the underlying d orbitals of the transition metal ions) and that occasionally a complex with 4 ligands actually has square-coplanar configuration, the general predictive utility is great.

PROBLEM:
Sketch the shape of the ions $ZnCl_4^=$ and $Co(NH_3)_6^{+++}$.

SOLUTION:
You must first recognize that Zn^{++} and Co^{+++} are transition metal ions and that $ZnCl_4^=$ and $Co(NH_3)_6^{+++}$ are complex ions. Once this is established you note that Zn^{++} has 4 ligands and therefore 4 electron pairs with tetrahedral geometry, and that Co^{+++} has 6 ligands and therefore 6 electron pairs with octahedral geometry. With Zn^{++} each Cl^- contributes one electron pair for binding,

and with Co^{+++} each H — $\overset{..}{\underset{H}{N}}$ — H contributes its lone pair for binding.

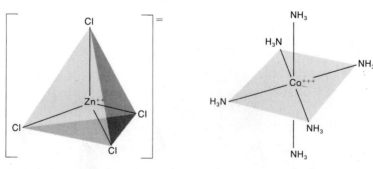

FIGURE 9-29
Molecular shapes of $ZnCl_4^=$ and $Co(NH_3)_6^{+++}$.

PROBLEMS A   Answers on page 340

1. Sketch each of the following molecules or ions, and give approximate values for bond angles, bond distances, and bond energies for those for which sufficient data are given in Table 9-1:

(a) CO_2	(f) NO_2	(k) NO_2Cl	(p) $S_2O_3^=$	(u) ICl_3
(b) $SiO_4^{\equiv}$	(g) NO_2^+	(l) $Mo(CN)_8^{\equiv}$	(q) $SbF_7^=$	(v) $AsS_3^=$
(c) $SO_3^=$	(h) NOF	(m) BrF_4^-	(r) N_2O_4	(w) SO_2Cl_2
(d) XeF_2	(i) $SeCl_4$	(n) $AlF_6^{\equiv}$	(s) CH_2O	(x) C_6H_6
(e) NO_2^-	(j) $SiF_6^=$	(o) OCN^-	(t) BrF_5	(y) C_2H_3Cl
		(z) $(SiO_3^=)_x$		

PROBLEMS B   No answers given

2. Sketch each of the following molecules or ions, and give approximate values for bond angles, bond distances and bond energies for those for which sufficient data are given in Table 9-1:

(a) F_2O	(f) ClO_2	(k) $PO_2F_2^-$	(p) BrF_3	(u) SO_3
(b) $SO_4^=$	(g) $SnS_3^=$	(l) XeF_4	(q) H_2O_2	(v) I_5^-
(c) IO_3^-	(h) $IF_2O_2^-$	(m) $TaF_8^{\equiv}$	(r) SCN^-	(w) C_2H_2
(d) NO_3^-	(i) $COCl_2$	(n) HCN	(s) I_2Br^-	(x) $ZrF_7^=$
(e) N_3^-	(j) POF_3	(o) H_5IO_6	(t) H_3BO_3	(y) Al_2Cl_6
		(z) $(PO_3^-)_x$		

Stoichiometry I.
Calculations Based on Formulas

A chemical formula tells the number and the kind of atoms that make up a molecule of a compound. Since each atom is an entity with a characteristic mass, a formula also provides a means for computing the relative weights of each kind of atom in a compound. Calculations based on the masses and and numbers of atoms in a compound, or the numbers and masses of molecules participating in a reaction, are designated *stoichiometric calculations*. We shall consider in this chapter the stoichiometry of chemical formulas, and in later chapters, the stoichiometric relations involved in reactions and in solutions.

Atomic and Molecular Weights

The basis for all stoichiometric calculations is the weights of the atoms involved. Weights of individual atoms can be determined experimentally and it is possible to assign absolute values for all atomic masses. Such numbers are, however, very small if expressed in grams. A single hydrogen atom, for example, weighs only 1.67×10^{-24} g. Consequently, it is more convenient to select relative values for atomic weights. Those in use since 1961 are based upon a value of 12.0000 for the most common isotope of carbon.

Once the relative atomic weights are established, it is simple to compute a molecular weight on the same relative basis. One need only add the weights of the atoms that compose a molecule. Thus the molecular weight of H_2SO_4 is $(2 \times 1.0) + (32.1) + (4 \times 16.0) = 98.1$.

The Mole

Since atomic weights are relative numbers, it is convenient to let the symbol of an element or the formula of a compound represent the number of grams indicated by the atomic or molecular weight. This quantity is known as a gram-atomic weight (g-atom) when applied to atoms, and as a gram-molecular weight (GMW) when applied to compounds; it is the weight in grams of 6.02×10^{23} atoms or molecules. The number 6.02×10^{23} is known as "Avogadro's Number." Thus a GMW of H_2SO_4 is 98.1 g, made up of 2.0 g H, 64.0 g O, and 32.1 g S. We customarily abbreviate the term gram-molecular weight to the single word, *mole* (the term is also often used for a gram atomic weight).

There is no special reason, other than convention, why the gram is the weight unit for the mole; we could just as well use any other weight unit. In fact, in many calculations it is convenient to express molecular or atomic weight in units other than grams, such as pounds, tons, and so on. Whenever we do this, the term mole is preceded by the weight unit in use. Thus a ton mole of H_2SO_4 is 98.1 tons. It is often convenient, in dealing with small weights of materials, to express the molecular weight in milligrams, and we denote this quantity as the millimole (m mol). A millimole therefore represents the same number of milligrams as a mole does for grams; a millimole of H_2SO_4 is 98.1 mg. Similarly, a micromole (μ mole) is the same number expressed as micrograms.

Determination of Atomic Weights

The relative atomic weights we use for the elements are quantities that have been found by experimental measurements. Like all such measurements, they are subject to experimental errors and some atomic weights are more precisely known than others. The International Atomic Weights given in your textbook are the values thought most reliable at the time a committee reviewed them, and their reliability is indicated by the number of significant figures. An abbreviated list of atomic weights is given inside the front cover of this book. These are rounded off to the first decimal figure, since the calculations of this course do not require high accuracy.

To illustrate how the values of atomic weights have been established, let us take a hypothetical case. We know that the formula of cupric oxide is CuO, that is, that 1 atom of copper combines with 1 atom of oxygen, and we know the atomic weight of oxygen to be 16. We know also that if copper is heated in oxygen it is converted to CuO. To find the atomic weight of copper, we therefore need only to determine the weight of copper that combines with 1 gram atom or 16 g of oxygen. This we can do by a simple experiment. We heat a weighed piece of copper in oxygen, then weigh the CuO formed. This gives the following data:

$$\text{weight of copper} = 0.32 \text{ g}$$
$$\text{weight of oxide formed} = 0.40 \text{ g}$$

The weight of oxygen combined with 0.32 g Cu is: $0.40 \text{ g} - 0.32 \text{ g} = 0.08 \text{ g}$. We can find the grams of copper that combined with 1 g of oxygen by dividing the total weight of copper by the total weight of oxygen with which it is combined:

$$\frac{0.32 \text{ g Cu}}{0.08 \text{ g O}} = 4.0 \frac{\text{g Cu}}{\text{g O}} \text{ or } 4.0 \text{ g Cu per gram O}$$

The weight of copper combined with 16 g (that is, 1 g-atom) of oxygen is 16 times this quantity.

$$4.0 \frac{\text{g Cu}}{\text{g O}} \times 16.0 \text{ g O} = 64 \text{ g Cu}$$

Methods for determining the atomic weight from the measured weight that combines with 16.0 g of oxygen when the formula of the compound is not previously known are discussed in Chapter 23, in connection with Dulong and Petit's rule for specific heats of elements.

Percentage Composition

A chemical formula may be used to compute the percentage composition of a compound; that is, the per cent by weight of each type of atom in the compound.

PROBLEM:
Calculate the percentages of oxygen and hydrogen in water (H_2O).

SOLUTION:
The formula shows that 1 mole of H_2O contains 2 g-atom of hydrogen and 1 g-atom of oxygen. The weight of 1 mole is 18 g. The weight of hydrogen in 1 mole is $2 \times 1 \text{ g} = 2 \text{ g}$. The percentage of hydrogen is

$$\%H = \frac{2.0 \text{ g}}{18.0 \text{ g}} \times 100 = 11.1\%$$

Similarly, since 1 mole H_2O contains 16 g of oxygen,

$$\%O = \frac{16.0 \text{ g}}{18.0 \text{ g}} \times 100 = 88.9\%$$

PROBLEM:
Compute the percentages of K, Fe, C, N, and H_2O in $K_4Fe(CN)_6 \cdot 3H_2O$ crystals.

SOLUTION:
The dot in the formula indicates that 3 moles of H_2O are combined with 1 mole of $K_4Fe(CN)_6$ in the crystalline compound. The weight of 1 mole is $(4 \times 39.1) + (55.8) + (6 \times 12.0) + (6 \times 14.0) + (3 \times 18.0) = 422.2$ g/mole.

$$\%K = \frac{4 \times 39.1 \text{ g}}{422.2 \text{ g}} \times 100 = 37.0\%$$

$$\%Fe = \frac{55.8 \text{ g}}{422.2 \text{ g}} \times 100 = 13.2\%$$

$$\%C = \frac{6 \times 12.0 \text{ g}}{422.2 \text{ g}} \times 100 = 17.1\%$$

$$\%N = \frac{6 \times 14.0 \text{ g}}{422.2 \text{ g}} \times 100 = 19.9\%$$

$$\%H_2O = \frac{3 \times 18.0 \text{ g}}{422.2 \text{ g}} \times 100 = 12.8\%$$

Calculation of Formulas from Chemical Analysis

When a new chemical compound is prepared, we do not know its formula. To establish it, we find by experiment the weights of the various atoms in the compound and from these weights compute the relative number of each kind of atom in the molecule. The formula so computed is the simplest formula, not necessarily the true one. It is therefore called the *empirical* formula. For example, we may find the empirical formula for benzene to be CH whereas the true formula is C_6H_6. To get the true formula from the empirical formula, we must also be able to determine the molecular weight. (This is accomplished by methods that will be discussed later.)

To illustrate the use of experiments in establishing an empirical formula, we will assume a simple experiment. A sample of copper weighing 0.160 g is heated in oxygen and converted to 0.200 g CuO. The gain in weight is the weight of oxygen taken up: 0.160 g Cu combines with 0.040 g O. If we divide each of these values by the atomic weight, we obtain the relative number of gram atoms of each element in the compound:

$$\text{g-atoms of Cu} = \frac{0.160 \text{ g Cu}}{63.6 \dfrac{\text{g Cu}}{\text{g-atom Cu}}} = 0.00251$$

$$\text{g-atoms of O} = \frac{0.040 \text{ g O}}{16.0 \dfrac{\text{g O}}{\text{g-atom O}}} = 0.00250$$

We might then write the formula as

$$Cu_{0.00251}O_{0.00250}$$

but it is better to express it in whole numbers of atoms. If we divide each of the figures by 0.00250, we get

$$Cu_1O_1 \text{ or } CuO$$

as the formula. (The slight deviation between 0.00251 and 0.00250 is due to experimental errors. The two figures are so nearly the same that we know the ratio is 1:1.)

PROBLEM:
A compound contains 90.6 per cent Pb and 9.4 per cent O. Find the empirical formula.

SOLUTION:
Usually the results of analysis are given, not in the amounts actually weighed, as in the preceding problem, but in percentages of the constituents. This permits comparison of results from different experiments. You recall that "per cent" means "per 100." We can say, then, that if we have 100 g of the compound, 90.6 g is Pb and 9.4 g is O:

$$\text{g-atoms of Pb} = \frac{90.6 \text{ g Pb}}{207.2 \dfrac{\text{g Pb}}{\text{g-atom Pb}}} = 0.437$$

$$\text{g-atoms of O} = \frac{9.4 \text{ g O}}{16.0 \dfrac{\text{g O}}{\text{g-atom O}}} = 0.583$$

At this stage a common student error is to say, "The numbers 0.437 and 0.583 are approximately the same, so we assume the formula to be Pb_1O_1 or PbO." This is not correct. If the analysis were so poor that results which should be identical were this far apart, we should find another analyst! To find the relative numbers of gram atoms, divide through by the smaller number:

$$Pb_{\frac{0.437}{0.437}}O_{\frac{0.583}{0.437}} = Pb_1O_{1.33}$$

Multiplying by 3, we convert this ratio into integers:

$$3(Pb_1O_{1.33}) = Pb_3O_4$$

PROBLEM:
Many crystalline compounds contain water of hydration that is driven off when the compound is heated. The loss of weight in heating can be used to determine the formula. For example, a hydrate of barium chloride, $BaCl_2 \cdot xH_2O$, weighing 1.222 g was heated until all the combined water was expelled. The dry powder remaining weighed 1.042 g. Compute the formula for the hydrate.

SOLUTION:

The residue after heating is anhydrous $BaCl_2$, and the loss of weight, $1.222 \text{ g} - 1.042 \text{ g} = 0.180 \text{ g}$, is H_2O. These figures give:

$$\text{moles of } BaCl_2 = \frac{1.042 \text{ g } BaCl_2}{208.4 \dfrac{\text{g } BaCl_2}{\text{mole of } BaCl_2}} = 0.005$$

$$\text{moles of } H_2O = \frac{0.180 \text{ g } H_2O}{18 \dfrac{\text{g } H_2O}{\text{mole of } H_2O}} = 0.010$$

Moles of H_2O per mole of $BaCl_2$ is obtained by dividing the moles of H_2O by the moles of $BaCl_2$ to which H_2O is attached; that is,

$$\frac{.010 \text{ mole of } H_2O}{.005 \text{ mole of } BaCl_2} = 2 \frac{\text{moles of water}}{\text{mole of } BaCl_2}$$

and the formula of the hydrate is $BaCl_2 \cdot 2H_2O$.

PROBLEMS* A   Answers on page 342

1. Find the percentage composition of (the percentage of each element in) each of the following compounds: (a) N_2O; (b) NO; (c) NO_2; (d) Na_2SO_4; (e) $Na_2S_2O_3$; (f) $Na_2SO_4 \cdot 10H_2O$; (g) $Na_2S_2O_3 \cdot 5H_2O$; (h) $Ca(CN)_2$; (i) $(NH_4)_2CO_3$; (j) $UO_2(NO_3)_2 \cdot 6H_2O$; (k) penicillin, $C_{16}H_{26}O_4N_2S$.

2. What is the weight of 1 mole of each compound in Problem 1?

3. How many moles are in 1 lb of each compound in Problem 1?

4. From the following analytical results, determine the empirical formulas for the compounds:
   (a) 77.7% Fe, 22.3% O
   (b) 70.0% Fe, 30.0% O
   (c) 72.4% Fe, 27.6% O
   (d) 40.2% K, 26.9% Cr, 32.9% O
   (e) 26.6% K, 35.4% Cr, 38.0% O
   (f) 92.4% C, 7.6% H
   (g) 75.0% C, 25.0% H
   (h) 21.8% Mg, 27.9% P, 50.3% O
   (i) 66.8% Ag, 15.9% V, 17.3% O
   (j) 52.8% Sn, 12.4% Fe, 16.0% C, 18.8% N

5. Weighed samples of the following hydrates were heated to drive off the water, and then the cooled residues were weighed. From the data given, find the formulas of the hydrates.
   (a) 0.695 g of $CuSO_4 \cdot xH_2O$ gave a residue of 0.445 g

---

*Problems concerning the determination of approximate atomic weights by the rule of Dulong and Petit may be found in Chapter 23.

(b) 0.573 g of $Hg(NO_3)_2 \cdot xH_2O$ gave a residue of 0.558 g

(c) 1.205 g of $Pb(C_2H_3O_2)_2 \cdot xH_2O$ gave a residue of 1.032 g

(d) 0.809 g of $CoCl_2 \cdot xH_2O$ gave a residue of 0.442 g

(e) 2.515 g of $CaSO_4 \cdot xH_2O$ gave a residue of 1.990 g

6. Weighed samples of the following metals were completely converted to other compounds by heating them in the presence of other elements, and then were reweighed to find the increase in weight. The excess of the non-metal is easily removed in each case. From the data given, find the formulas of the compounds formed.

(a) 0.527 g of Cu gave a 0.659-g residue with S

(b) 0.273 g of Mg gave a 0.378-g residue with N_2

(c) 0.406 g of Li gave a 0.465-g residue with H_2

(d) 0.875 g of Al gave a 4.325-g residue with Cl_2

(e) 0.219 g of La gave a 0.256-g residue with O_2

7. It is found experimentally that when a metal M is heated in chlorine gas, 0.54 g of M gives 2.67 g of metal chloride. The formula of the chloride is not known. (a) Compute possible values of the atomic weight of M, for each of the following formulas: MCl, MCl_2, MCl_3, MCl_4. (b) It is found by other methods that the atomic weight of M is about 27. Which of the above formulas is the correct one?

8. A metal forms two different chlorides. Analysis shows one to be 54.7% Cl and the other to be 64.4% Cl. What are the possible values of the atomic weight of the metal? (Assume that the atomic weight of Cl is 35.5.)

9. In Wöhler's *Grundriss der Chemie,* published in 1823, the gram-atomic weight of oxygen is given as 100. On this basis calculate the gram-molecular weight of NH_4Cl.

10. An organic compound containing C, H, O, and S was subjected to two analytical procedures. When a 9.33-mg sample was burned, it gave 19.50 mg of CO_2 and 3.99 mg of H_2O. A separate 11.05-mg sample was fused with Na_2O_2, and the resulting sulfate was precipitated as $BaSO_4$, which, when washed and dried, weighed 20.4 mg. The amount of oxygen in the original sample is obtained by difference. Determine the empirical formula of this compound.

PROBLEMS B   No answers given

11. Find the percentage composition of (the percentage of each element in) each of the following compounds: (a) NH_3; (b) N_2H_4; (c) HN_3; (d) $Zn(NO_2)_2$; (e) $Zn(NO_3)_2$; (f) $Zn(NO_3)_2 \cdot 6H_2O$; (g) $(NH_4)_2CrO_4$; (h) $CaCN_2$; (i) PtP_2O_7; (j) $BiONO_3 \cdot H_2O$; (k) streptomycin, $C_{21}H_{39}O_{12}N_7$.

12. What is the weight of 1 mole of each compound in Problem 11?

13. How many moles are in 1 lb of each compound in Problem 11?

14. From the following analytical results determine the empirical formulas for the compounds:
    (a) 42.9% C, 57.1% O
    (b) 27.3% C, 72.7% O
    (c) 53.0% C, 47.0% O
    (d) 19.3% Na, 26.8% S, 53.9% O·
    (e) 29.1% Na, 40.5% S, 30.4% O
    (f) 32.4% Na, 22.6% S, 45.0% O
    (g) 79.3% Tl, 9.9% V, 10.8% O
    (h) 25.8% P, 26.7% S, 47.5% F
    (i) 19.2% P, 2.5% H, 78.3% I
    (j) 14.2% Ni, 61.3% I, 20.2% N, 4.3% H

15. Weighed samples of the following hydrates were heated to drive off the water, and then the cooled residues were weighed. From the data given, find the formulas of the hydrates.
    (a) 0.520 g of $NiSO_4 \cdot xH_2O$ gave a residue of 0.304 g
    (b) 0.895 g of $MnI_2 \cdot xH_2O$ gave a residue of 0.725 g
    (c) 0.654 g of $MgSO_4 \cdot xH_2O$ gave a residue of 0.320 g·
    (d) 1.216 g of $CdSO_4 \cdot xH_2O$ gave a residue of 0.990 g
    (e) 0.783 g of $KAl(SO_4)_2 \cdot xH_2O$ gave a residue of 0.427 g

16. Weighed samples of the following metals were completely converted to compounds by heating them in the presence of the specified elements, and then were reweighed to find the increase in weight. The excess of the non-metal is easily removed in each case. From the data given find the formulas of the compounds formed.
    (a) 0.753 g of Ca gave a 0.792-g residue with H_2
    (b) 0.631 g of Al gave a 1.750-g residue with S
    (c) 0.137 g of Pb gave a 0.243-g residue with Br_2
    (d) 0.211 g of U gave a 0.248-g residue with O_2
    (e) 0.367 g of Co gave a 0.463-g residue with P

17. If a container holds 5.0 g CCl_4 vapor, how many CCl_4 molecules are present?

18. A metal forms two different chlorides. Analysis shows one to be 40.3% metal and the other to be 47.4% metal. What are the possible values of the atomic weight of the metal?

19. A sample of an organic compound containing C, H, and O, which weighed 12.13 mg, gave 30.6 mg of CO_2 and 5.36 mg of H_2O on combustion. The amount of oxygen in the original sample is obtained by difference. Determine the empirical formula of this compound.

20. A 5.135-g sample of impure limestone ($CaCO_3$) yielded 2.050 g of CO_2 (which was absorbed in a soda-lime tube) when treated with an excess of acid. Assuming that limestone was the only component that would yield CO_2, calculate the percentage purity of the limestone sample.

Gases

In order for chemists to prepare and handle gases under a variety of conditions, they must understand the relationships between the weight, volume, temperature, and pressure of a gas sample. Measurement of the first three of these quantities is relatively straightforward, but because the measurement of pressure can present some complications, we will discuss it before we consider the interrelationship of all four variables.

Measurement of Gas Pressure

The most common laboratory instrument used to measure gas pressure is a manometer, a glass U-tube partially filled with a liquid, as shown in Figure 11-1. Mercury is the most commonly used liquid, because it is fairly non-volatile, chemically inactive, and dense, does not dissolve gases, and does not wet (adhere to) glass. Since mercury does not wet glass, its meniscus will curve upward instead of down, and its position is recorded as that of the horizontal plane at the top of the meniscus.

Figure 11-1 illustrates the measurement of gas pressure by a manometer. In (*a*) both sides of the mercury column are at the same pressure, and the mercury level is the same in both tubes. In (*b*) the righthand tube has been evacuated and, as a result, the mercury has risen until the weight of the mercury column of height *h* exactly balances the gas pressure in the lefthand tube. When one side of the manometer is open to the atmosphere and the other side is evacuated, the instrument functions as a barometer; this could be the arrangement in Figure 11-1(b), with stopcock 2 opened to the atmosphere. At sea level and a temperature of 0°C, the average barometric pres-

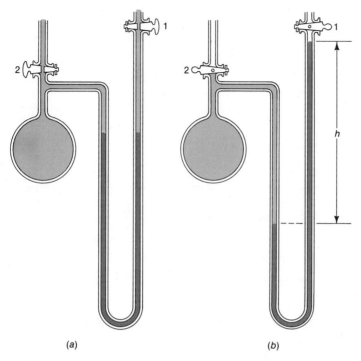

FIGURE 11-1.
Mercury manometer: (a) atmospheric pressure in both tubes,
(b) gas pressure in flask and vacuum in righthand tube.

sure corresponds to a 760.0 mm column of mercury. Such a pressure is designated as 1 atmosphere (atm). It is necessary to specify the temperature when using a mercury manometer because mercury expands with an increase in temperature. Because of the mercury expansion, a longer column of mercury will be needed at a higher temperature in order to provide the mass of mercury required to balance a given gas pressure.

Various units are used to express gas pressures. Since pressure is defined as force per unit area, one should employ units of force in measurements and calculations. In practice, however, it is convenient to use units more directly related to the measurements (such as the height of the mercury manometer column), or to use weight units per unit area (such as g/cm^2 or $lb/in.^2$). In scientific work the commonly used unit is the *torr*, defined as the pressure that will support a column of mercury exactly 1 mm in height at a temperature of 0°C. In order to correct mercury manometer readings (H, in mm) taken at normal laboratory temperatures (t, in °C) to values (P, in torr) that would have been observed had the mercury temperature been 0°C, one can use the simple formula

$$P = H[1 - \alpha t]$$

where $\alpha = 1.63 \times 10^{-4}$ deg^{-1} if the scale is made of brass, or $\alpha = 1.72 \times 10^{-4}$ deg^{-1} if the scale is etched on glass. Conversions from one set of pressure units to others are illustrated in the following problems.

PROBLEM:

Express a pressure of 1 atm in terms of the following units: (a) grams per cm^2; (b) pounds per in.^{2}; and (c) dynes per cm^2.

SOLUTION:

(a) We shall assume that the mercury column of the barometer has a diameter of 0.60 cm. At a temperature of 0°C, the density of mercury is 13.60 g/ml and the column height corresponding to 1 atm is 76.0 cm.

$$\text{cross-sectional area of column} = (\pi)(0.30 \text{ cm})^2$$

$$\text{volume of mercury} = (\pi)(0.30 \text{ cm})^2(76.0 \text{ cm})$$

$$\text{mass of mercury} = (\pi)(0.30 \text{ cm})^2(76.0 \text{ cm})(13.60 \text{ g/cm}^3)$$

$$1 \text{ atm} = \frac{\text{mass}}{\text{area}} = \frac{(\pi)(0.30 \text{ cm})^2(76.0 \text{ cm})(13.60 \text{ g/cm}^3)}{(\pi)(0.30 \text{ cm})^2}$$

$$= 1,033 \text{ g/cm}^2$$

Note that the result does *not* depend on the cross-sectional area of the mercury column, since this cancels in the computation.

(b) To convert to pounds per in.^{2}, we use the approximate factors of 454 g/lb and 2.54 cm/in. This gives:

$$\text{pounds per in.}^2 = \left(1,033 \frac{g}{cm^2}\right)\left(2.54 \frac{cm}{in.}\right)^2\left(\frac{1 \text{ lb}}{454 \text{ g}}\right)$$

$$1 \text{ atm} = 14.7 \text{ lb/in.}^2$$

(c) To express 1 atm in dynes/cm^2, multiply the mass in grams by the acceleration due to gravity:

$$\text{dynes per cm}^2 = \left(1,033 \frac{g}{cm^2}\right)\left(980.7 \frac{\text{dynes}}{g}\right)$$

$$1 \text{ atm} = 1.013 \times 10^6 \text{ dynes/cm}^2$$

The Ideal Gas Law

As is explained in every chemistry textbook, the volume of a gas depends on three factors; the absolute temperature (T), the pressure (P), and the number of moles (n). Mathematically,

$$V = f(T, P, n) \tag{11-1}$$

For a given quantity of gas it may be shown, by combining Boyle's and Charles' Laws, that

$$\frac{PV}{T} = \text{a constant, } k \tag{11-2}$$

or that

$$\frac{P_1 V_1}{T_1} = \frac{P_2 V_2}{T_2} = k \tag{11-3}$$

These equations predict the behavior of gases very well except at relatively low temperatures and high pressures. Just how low the temperature must be and how high the pressure must be before serious deviations from these equations are observed will vary from one gas to another. Under extreme conditions these equations cannot be used without correction for the volume occupied by the molecules themselves or for the attractive forces between neighboring molecules. For every gas, conditions exist under which the molecules condense to a liquid that occupies a fairly incompressible volume. A hypothetical gas, called an *ideal gas,* would obey equation 11-3 under all conditions, and possess zero volume at a temperature of absolute zero. Most gases obey the ideal gas law (equation 11-3) at normal temperatures and pressures.

Experiments have shown that one mole of any gas (behaving ideally) at the *standard conditions* of 1 atm (760 torr) and 273°K occupies a volume of 22.4 liters. These values make it possible to evaluate the constant k in equation 11-2:

$$\frac{PV}{T} = \frac{(1 \text{ atm})(22.4 \text{ liter/mole})}{273 \text{ deg}} = 0.0820 \frac{\text{liter atm}}{\text{mole deg}}$$

This constant, known as the "ideal gas constant" is given the special symbol R.

When pressures are measured in torr, it is convenient to have the value of R in the same units. This may be computed as above, substituting 760 torr for P in equation 11-2:

$$\frac{PV}{T} = \frac{(760 \text{ torr})(22.4 \text{ liter/mole})}{273 \text{ deg}} = 62.36 \frac{\text{torr liter}}{\text{mole deg}}$$

Some other frequently used values for R are

$$8.317 \times 10^7 \text{ ergs/mole deg, and}$$

$$1.987 \text{ cal/mole deg.}$$

Many other values of R can be derived, depending on which units are used for pressure and volume and for mass on the atomic weight scale.

Equation 11-2 is written as $PV = RT$ for one mole of gas. For n moles it follows that

$$PV = nRT \tag{11-4}$$

which, for convenience, may be used in the form

$$PV = \frac{wRT}{M} \tag{11-5}$$

where w is the weight of gas in grams and M is its molecular weight. The following examples illustrate some applications of equations 11-4 and 11-5.

PROBLEM:

What is the volume of one mole of *any* gas at room conditions, 740 torr and 27°C?

SOLUTION:

$$V = \frac{nRT}{P}$$

$$= \frac{(1 \text{ mole})\left(62.4 \dfrac{\text{torr liter}}{\text{mole deg}}\right)(300 \text{ deg})}{740 \text{ torr}}$$

$$= 25.3 \text{ liters}$$

PROBLEM:

What is the density of NH_3 gas at 67°C and 800 torr?

SOLUTION:

Density (D) is defined as mass per unit volume:

$$D = \frac{g}{ml} = \frac{w}{V}$$

By rearranging equation 11-5,

$$D = \frac{w}{V} = \frac{MP}{RT} = \frac{\left(17.0 \dfrac{g}{\text{mole}}\right)(800 \text{ torr})}{\left(62.4 \dfrac{\text{torr liter}}{\text{mole deg}}\right)(340 \text{ deg})}$$

$$= 0.641 \frac{g}{\text{liter}}$$

PROBLEM:

What is the molecular weight of a gas if a 0.894-g sample in a 60.0-ml bulb has a pressure of 400 torr at 20°C?

SOLUTION:
Rearrange equation 11-5 and solve for M.

$$M = \frac{wRT}{PV} = \frac{(0.894 \text{ g})\left(62.4 \, \frac{\text{torr liter}}{\text{mole deg}}\right)(293 \text{ deg})}{(400 \text{ torr})(0.0600 \text{ liter})} = 68.1 \, \frac{\text{g}}{\text{mole}}$$

If the empirical formula of this compound had been found to be $(CH_3F)_x$, this molecular-weight determination could be used to find the *true* formula. The gram-molecular weight of the CH_3F unit is 34.0. The true molecular weight, 68.1, must be some integral multiple (x) of 34.0; i.e.,

$$x = \frac{68.1}{34.0} = 2$$

True formula $= C_2H_6F_2$

PROBLEM:
What is the apparent molecular weight of air, assuming that it contains 78% nitrogen, 21% oxygen, and 1% argon by volume?

SOLUTION:
The apparent molecular weight of a gas mixture will be the total weight of the mixture divided by the total number of moles in the mixture (i.e., the average weight of one mole). Since the ideal gas law states that equal volumes of gases have the same number of moles at the same conditions of temperature and pressure, it follows that the percentage by volume of components in a gas mixture is the same as their percentage by moles. If we arbitrarily take 100 moles of air, the percentages by volume indicate that we have 78 moles of N_2, 21 moles of O_2 and 1 mole of Ar. We know the molecular weights of the individual components, so we calculate the total weight as follows.

$$78 \text{ moles of } N_2 \text{ weigh} \quad (78 \text{ moles})\left(28.0 \, \frac{\text{g}}{\text{mole}}\right) = 2184 \text{ g}$$

$$21 \text{ moles of } O_2 \text{ weigh} \quad (21 \text{ moles})\left(32.0 \, \frac{\text{g}}{\text{mole}}\right) = 672 \text{ g}$$

$$1 \text{ mole of Ar weighs} \quad (1 \text{ mole})\left(40.0 \, \frac{\text{g}}{\text{mole}}\right) = 40 \text{ g}$$

Total weight of 100 moles $= 2896$ g

$$\text{Average weight of one mole} = \frac{2896 \text{ g}}{100 \text{ moles}} = 29.0 \, \frac{\text{g}}{\text{mole}}$$

$$= \text{apparent molecular weight of air}$$

PROBLEM:
How many molecules are there in 3.00 liters of a gas at a temperature of 500°C and a pressure of 50.0 torr?

SOLUTION:
Substituting in the ideal gas equation

$$n = \frac{PV}{RT}$$

we have

$$n = \frac{(50.0 \text{ torr})(3.00 \text{ liters})}{\left(62.4 \dfrac{\text{torr liter}}{\text{mole deg}}\right)(773 \text{ deg})}$$

$$= 0.00312 \text{ mole}$$

Since a mole of gas contains 6.02×10^{23} molecules, the number of molecules in the sample is given by

$$\text{no. of molecules} = (0.0312 \text{ moles})\left(6.02 \times 10^{23} \frac{\text{molecules}}{\text{mole}}\right)$$

$$= 1.87 \times 10^{21} \text{ molecules}$$

Partial Pressure of Gases

Since the pressure of a gas is due to the impacts of the molecules on the walls of the container, the pressure depends on the number of molecules present. If we have two gases, A and B, each contributes to the pressure on the walls, and the total pressure is $P_A + P_B$. In general terms,

$$P = P_A + P_B + \text{etc.}$$

Thus, if we have air at a total pressure of 760 torr, the partial pressure of oxygen is $21\% \times 760$ or 160 torr, and the partial pressure of nitrogen is 600 torr.

In the laboratory work of general chemistry, we have important applications of partial pressures. Gases (such as oxygen) that are not very soluble in water are collected in bottles by displacement of water. As the gas bubbles rise through the water, they become saturated with vapor, and the collected gas is a mixture of water vapor and the original gas. When the bottle is filled, it is at atmospheric pressure, or

$$P_{\text{gas}} + P_{\text{H}_2\text{O}} = \text{barometric pressure}$$

TABLE 11-1.
Vapor Pressures of Water at Various Temperatures

t, °C	p, torr	t, °C	p, torr	t, °C	p, torr
0	4.58	15	12.8	29	30.0
1	4.93	16	13.6	30	31.8
2	5.29	17	14.5	31	33.7
3	5.69	18	15.5	32	35.7
4	6.10	19	16.5	33	37.7
5	6.54	20	17.5	34	39.9
6	7.01	21	18.7	35	42.2
7	7.51	22	19.8	40	55.3
8	8.05	23	21.1	50	92.5
9	8.61	24	22.4	60	149.4
10	9.21	25	23.8	70	233.7
11	9.84	26	25.2	80	355.1
12	10.5	27	26.7	90	525.8
13	11.2	28	28.3	100	760.0
14	12.0				

To obtain the partial pressure of the gas, we must subtract from the barometric pressure the water-vapor pressure, which is given as a function of temperature in Table 11-1.

PROBLEM:
A 250-ml flask is filled with oxygen, collected over water at a barometric pressure of 730 torr and a temperature of 25°C. What will be the volume of this oxygen sample, dry, at standard conditions?

SOLUTION:
Equation 11-3 is conveniently used for problems of this type, where we wish to convert a gas volume at a given temperature and pressure to its volume at other conditions. From Table 11-1 we find that the partial pressure of water vapor at 25°C is 24 torr. The partial pressure of O_2 is therefore $730 - 24 = 706$ torr.

$$V = (250 \text{ ml}) \left(\frac{706 \text{ torr}}{760 \text{ torr}}\right)\left(\frac{273 \text{ deg}}{298 \text{ deg}}\right)$$

$$= 213 \text{ ml at standard conditions}$$

PROBLEM:

What is the volume of one mole of gas measured over water at 730 torr and 30°C?

SOLUTION:

From Table 11-1 we find that the partial pressure of water at 30°C is 32 torr. Thus, the partial pressure of the gas is $730 - 32 = 698$ torr. Applying the ideal gas equation,

$$V = \frac{(1 \text{ mole})\left(62.4 \dfrac{\text{torr liter}}{\text{mole deg}}\right)(303 \text{ deg})}{698 \text{ torr}}$$

$$= 27.0 \text{ liter}$$

Graham's Law of Diffusion

When we put two gases together, the molecules diffuse throughout the container so that within a short time the mixture is homogeneous, or of uniform concentration throughout. Not all gases diffuse at the same rate, however: the lighter the molecule, the more rapid the diffusion process.

In 1829 Graham discovered the relation between the rate of diffusion and the density of gaseous molecules. More recently the relation has been derived theoretically from the kinetic theory of gases. The law may be stated mathematically as

$$\frac{r_1}{r_2} = \sqrt{\frac{M_2}{M_1}}$$

where r_1 and r_2 are the rates of diffusion of gases having the respective molecular weights, M_1 and M_2.

PROBLEM:

The molecular weight of an unknown gas is found by measuring the time required for a known volume of the gas to diffuse through a small pinhole, under constant pressure. The apparatus is calibrated by measuring the time needed for the same volume of O_2 (mol weight $= 32$) to diffuse through the same pinhole, under the same conditions. It is found that the time for O_2 is 60 sec and for the unknown gas is 120 sec. Compute the molecular weight of the unknown gas.

SOLUTION:

Graham's Law states

$$\frac{\text{rate of diffusion of } O_2}{\text{rate of diffusion of gas}} = \sqrt{\frac{\text{mol wt of gas}}{32}}$$

We do not measure the rate, however; we measure the time required for a known volume to diffuse. We see that this time is inversely proportional to the rate. That is, the faster the gas diffuses, or the higher its rate, the shorter is the time required. We therefore substitute the measured times, with proper change in the equation:

$$\frac{\text{time for O}_2}{\text{time for gas}} = \sqrt{\frac{32}{\text{mol wt of gas}}}$$

Substituting the measured times and solving,

$$\frac{60 \text{ sec}}{120 \text{ sec}} = \sqrt{\frac{32}{\text{mol wt of gas}}}$$

$$\frac{1}{2} = \sqrt{\frac{32}{\text{mol wt of gas}}}$$

To solve, square both sides of the equation:

$$\frac{1}{4} = \frac{32}{\text{mol wt of gas}}$$

$$\text{mol wt of gas} = 4 \times 32 = 128$$

PROBLEMS A   Answers on page 343

1. A student collects 265 ml of a gas over Hg at 25°C and 750 torr. What is the volume at standard conditions?

2. What is the volume at room conditions (740 torr and 25°C) of 750 ml of a gas at standard conditions?

3. A sample of nitrous oxide is collected over water at 24°C and 735 torr. The volume is 235 ml. What is the volume at standard conditions?

4. (a) What volume will 0.500 g of O_2 occupy at 750 torr and 26°C over water? (b) What volume will it occupy if collected over mercury at the same conditions?

5. What is the weight of 250 ml of N_2 measured at 740 torr and 25°C?

6. What volume will be occupied by 1.00 g of O_2 measured over water at 27°C and 730 torr?

7. What is the molecular weight of a gas if 250 ml measured over water at 735 torr and 28°C weighs 1.25 g?

8. What is the density of chlorine gas (Cl_2) at 83°F and 723 torr?

9. Calculate the density of N_2O (a) at standard conditions; (b) at 730 torr and 25°C, dry.

10. It is found experimentally that 0.563 g of a vapor at 100°C and 725 torr has a volume of 265 ml. Find the molecular weight.

11. A compound has the formula C_8H_{18}. What volume will 1.00 g of this material have at 735 torr and 99°C?

12. A sample of $NaNO_2$ was tested for purity by heating it with an excess of NH_4Cl and collecting the evolved N_2 over water. The volume collected was 567.3 ml at a barometric pressure of 741 torr. The temperature was 22°C. What volume would the nitrogen, dry, occupy at standard conditions?

13. A graduated tube, sealed at the upper end, has a mercury-leveling bulb connected to the lower end. The gas volume is 25.0 ml when the mercury level is the same in both tubes. The barometric pressure is 732 torr. What is the volume when the level on the open side is 10.0 cm above the level of the closed side? (Remember that the pressure is 732 torr plus that due to the mercury column.)

14. If a barometer were filled with a liquid of density 1.6 g/ml, what would be the reading when the mercury barometer read 730 torr? The density of mercury is 13.56 g/ml.

15. What is the atmospheric pressure in lbs/in.² when the barometer reading is 720 torr?

16. A sealed vessel containing methane, CH_4, at 730 torr and 27°C is put into a box cooled with "dry ice" (−78°C). What pressure will the CH_4 exert under these conditions?

17. Liquid nitrogen (boiling point, −195.8°C) is commonly used as a cooling agent. A vessel containing helium at 10 lb/in.² at the temperature of boiling water is sealed off and then cooled with boiling liquid nitrogen. What will be its pressure expressed in torr?

18. What is the apparent mole weight of a gas mixture composed of 20% H_2, 70% CO_2, and 10% N_2O? (Composition given is per cent by volume.)

19. An oxygen-containing gas mixture at 1 atm was subjected to the action of yellow phosphorus, which removed the oxygen. In this way it was found that there was 35% oxygen by volume in the mixture. What was the partial pressure of O_2 in the mixture?

20. A mixture of gases contained in a vessel at 0.5 atm is found to comprise 15% N_2, 50% N_2O, and 35% CO_2 by volume. (a) What is the partial pressure of each gas? (b) A bit of solid KOH is added to remove the CO_2. Calculate the resulting total pressure and the partial pressures of the remaining gases.

21. Two liters of N_2 at 1 atm, 5 liters of H_2 at 5 atm, and 3 liters of CH_4 at 2 atm are mixed and transferred to a 10-liter vessel. What is the resulting pressure?

22. A vessel whose volume is 235.0 ml, and whose weight evacuated is 13.5217 g + a tare vessel, is filled with an unknown gas at a pressure of 725 torr and a temperature of 19°C. It is then closed, wiped with a damp cloth, and hung in the balance case to come to equilibrium with the tare

vessel. The tare vessel has about the same surface area and is needed to minimize surface moisture effects. This second weighing is 13.6109 g + the tare vessel. What is the mole weight of the gas?

23. Two or three milliliters of a liquid that boils at about 50°C were put into an Erlenmeyer flask. The flask was closed with a polystyrene stopper that had a fine glass capillary running through it. The gas-containing part of the flask was then completely immersed in a bath of boiling water, which, at the elevation of the experiment, boiled at 99.2°C. After a short time the air had been completely swept out through the capillary, and the excess liquid had boiled away, leaving the flask filled only with the vapor of the liquid. At this point the flask was removed from the boiling water and cooled. The vapor condensed to liquid, and air rushed in to fill the flask again. Whereas the flask, when empty and dry, weighed 45.3201 g, after the experiment it weighed 46.0513 g. The barometric pressure during the experiment was 735 torr. The volume of the flask was determined by filling the flask with water, inserting the stopper to its previous position, and squeezing out the excess water through the capillary. The volume of water so held was 263.2 ml. What is the mole weight of the liquid?

24. The liquid used in Problem 23 was analyzed and found to be 54.5%C, 9.10% H, and 36.4% O. What is the true molecular formula of this liquid?

25. An automobile tire has a gauge pressure of 32 lb/in.² at 20°C when the prevailing atmospheric pressure is 14.7 lb/in.². What is the gauge pressure if the temperature rises to 50°C?

26. It took 1 min and 37 sec for a given volume of chlorine (Cl_2) to effuse through a pinhole under given conditions of temperature and pressure. How long will it take for the same volume of water vapor to effuse through the same hole under the same conditions?

27. Argon effuses through a hole (under prescribed conditions of temperature and pressure) at the rate of 3.0 ml/min. At what velocity will xenon effuse through the same hole under the same conditions?

28. A low pressure easily achieved with a diffusion pump and a mechanical vacuum pump is 10^{-6} torr. Calculate the number of molecules still present in 1 ml of gas at this pressure and at 0°C.

29. Oxygen is commonly sold in 6 ft³ steel cylinders at a pressure of 2,000 lb/in.² (at 70°F). What weight of oxygen does such a cylinder contain? (Assume oxygen to be an ideal·gas under these conditions.)

30. An organic compound containing C, H, O, and N was analyzed. When a sample weighing 0.01230 g was burned, it produced 18.62 mg of CO_2 (absorbed in a soda-lime tube) and 7.62 mg of H_2O [absorbed in a tube containing $Mg(ClO_4)_2$]. When another sample, weighing 0.00510 g, was burned, the CO_2 and H_2O were absorbed, and the N_2 formed was collected in a measuring tube. At 730 torr and 22°C the N_2 gas displaced an equal

volume of mercury, which was weighed and found to be 15.000 g. The density of mercury is 13.56 g/ml. Calculate the empirical formula of the compound.

31. In gas analysis it is common to make measurements at constant temperature and pressure. If one component is removed, the decrease in volume is therefore a direct measure of the amount of that component present. A typical gas analysis performed on a Blacet-Leighton micro gas-analysis apparatus follows. It assumes that the original gas sample is dry. The original volume is 35.25 mm³; a bead of KOH removes CO_2 and leaves a residual volume of 27.80 mm³; 32.15 mm³ of oxygen is added, H_2 and CO are burned to H_2O and CO_2, and the resulting water vapor is removed with a bead of P_4O_{10}. The remaining volume is now determined to be 48.00 mm³. Another KOH bead is added to remove the CO_2 formed on combustion of CO, and the residual volume is 39.40 mm³. All that now remains is the excess of the oxygen used for combustion and the nitrogen present in the original sample. With these data, calculate the percentages by volume of CO_2, H_2, CO, and N_2 in the original sample.

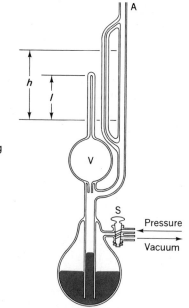

FIGURE 11-2
McLeod gauge for measuring low gas pressures.

32. A McLeod gauge (Figure 11-2) is frequently used to measure very low pressures of noncondensable gases. The gauge is connected at A to the system having the low pressure. To measure the low pressure, stopcock S is turned to admit air, which forces mercury from the reservoir up the central tube into V and into the tube leading to A. Just as vessel V is closed off, it contains a gas volume V at the low pressure P_1. As the mercury rises, the gas is compressed in V, and its pressure rises. It is finally

compressed into the capillary of radius r. To make a reading, suppose that one mercury level is l cm below the top of the closed capillary and that the mercury level in the other capillary arm is h cm above the level of the mercury in the first. The compressed gas is now under a pressure of h cm of mercury plus the pressure (P_1) of the gas in the system at A. (a) Derive a formula that shows P_1 as a function of V, r, l, and h. (b) If $V = 250$ ml, $r = 0.5$ mm, $l = 10$ cm, and $h = 30$ cm, calculate the low pressure in the system.

33. The average breath that an 18-year-old takes when not exercising is about 300 ml at 20°C and 750 torr. His respiratory rate is about 20 breaths/min. (a) What volume of air, corrected to standard conditions, does an average 18-year-old breathe each day? (b) What weight of air does he breathe each day? (Assume that air is 21% oxygen and 79% nitrogen by volume.)

34. The percentage of CO_2 in normal air is 0.04% by volume, and that in the exhaled air of the average person is about 4%. (a) What volume of CO_2, at standard conditions, does the average 18-year-old make each day? (b) What weight of CO_2 does he make each day?

35. The sample of impure $NaNO_2$ used in Problem 12 weighed 2.053 g. The reaction involved is

$$NaNO_2 + NH_4Cl \rightarrow N_2 + NaCl + 2H_2O.$$

Calculate the percentage of $NaNO_2$ in the original sample.

36. A rubber balloon weighing 5.0 g is 12 inches in diameter when filled with hydrogen at 730 torr and 25°C. How much will the balloon lift in addition to its own weight? (Assume the density of air to be 1.2 g/liter under these conditions.)

37. (a) If the balloon of Problem 36 were filled with ammonia gas (NH_3) under the same conditions, would it rise? (b) If so, how much weight would it lift?

38. What must be the composition of a mixture of H_2 and O_2 if it inflates a balloon to a diameter of 15 inches and yet the balloon *just barely* rises from a table top? (The balloon weighs 6 g, and the pressure and temperature are 740 torr and 30°C.)

PROBLEMS B   No answers given

39. What is the volume at 730 torr and 27°C of 350 ml of H_2S as standard conditions?

40. A 50-ml quartz vessel is filled with O_2 at 300 torr and at 35°C. It is then heated to 1,400°C in an electric furnace. What will be the oxygen pressure at the higher temperature?

41. A sample of NH_3 gas collected over mercury measures 595 ml at 19°C and 755 torr. What will its volume be at standard conditions?

42. An automobile tire has a gauge pressure of 35 lb/in.² when the atmospheric pressure is 14.7 lb/in.² and the temperature is 40°F. What will its gauge pressure be if its temperature rises to 120°F?

43. A graduated gas tube, sealed at the upper end, has a mercury-leveling bulb connected to the lower end. The gas volume is 17.2 ml when the leveling bulb is 8 cm above the other mercury level. What will be the gas volume when the leveling bulb is 8 cm below the other mercury level? (The barometric pressure is 738 torr.)

44. If a barometer were filled with a silicone fluid whose vapor pressure is very low but whose density is 1.15 g/ml, what would be the barometer reading when the atmospheric pressure was 710 torr? (The density of mercury is 13.56 g/ml.)

45. A mixture of gases contained in a vessel at 1.3 atm is found to be 60% NH_3, 25% NO, and 15% N_2 by volume. (a) What is the partial pressure of each gas? (b) A bit of solid P_4O_{10} is added to remove the NH_3. Calculate the resulting total pressure and the partial pressures of the remaining gases.

46. A gas mixture containing CO_2 was subjected, at 1 atm, to the action of KOH, which removed the CO_2. In this way it was found that there was 27% CO_2 by volume in the mixture. What was the partial pressure of CO_2 in the mixture?

47. A 375-ml sample of hydrogen is collected over water at 18°C and 720 torr. What is its volume at standard conditions?

48. What will be the difference in volume occupied by 0.100 g of hydrogen at 740 torr and 19°C if it is collected over water instead of mercury?

49. In some recent work at the Bell Telephone Research Laboratory a low pressure of 10^{-11} torr of mercury was used. This is an unusually low pressure, one not easily obtained. Calculate the number of molecules still remaining in 1 ml of gas at this pressure at 27°C.

50. Hydrogen chloride effuses through a hole (under prescribed conditions of temperature and pressure) at the rate of 2.7 ml/min. At what velocity will helium effuse through the same hole under the same conditions?

51. It took 45 sec for a given volume of CO to effuse through a pinhole under given conditions of temperature and pressure. How long will it take for the same volume of Br_2 vapor to effuse through the same hole under the same conditions?

52. A 1-g sample of helium occupies 5.6 liters at standard conditions. What will be its weight when expanded to a pressure of 0.1 atm?

53. What volume will 5.0 g of methyl alcohol, CH_3OH, occupy at 720 torr and 98°C?

54. What is the weight of 420 ml of NH_3 measured at 735 torr and 27°C?

55. What volume will be occupied by 2.5 g of CO measured over water at 27°C and 725 torr?

56. What is the molecular weight of a gas if 365 ml of it measured over water at 727 torr and 30°C weighs 1.42 g?

57. What is the density of NH_3 gas at 78°F and 741 torr?

58. Calculate the density of C_2H_6 (a) at standard conditions; (b) at 725 torr and 27°C, dry.

59. What is the apparent mole weight of a gas mixture whose composition by volume is 60% NH_3, 25% NO, and 15% N_2?

60. If 0.670 g of a vapor at 100°C and 735 torr has a volume of 249 ml, what must its mole weight be?

61. A vessel, whose volume is 205.3 ml and whose weight evacuated is 5.3095 g + a tare flask, is filled with an unknown gas to a pressure of 750 torr at a temperature of 27°C. It is then cleaned, wiped with a damp cloth, and hung in the balance case to come to equilibrium with the tare vessel. The tare vessel has about the same surface area and is needed to minimize effects of surface moisture. The second weighing is 5.6107 g + the tare flask. What is the mole weight of the gas?

62. Two or three milliliters of a liquid that boils at about 60°C were put into an Erlenmeyer flask. The flask was closed with a polystyrene stopper that had a fine glass capillary running through it. The gas-containing part of the flask was then completely immersed in a bath of boiling water, which at the elevation of the experiment boils at 98.8°C. After a short time the air had been completely swept out through the capillary, and the excess liquid had boiled away, leaving the flask filled only with the vapor of the liquid. At this point the flask was removed from the boiling water and cooled. The vapor condensed to liquid, and air rushed in to fill the flask again. The flask, when dry and empty, had weighed 39.5762 g; after the experiment it weighed 40.3183 g. The barometric pressure during the experiment was 730 torr. The volume of the flask was determined by filling the flask with water, inserting the stopper to its previous position, and squeezing out the excess water through the capillary. The volume of water so held was 239.6 ml. What is the mole weight of the liquid?

63. The liquid used in Problem 62 was analyzed and found to be 24.2% C, 4.05% H, and 71.7% Cl. What is the true molecular formula of this liquid?

64. Under certain prescribed conditions, O_2 effuses through a pinhole at the rate of 3.65 ml/min. A mixture of CO and CO_2 effuses through the same pinhole under the same conditions at the rate of 3.21 ml/min. Calculate the percentage of CO in the gas mixture.

65. A mixture of copper and zinc was analyzed for the percentage of zinc by adding an excess of HCl and collecting the evolved H_2 over water. (Copper will not react with HCl.) The volume of H_2 collected was 229.5 ml at a

barometric pressure of 732 torr. The temperature was 29°C. What volume would the H_2 occupy at standard conditions, dry?

66. The sample of copper and zinc in Problem 65 weighed 1.500 g. The equation for the reaction by which the H_2 was evolved is

$$Zn + 2HCl \rightarrow ZnCl_2 + H_2$$

Calculate the percentage of zinc in the original sample.

67. When very accurate weighings are desired, correction must be made for buoyancy. This is caused by the unequal displacement of air by the object and the weights. The object is buoyed up by the weight of the air it displaces, and the weights are buoyed up by the weight of the air they displace. A sealed flask having a volume of 145 ml weighs 37.6215 g in air, using brass weights whose density is 8.3 g/ml. Calculate the true weight of the flask; that is, calculate what the flask would weigh in a vacuum. (The density of air at the conditions of the weighing was 0.0012 g/ml.)

68. A 16-gal fuel tank on an automobile is half full of gasoline, which for simplicity we shall assume to be C_8H_{18}. If the tank temperature goes from 45°F at night to 80°F in the daytime, the tank will "breathe out" gasoline vapor. (a) What weight of gasoline will be lost per day in this way? (b) If the loss continues each day, how long will it take to lose 1 gal of gasoline? (The density of liquid C_8H_{18} is 0.692 g/ml. Assume that the vapor volume of the tank remains unchanged during this period of time, and that the average partial pressure of C_8H_{18} is 16 torr.)

69. A rubber balloon weighing 10 g is 15 inches in diameter when inflated with helium at 735 torr and 75°F. How much will the balloon lift in addition to its own weight? (Assume the density of air to be 1.20 g/liter under these conditions.)

70. If the balloon of Problem 69 were filled with methane, CH_4, at 735 torr and 75°F, would it rise? If so, how much additional weight would it lift?

Stoichiometry II. Calculations Based on Chemical Equations

A chemical equation is a statement of experimental fact. It gives on the left side the reactants and on the right side the products of the reaction. Since no atoms are produced or destroyed in a nonnuclear chemical reaction, the equation must be so balanced that every atom originally present in the reactants is accounted for in the products. This means that the combined weight of the reaction products is exactly equal to the combined weight of the original reactants.

Most of the stoichiometric calculations we make from equations involve three simple steps:

1. Find how many moles correspond to the given quantity of some substance in the reaction.

2. Use the balanced chemical equation to find, from the number of moles of the given substance, the number of moles of the substance sought in the calculation.

3. Convert the number of moles of the substance sought to the units requested in the statement of the problem.

These three steps are illustrated in the following problems.

PROBLEM:

Oxygen is prepared by heating $KClO_3$.

    (a) What weight of O_2 is obtained from 3.0 g $KClO_3$?

    (b) What is the volume of O_2, measured at standard conditions?

    (c) What volume does the O_2 occupy if collected over water at 730 torr and 25°C?

SOLUTION:

The first step is to write the balanced equation for the reaction (if it is not given). This requires knowledge of the experimental facts. We note in the text that when $KClO_3$ is decomposed by heating, the products are KCl and O_2, so we start with the unbalanced equation

$$KClO_3 \rightarrow KCl + O_2$$

To account for the three oxygen atoms of $KClO_3$ we need $\frac{3}{2}$ moles of O_2 in the products

$$KClO_3 \nrightarrow KCl + \tfrac{3}{2}O_2$$

This is now a balanced equation, but we prefer to eliminate fractional numbers of moles, so we multiply all terms by 2, getting the final equation

$$2KClO_3 \rightarrow 2KCl + 3O_2$$

We now examine the problem, asking two questions: (1) what is given, and (2) what is sought? We see that (1) the weight of $KClO_3$ used is given and (2) that we seek the amount of O_2 produced. We then proceed with the three steps listed above:

$$\text{moles } KClO_3 \text{ given} = \frac{3.0 \text{ g } KClO_3}{122.6 \dfrac{\text{g } KClO_3}{\text{mole } KClO_3}} = 0.0244 \text{ moles } KClO_3$$

We next use this value to compute the number of moles of O_2 produced. According to the equations, 2 moles of $KClO_3$ produce 3 moles of O_2. Therefore,

$$\text{moles } O_2 \text{ produced} = \frac{3 \text{ moles } O_2}{2 \text{ moles } KClO_3} \times 0.0244 \text{ moles } KClO_3$$

$$= 0.0366 \text{ moles } O_2$$

We now compute the amount of O_2 in the units specified in the statement of the problem.

(a) $$\text{weight of } O_2 = 0.0366 \text{ moles } O_2 \times 32.0 \frac{\text{g } O_2}{\text{mole } O_2}$$

$$= 1.17 \text{ g } O_2$$

(b) We know that a mole of O_2 gas at standard conditions occupies the GMV of 22.4 liters. Consequently,

$$\text{volume of } O_2 = 0.0366 \text{ moles } O_2 \times \frac{22.4 \text{ liters } O_2}{\text{mole } O_2}$$

$$= 0.82 \text{ liters } O_2$$

(c) The volume of O_2 at the stated conditions, over water at 25°C and 730 torr, is computed from the ideal gas law. First, the partial pressure of O_2 must be calculated, knowing the vapor pressure of water at 25°C is 24 torr (see page 131).

$$P_{O_2} = 730 \text{ torr} - 24 \text{ torr} = 706 \text{ torr}$$

$$V = \frac{nRT}{P} = \frac{(0.0366 \text{ moles})\left(62.4 \dfrac{\text{torr liter}}{\text{mole deg}}\right)(298 \text{ deg})}{706 \text{ torr}}$$

$$= 0.960 \text{ liter } O_2$$

PROBLEM:

Chlorine is prepared by the reaction

$$2NaCl + MnO_2 + 3H_2SO_4 \rightarrow MnSO_4 + 2NaHSO_4 + Cl_2 + 2H_2O$$

What weights of MnO_2 and NaCl are needed to prepare 500 ml of Cl_2 gas, measured dry at 730 torr and 25°C?

SOLUTION:

The given quantity in this problem is the volume of chlorine gas measured at room conditions. We must first compute the number of moles of Cl_2 produced, using the ideal gas equation:

$$n = \frac{PV}{RT} = \frac{(730 \text{ torr})(0.500 \text{ liter})}{\left(62.4 \dfrac{\text{torr liter}}{\text{mole deg}}\right)(298 \text{ deg})}$$

$$= 0.0196 \text{ moles } Cl_2$$

We see from the chemical equation that 2 moles of NaCl and 1 mole of MnO_2 are required to produce 1 mole of Cl_2. Consequently:

$$\text{moles } MnO_2 = \frac{1 \text{ mole } MnO_2}{1 \text{ mole } Cl_2} \times 1.96 \times 10^{-2} \text{ moles } Cl_2$$

$$= 1.96 \times 10^{-2} \text{ moles } MnO_2$$

$$\text{moles } NaCl = \frac{2 \text{ moles } NaCl}{1 \text{ mole } Cl_2} \times 1.96 \times 10^{-2} \text{ moles } Cl_2$$

$$= 3.92 \times 10^{-2} \text{ moles } NaCl$$

We use these values to compute the weights of MnO_2 and $NaCl$:

$$\text{weight of } NaCl = 3.92 \times 10^{-2} \text{ moles } NaCl \times 58.5 \frac{\text{g } NaCl}{\text{mole } NaCl}$$

$$= 2.29 \text{ g } NaCl$$

$$\text{weight of } MnO_2 = 1.96 \times 10^{-2} \text{ moles } MnO_2 \times 86.9 \frac{\text{g } MnO_2}{\text{mole } MnO_2}$$

$$= 1.70 \text{ g } MnO_2$$

These two problems illustrate the general procedure for any stoichiometric calculation involving weights of reagents and volumes of gases. It may be helpful, however, to consider two additional problems that on first sight might appear to be different. Actually, as we shall see, they are not.

PROBLEM

Sulfur dioxide is prepared by heating iron pyrites, FeS_2, in presence of air (a process known as roasting). The reaction is

$$4FeS_2 + 11O_2 \xrightarrow{\Delta} 2Fe_2O_3 + 8SO_2$$

(a) What weight of SO_2 is obtained from 20 tons of FeS_2?
(b) What volume of air, in cubic feet at 1 atm and 75°F, is required to roast 20 tons of FeS_2? Assume that air is 21 per cent O_2 by volume.

SOLUTION:

(a) The equation shows that 8 moles of SO_2 are obtained from 4 moles of FeS_2. We could work the problem by converting the given 20 tons of FeS_2 to moles and use this figure to compute the moles of SO_2 obtained, but this would require a further conversion back to tons to get the weight of SO_2. When the data are given in units other than grams, as in this problem, it is simpler to express the moles in the given weight unit (in this case, tons). Since 4 moles FeS_2 yield 8 moles SO_2 it follows that 4 ton-moles FeS_2 yield 8 ton-moles SO_2.

$$\text{ton-moles of } FeS_2 = \frac{20 \text{ tons } FeS_2}{120 \frac{\text{tons } FeS_2}{\text{ton-mole } FeS_2}}$$

$$= 0.167 \text{ ton-moles } FeS_2$$

$$\text{ton-moles of } SO_2 = \frac{8 \text{ ton-moles } SO_2}{4 \text{ ton-moles } FeS_2} \times 0.167 \text{ ton-moles } FeS_2$$

$$= 0.334 \text{ ton-moles } SO_2$$

$$\text{weight of } SO_2 = 0.334 \text{ ton-moles } SO_2 \times \frac{64.1 \text{ tons } SO_2}{\text{ton-mole } SO_2}$$

$$= 21.4 \text{ tons } SO_2$$

(b) The equation shows that 4 moles of FeS_2 require 11 moles of O_2. Consequently,

$$\text{ton moles of } O_2 \text{ needed} = \frac{11 \text{ ton-moles } O_2}{4 \text{ ton-moles } FeS_2} \times 0.167 \text{ ton-moles } FeS_2$$

$$= 0.460 \text{ ton-moles } O_2$$

It now remains to use this weight of oxygen to find the volume of air (in cubic feet) at 75°F and 1 atm that are required to roast 20 tons of FeS_2. Since we have not been given the gram-molecular volume in cubic feet for a ton-mole of gas at standard conditions, it is necessary to start with what we know, that is, that a mole of gas occupies a volume of 22.4 liters at standard conditions. Using the conversion factors of page 28, we have:

volume of ton-mole of gas in cubic feet =

$$\frac{22.4 \text{ liter}}{\text{mole}} \times \frac{2,000 \text{ lb}}{\text{ton}} \times \frac{454 \text{ g}}{\text{lb}} \times \frac{61.0 \text{ in.}^3}{\text{liter}} \times \left(\frac{1 \text{ ft}}{12 \text{ in.}}\right)^3 = 7.17 \times 10^5 \frac{\text{ft}^3}{\text{ton-mole}}$$

$$\text{volume of } O_2 = 0.460 \text{ ton-mole } O_2 \times 7.17 \times 10^5 \frac{\text{ft}^3 \, O_2}{\text{ton-mole } O_2}$$

$$= 3.30 \times 10^5 \text{ ft}^3 \, O_2$$

Since air contains 21 per cent O_2 by volume,

$$\text{volume of air} = 3.30 \times 10^5 \text{ ft}^3 \, O_2 \times \frac{100 \text{ ft}^3 \text{ air}}{21 \text{ ft}^3 \, O_2}$$

$$= 15.7 \times 10^5 \text{ ft}^3 \text{ air}$$

Finally we convert to the volume at the given conditions, 75°F and 1 atm. Converting 75°F to C degrees,

$$t(\text{Celsius}) = (75°F - 32°F)\left(\frac{5°C}{9°F}\right) = 24°C$$

The absolute temperature is 297°K. Using this value we have

$$\text{volume of air} = 15.7 \times 10^5 \text{ ft}^3 \text{ air} \times \frac{297°K}{273°K}$$

$$= 17.1 \times 10^5 \text{ ft}^3 \text{ air}$$

PROBLEM:
What volume of oxygen at 20°C and 730 torr is needed to burn 3 liters propane, C_3H_8, at 20°C and 730 torr? The equation is

$$C_3H_8 + 5O_2 \xrightarrow{\Delta} 3CO_2 + 4H_2O$$

SOLUTION:
This type of problem may be deceptive when first encountered. Many students assume that such problems are to be solved in exactly the same way as the preceding exercises, and they go through the steps of converting the propane

volume to standard conditions, computing the O_2 volume at standard conditions and converting to the given conditions. A little study shows that all these steps are not necessary, since the volumes for both gases are measured at the same conditions of temperature and pressure. One need only apply Avogadro's Principle, that equal volumes of gases at the same temperature and pressure contain the same number of molecules. Using this relation,

$$\text{volume of } O_2 = 3 \text{ liters C}_3\text{H}_8 \times \frac{5 \text{ liters O}_2}{1 \text{ liter C}_3\text{H}_8}$$

$$= 15 \text{ liters O}_2$$

Note, however, that this method may be used only when both of the volumes involved are measured at the same conditions.

PROBLEMS A   Answers on page 343

1. Balance the following equations, which show the starting materials and the reaction products. It is not necessary to supply any additional reactants or products.
   (a) $KNO_3 \rightarrow KNO_2 + O_2$
   (b) $Pb(NO_3)_2 \rightarrow PbO + NO_2 + O_2$
   (c) $Na + H_2O \rightarrow NaOH + H_2$
   (d) $Fe + H_2O \rightarrow Fe_3O_4 + H_2$
   (e) $C_2H_5OH + O_2 \rightarrow CO_2 + H_2O$
   (f) $Fe_3O_4 + H_2 \rightarrow Fe + H_2O$
   (g) $CO_2 + NaOH \rightarrow NaHCO_3$
   (h) $MnO_2 + HCl \rightarrow H_2O + MnCl_2 + Cl_2$
   (i) $Zn + KOH \rightarrow K_2ZnO_2 + H_2$
   (j) $Cu + H_2SO_4 \rightarrow H_2O + SO_2 + CuSO_4$
   (k) $Al(NO_3)_3 + NH_3 + H_2O \rightarrow Al(OH)_3 + NH_4NO_3$
   (l) $Al(NO_3)_3 + NaOH \rightarrow NaAlO_2 + NaNO_3 + H_2O$

2. Some common gases may be prepared in the laboratory by reactions represented by the following equations. A Δ sign indicates that heating is necessary. For purity, air must be swept out of the apparatus before the gas is collected, and some gases must be dried with a suitable desiccant.
   (1) $FeS + 2HCl \rightarrow H_2S + FeCl_2$
   (2) $CaCO_3 + 2HCl \rightarrow CO_2 + CaCl_2 + H_2O$
   (3) $2NH_4Cl + CaO \xrightarrow{\Delta} 2NH_3 + CaCl_2 + H_2O$
   (4) $NaCl + H_2SO_4 \xrightarrow{\Delta} HCl + NaHSO_4$
   (5) $NH_4Cl + NaNO_3 \xrightarrow{\Delta} N_2O + NaCl + 2H_2O$
   (6) $2Al + 3H_2SO_4 \rightarrow Al_2(SO_4)_3 + 3H_2$
   (7) $CaC_2 + 2H_2O \rightarrow C_2H_2 + Ca(OH)_2$
   (a) What weight of FeS is needed to prepare (i) 4.5 moles of H_2S? (ii) 1 lb of H_2S?

(b) How many tons of limestone ($CaCO_3$) are needed to prepare 5 tons of "dry ice" (CO_2), assuming that 30% of the CO_2 produced is wasted in converting it to the solid?

(c) How many grams of NH_4Cl and CaO are needed to make 0.1 mole of NH_3?

(d) How many grams of 95%-pure NaCl are needed to produce 2 lb of HCl?

(e) What volume of commercial HCl (36% HCl by weight, density $= 1.18$ g/ml) and what weight of limestone (90% pure) are needed to produce 2 kg of CO_2?

(f) Commercial sulfuric acid that has a density of 1.84 g/ml and is 95% H_2SO_4 by weight is used for the production of HCl. (i) What weight of commercial acid is needed for the production of 365 g of HCl? (ii) What volume of acid is needed for the production of 365 g of HCl?

(g) Commercial sulfuric acid that has a density of 1.45 g/ml and is 55.1% H_2SO_4 by weight is used for the production of H_2. What (i) weight and (ii) volume of this commercial acid are needed for the production of 50 g of H_2 gas?

(h) A manufacturer supplies 1-lb cans of calcium carbide whose purity is labeled as 85%. How many grams of acetylene can be prepared from 1 lb of this product if the label is correct?

(i) A whipped-cream manufacturer wished to produce 500 lb of N_2O for his chain of soda fountains. What is the cost of the NH_4Cl and $NaNO_3$ if they are $120 and $47.50 per ton, respectively?

(j) A 0.795-g sample of impure limestone was tested for purity by adding H_2SO_4 [instead of HCl as shown in equation (2)]. After the generated gas was passed over $Mg(ClO_4)_2$ to dry it, it was passed over soda lime (a mixture of sodium and calcium hydroxides), which absorbs the CO_2. The soda-lime tube increased in weight by 0.301 g. What was the percentage of limestone in the original sample?

(k) The purity of a 0.617-g sample of impure FeS was tested by passing the H_2S produced by the HCl [as in equation (1)] into a dilute solution of $AgNO_3$. The precipitate of Ag_2S was filtered off, washed, and gently dried. The weight of the Ag_2S produced was 1.322 g. How pure was the original sample of FeS?

3. How many grams of zinc are needed to prepare 3 liters of hydrogen at standard conditions? The reaction is

$$Zn + H_2SO_4 \rightarrow ZnSO_4 + H_2$$

4. (a) How many grams of zinc are needed to prepare 3 liters of H_2 collected over water at 750 torr and 26°C? (b) How many moles of H_2SO_4 are used?

5. What volume of O_2 at 730 torr and 25°C will react with 3 liters of H_2 at the same conditions?

6. What volume of steam at 1,000°C and 1 atm is needed to produce 10^6 ft^3 of H_2, under the same conditions, by the reaction

$$4H_2O + 3Fe \rightarrow Fe_3O_4 + 4H_2$$

7. What volume of Cl_2 at 730 torr and 27°C is needed to react with 7 g of sodium metal by the reaction

$$2Na + Cl_2 \rightarrow 2NaCl$$

8. (a) How many grams of MnO_2 are needed to prepare 5 liters of Cl_2 at 750 torr and 27°C? (b) How many moles of HCl are needed for the reaction? The reaction is

$$MnO_2 + 4HCl \rightarrow MnCl_2 + Cl_2 + 2H_2O$$

9. How much H_2S gas at 725 torr and 25°C is needed to react with the copper in 1.5 g of $CuSO_4$? The reaction is

$$CuSO_4 + H_2S \rightarrow CuS + H_2SO_4$$

10. (a) What volume of O_2 at 730 torr and 60°F is needed to burn 500 g of octane, C_8H_{18}? (b) What volume of air (21% O_2 by volume) is needed to provide this amount of O_2? Balance the equation before working the problem:

$$C_8H_{18} + O_2 \rightarrow CO_2 + H_2O$$

11. Chlorine is prepared by the reaction

$$2KMnO_4 + 16HCl \rightarrow 2KCl + 2MnCl_2 + 5Cl_2 + 8H_2O$$

(a) What weight of $KMnO_4$ is needed to prepare 2.5 liters of Cl_2 at SC? (b) How many moles of HCl are used? (c) What volume of solution is needed if there are 12 moles of HCl per liter? (d) What weight of $MnCl_2$ is obtained from the reaction?

12. Nitric oxide is prepared by the reaction

$$3Cu + 8HNO_3 \rightarrow 3Cu(NO_3)_2 + 2NO + 4H_2O$$

(a) What weight of $KMnO_4$ is needed to prepare 2.5 liters of Cl_2 at standard conditions? (b) How many moles of HCl are used? (c) What volume of solution is needed if there are 12 moles of HCl per liter? (d) What weight of $MnCl_2$ is obtained from the reaction?

13. Arsenic compounds may be easily detected by the Marsh test. In this test some metallic zinc is added to an acid solution of the material to be tested, and the mixture heated. The arsenic is liberated as arsine, AsH_3, which may be decomposed by heat to give an "arsenic mirror." The reaction is

$$4Zn + H_3AsO_4 + 8HCl \rightarrow 4ZnCl_2 + AsH_3 + 4H_2O$$

What volume of AsH_3 at 720 torr and 25°C is evolved by 7×10^{-7} g of arsenic, the smallest amount of arsenic that can be detected with certainty by this method?

14. H_2S gas will cause immediate unconsciousness at a concentration of 1 part per 1,000 by volume. What weight of FeS is needed to fill a room 20 ft × 15 ft × 9 ft with H_2S at this concentration? Barometric pressure is 740 torr, and the temperature is 80°F. The reaction is

$$FeS + 2HCl \rightarrow FeCl_2 + H_2S$$

15. The Mond process separates nickel from other metals by passing CO over the hot metal mixture. The nickel reacts to form a volatile compound (called nickel carbonyl), which is then swept away by the gas stream. The reaction is

$$Ni + 4CO \rightarrow Ni(CO)_4$$

How many cubic feet of CO at 3 atm and 65°F are needed to react with 1 ton of nickel?

16. A cement company produces 100 tons of cement per day. Its product contains 62% CaO, which is prepared by calcining limestone by the reaction

$$CaCO_3 \rightarrow CaO + CO_2$$

What volume of CO_2 at 735 torr and 68°F is sent into the air around the plant each day as a result of this calcination?

17. The Ostwald process of making HNO_3 involves the air oxidation of NH_3 over a platinum catalyst. The first two steps in this process are

$$4NH_3 + 5O_2 \rightarrow 6H_2O + 4NO$$
$$2NO + O_2 \rightarrow 2NO_2$$

How many cubic feet of air (21% O_2 by volume) at 27°C and 1 atm are needed for the conversion of 50 tons of NH_3 to NO_2 by this process?

18. How many cubic feet of air (21% O_2 by volume) are needed for the production of 10^5 ft^3 NO_2 at the same conditions as used to measure the air? See Problem 17 for the equations involved.

19. The du Pont company has developed a nitrometer, an apparatus for the rapid routine analysis of nitrates, which measures the volume of NO liberated by the reaction of concentrated H_2SO_4 with nitrates in the presence of metallic mercury, by the reaction

$$2KNO_3 + 4H_2SO_4 + 3Hg \rightarrow K_2SO_4 + 3HgSO_4 + 4H_2O + 2NO$$

In a simple form of this apparatus, the NO is collected over water in a graduated tube and its volume, temperature, and pressure measured. A 1-g sample containing a mixture of KNO_3 and K_2SO_4 was treated in this manner, and 37.5 ml of NO was collected over water at a temperature of 23°C and a pressure of 732 torr. Calculate the percentage of KNO_3 in the original sample.

20. A commercial laboratory wished to speed up its routine analyses for HNO_3 in a mixture of acids, using the nitrometer mentioned in Problem

19. To do this, it collected the NO over mercury, used enough concentrated H_2SO_4 to make correction for water vapor unnecessary, thermostated its graduated tube at 25.0°C, and took all pressure measurements at 730 torr. The tube is graduated to 100 ml. What weight of original acid sample should always be taken so that the buret reading under these conditions also indicates directly the percentage of HNO_3 in the original sample?

21. In the Dumas method for measuring the total nitrogen in an organic compound, the compound is mixed with CuO and heated in a stream of pure CO_2. All the gaseous products are passed through a heated tube of Cu turnings, to reduce any oxides of nitrogen to N_2, and then through a 50% solution of KOH to remove the CO_2 and water. The N_2 is not absorbed, and its volume is measured by weighing the mercury (density = 13.56 g/ml) that the N_2 displaces from the apparatus. (a) A 20.1-mg sample of a mixture of glycine, $CH_2(NH_2)COOH$, and benzoic acid, $C_7H_6O_2$, yields N_2 at 730 torr and 21°C. This N_2 displaces 5.235 g of mercury. Calculate the percentage of glycine in the original mixture. (b) A 4.71-mg sample of a compound containing C, H, O, and N was subjected to a Dumas nitrogen determination. The N_2, at 735 torr and 27°C, displaced 10.5532 g of mercury. A carbon-hydrogen analysis showed that this compound contained 3.90% H and 46.78% C. Determine the empirical formula of this compound.

PROBLEMS B   No answers given

22. Balance the following equations, which show the starting materials and the reaction products. It is not necessary to supply any additional reactants or products.
   (a) $H_3BO_3 \rightarrow H_4B_6O_{11} + H_2O$
   (b) $C_6H_{12}O_6 \rightarrow C_2H_5OH + CO_2$
   (c) $CaC_2 + N_2 \rightarrow CaCN_2 + C$
   (d) $CaCN_2 + H_2O \rightarrow CaCO_3 + NH_3$
   (e) $BaO + C + N_2 \rightarrow Ba(CN)_2 + CO$
   (f) $C_6H_6 + O_2 \rightarrow CO_2 + H_2O$
   (g) $C_7H_{16} + O_2 \rightarrow CO_2 + H_2O$
   (h) $H_3PO_3 \rightarrow H_3PO_4 + PH_3$
   (i) $MnO_2 + KOH + O_2 \rightarrow K_2MnO_4 + H_2O$
   (j) $KMnO_4 + H_2SO_4 \rightarrow K_2SO_4 + Mn_2O_7 + H_2O$
   (k) $CO + Fe_3O_4 \rightarrow FeO + CO_2$
   (l) $ZnS + O_2 \rightarrow ZnO + SO_2$

23. Some common gases may be prepared in the laboratory by reactions represented by the following equations. A Δ sign indicates that heating is

necessary. For purity, air must be swept out of the apparatus before the gas is collected, and some gases must be dried with a suitable desiccant.

(1) $2NaHSO_3 + H_2SO_4 \rightarrow 2SO_2 + 2H_2O + Na_2SO_4$

(2) $MnO_2 + 4HCl \xrightarrow{\Delta} Cl_2 + MnCl_2 + 2H_2O$

(3) $2Na_2O_2 + 2H_2O \rightarrow 4NaOH + O_2$

(4) $SiO_2 + 2H_2F_2 \rightarrow SiF_4 + 2H_2O$

(5) $2HCOONa + H_2SO_4 \xrightarrow{\Delta} 2CO + 2H_2O + Na_2SO_4$

(6) $NH_4Cl + NaNO_2 \xrightarrow{\Delta} N_2 + NaCl + 2H_2O$

(7) $2Al + 2NaOH + 2H_2O \rightarrow 2NaAlO_2 + 3H_2$

(8) $Al_4C_3 + 12H_2O \rightarrow 3CH_4 + 4Al(OH)_3$

(a) What weight of $NaHSO_3$ is needed to prepare (i) 1.3 moles of SO_2 and (ii) 2 lb of SO_2?

(b) How many moles of pyrolusite, MnO_2, are needed to prepare (i) 100 g of Cl_2 and (ii) 2.6 moles of Cl_2?

(c) How many pounds of sand, SiO_2, are needed to prepare 10 lb of SiF_4, assuming that 25% of the sand is inert material and does not produce SiF_4?

(d) How many grams of sodium formate, $HCOONa$, are needed to make 0.25 mole of CO?

(e) How many pounds of aluminum hydroxide are produced along with 12 moles of CH_4 (methane)?

(f) How many moles of NH_4Cl are needed to prepare 1.33 moles of N_2 by equation (6)?

(g) How many grams of 90%-pure Na_2O_2 (sodium peroxide) are needed to prepare 2.5 lb of O_2?

(h) Commercial sulfuric acid that has a density of 1.84 g/ml and is 95% H_2SO_4 by weight is used for the production of CO by equation (5). (i) What weight of commercial acid is needed for the production of 560 g of CO? (ii) What volume of acid is needed for the production of 560 g of CO?

(i) What volume of commercial HCl (36% HCl by weight, density = 1.18 g/ml) and weight of pyrolusite (85% MnO_2) are needed to produce 5 kg of Cl_2 by equation (2)?

(j) Commercial sulfuric acid that has a density of 1.52 g/ml and is 62% H_2SO_4 by weight is used for the production of SO_2 by equation (1). What (i) weight and (ii) volume of this commercial acid are needed for the production of 720 g of SO_2 gas?

(k) Assume that an excess of metallic aluminum is added to a solution containing 100 g NaOH and that as a result all the NaOH is used up. When the reaction is complete, the excess aluminum metal is filtered off and the excess water evaporated. How many grams of $NaAlO_2$ are obtained?

24. How many grams of aluminum are needed for the preparation of 5.5 liters of H_2 at standard conditions? The reaction is

$$2Al + 3H_2SO_4 \rightarrow Al_2(SO_4)_3 + 3H_2$$

25. What weight of $(NH_4)_2SO_4$ is needed for the preparation of 5 moles of NH_3 gas? The reaction is

$$(NH_4)_2SO_4 + Ca(OH)_2 \rightarrow CaSO_4 + 2NH_3 + 2H_2O$$

26. What volume of O_2 at 0.9 atm and 75°F is needed to burn 21 liters of propane gas, C_3H_8, under the same conditions? The reaction is

$$C_3H_8 + 5O_2 \rightarrow 3CO_2 + 4H_2O$$

27. What volume of NO can react with 100 liters of air (21% O_2 by volume) at the same conditions of temperature and pressure? The reaction is

$$2NO + O_2 \rightarrow 2NO_2$$

28. An interesting lecture demonstration is the "Vesuvius" experiment, in which a small mound of $(NH_4)_2Cr_2O_7$ is heated to commence decomposition. It then continues its decomposition unaided, gives off heat, light, and sparks, and leaves a "mountain" of Cr_2O_3. The reaction is

$$(NH_4)_2Cr_2O_7 \rightarrow N_2 + 4H_2O + Cr_2O_3$$

What volume of N_2 at 730 torr and 31°C is produced from 1.6 moles of $(NH_4)_2Cr_2O_7$?

29. What volume of H_2S at 720 torr and 85°F is needed for the precipitation of the bismuth in 50 g of $BiCl_3$? The reaction is

$$2BiCl_3 + 3H_2S \rightarrow Bi_2S_3 + 6HCl$$

30. What volume of H_2S at 740 torr and 20°C is needed to precipitate the nickel from 50 g of $Ni_3(PO_4)_2 \cdot 7H_2O$ as NiS?

31. HCN gas is fatal at a concentration of 1 part per 500 by volume and is very dangerous within one hour at a concentration of 1 part per 10,000 by volume. What weight of NaCN is needed to fill a classroom 20 ft × 15 ft × 9 ft with HCN at a concentration of 1 part per 10,000? (The barometric pressure is 740 torr, and the temperature is 80°F.) The reaction is

$$2NaCN + H_2SO_4 \rightarrow Na_2SO_4 + 2HCN$$

32. (a) How many grams of aluminum are needed for the production of 5.5 liters of H_2 over H_2O at 730 torr and 18°C? (b) How many moles of H_2SO_4 are needed? The reaction is given in Problem 24.

33. What volume of O_2, collected over water at 735 torr and 20°C, can be produced from 4 lb of sodium peroxide, Na_2O_2, according to the reaction

$$2Na_2O_2 + 2H_2O \rightarrow 4NaOH + O_2$$

34. (a) What weight of $NaNO_3$ is needed to produce 5 liters of N_2O collected over water at 737 torr and 27°C? The reaction is

$$NH_4Cl + NaNO_3 \rightarrow NaCl + N_2O + 2H_2O$$

(b) How many moles of NH_4Cl are needed?

35. (a) What volume of NH_3 at 700 torr and 50°C is needed to make 5 lb of $(NH_4)_3PO_4$? The reaction is

$$3NH_3 + H_3PO_4 \rightarrow (NH_4)_3PO_4$$

(b) What volume of H_3PO_4 (density $= 1.69$ g/ml, 85% H_3PO_4 by weight) is needed for this preparation?

36. (a) What volume of H_2SO_4 (density $= 1.84$ g/ml, 95% H_2SO_4 by weight) is needed to produce 8.3 liters of H_2 collected over water at 740 torr and 18°C? The reaction is

$$Mg + H_2SO_4 \rightarrow MgSO_4 + H_2$$

(b) How many moles of Mg are used?

37. What volume of air (21% O_2 by volume) at 720 torr and 68°F is needed to burn 1 lb of butane gas, C_4H_{10}? Balance the equation before solving the problem:

$$C_4H_{10} + O_2 \rightarrow CO_2 + H_2O$$

38. In the process of photosynthesis, plants use CO_2 and water to produce sugars according to the over-all reaction

$$11H_2O + 12CO_2 \rightarrow C_{12}H_{22}O_{11} + 12O_2$$

(a) What volume of CO_2 at 30°C and 730 torr is used by a plant in making 1 lb of sucrose, $C_{12}H_{22}O_{11}$? (b) What volume of air is deprived of its normal amount of CO_2 in this process? (Air contains 0.04% CO_2 by volume. Assume a barometric pressure of 750 torr and a temperature of 70°F.)

39. A big national industry produces ethyl alcohol, C_2H_5OH, by the enzymatic action of yeast on sugar, by the reaction

$$C_6H_{12}O_6 \xrightarrow{\text{zymase}} 2C_2H_5OH + 2CO_2$$

What volume of CO_2 at 720 torr and 30°C is formed during the production of 100 gal of alcohol that is 95% C_2H_5OH by weight? (The density of 95% C_2H_5OH is 0.80 g/ml.)

40. The adsorbing capacity of charcoal may be greatly increased by "steam activation." The principal action of the steam passing over the very hot charcoal is to widen the pores as the result of the reaction

$$H_2O + C \rightarrow H_2 + CO$$

The mixture of CO and H_2, known as water gas, can be used as a fuel for other manufacturing concerns, as well as to heat up the coke for this process. (a) How many cubic feet of water gas at 730 torr and 68°F are obtained from the activation of 50 tons of charcoal if it is activated to a 60% weight loss? (b) How many gallons of water (density $= 1$ g/ml) will this take?

41. What volume of water gas ($CO + H_2$) can be prepared from 1,500 liters of steam at 900°C and 1 atm? (The steam and water gas are measured under the same conditions. See Problem 40 for the equation.)

42. In the contact process for making H_2SO_4, S is burned with air to SO_2, and the SO_2-air mixture is then passed over a V_2O_5 catalyst, which converts it to SO_3:

$$S + O_2 \rightarrow SO_2$$
$$2SO_2 + O_2 \rightarrow 2SO_3$$

(a) What volume of air (21% O_2) at 27°C and 1 atm is needed for the conversion of 1 ton of sulfur to SO_3? (Use a 15% excess of oxygen.) (b) How many gallons of H_2SO_4 (95% H_2SO_4 by weight, density = 1.84 g/ml) can be produced from 1 ton of sulfur?

43. A common biological determination is for amino-acid nitrogen. This determination is made by the van Slyke method, in which the amino groups ($-NH_2$) in protein material react with HNO_2 to produce N_2 gas, the volume of which is measured. A 0.53-g sample of a biological material containing glycine, $CH_2(NH_2)COOH$, yielded 37.2 ml of N_2 gas collected over water at a pressure of 737 torr and 27°C. What is the percentage of glycine in the original sample? The reaction is

$$CH_2(NH_2)COOH + HNO_2 \rightarrow CH_2(OH)COOH + H_2O + N_2$$

Stoichiometry III.
Calculations Based on
Concentrations of Solutions

When solutions are involved in reactions, the stoichiometric calculations must take into account two quantities not previously discussed: (1) the concentration and (2) the volume of the solution.

CONCENTRATIONS OF SOLUTIONS

The concentration of solute in a solution may be expressed in many different ways. The most convenient way in chemical calculations is to relate the concentrations and the volumes of solutions to the chemical unit, the mole.

Molarity

A solution containing 1 mole of solute per liter of solution is known as a molar or 1 M solution. Concentration values other than unity are indicated by a numerical value preceding the symbol M. Thus, $2M$ H_2SO_4 indicates a solution that contains 2 moles or 2×98.1 g H_2SO_4 per liter of solution.

This definition for molarity of solutions may be expressed by the equation

$$\text{molarity of solution} = \frac{\text{moles of solute}}{\text{liters of solution}}$$

From this it follows that

number of moles of solute = molarity (M) × volume (V) in liters

and that the weight of solute in a solution is given by the relation

$$\text{grams of solute} = M \times V \times \frac{\text{grams}}{\text{mole}}$$

Millimoles

In many measurements it is convenient to express the volume of a solution in milliliters rather than in liters. For example, when we withdraw a solution from a buret the graduation marks on the buret are in milliliters. If we use a 45.00-ml portion, we record the value as 45.00 ml rather than as 0.04500 liter (just as we express the price of apples in cents per pound rather than in dollars).

Since a 1 M solution contains one mole of solute per liter of solution, a milliliter of solution contains one-thousandth of a mole or a millimole (m mol) of solute. This is a formula weight of solute expressed in milligrams. It follows that

$$\text{molarity of solution} = \frac{\text{millimoles of solute}}{\text{milliliters of solution}}$$

and

number of millimoles of solute = molarity (M) × volume in milliliters

Thus, the weight of solute is given by

$$\text{mg solute} = M \times V \text{ (in ml)} \times \frac{\text{mg}}{\text{m mol}}$$

In our calculations involving molarities and volumes of solutions, we choose the most convenient units for the calculation. If volumes are in liters we usually prefer to use moles, grams, and liters, but if the measurement involves milliliters it is often convenient to use millimoles, milligrams, and milliliters.

Molality

In stoichiometric calculations it is usually most convenient to relate the total number of moles of a solute to the volume of the solution. In certain thermodynamic calculations, on the contrary, it is convenient to relate the number of moles of solute to the amount of solvent present. The term *molal* indicates a solution that contains a mole of solute per kilogram of solvent. The designation for such a solution is a 1 m solution.

Mol Fraction

The mol fraction of a constituent is defined by the relation

$$\text{mol fraction } A = \frac{\text{number of moles of } A}{\text{total number of moles in the solution}}$$

It follows from this definition that the sum of the mol fractions of all the substances in a solution must be unity.

Illustrative Problems

Preparation of solutions and conversions from one set of concentration units to another are illustrated in the following problems.

PROBLEM:
A sulfuric acid solution of density 1.802 g/ml contains 88.0% H_2SO_4 by weight. Find the weight of H_2SO_4 per liter of the solution.

SOLUTION:
The density of a solution is a function of the concentration of solute. Since density is a readily measured property, it is quite common to show on a bottle label the density (or specific gravity) of the solution and the corresponding percentage of solute by weight in the solution.

$$\text{weight of 1 liter of solution} = 1.802 \, \frac{\text{g solution}}{\text{ml}} \times 1{,}000 \, \frac{\text{ml}}{\text{liter}}$$

$$= 1{,}802 \text{ g solution per liter}$$

$$\text{weight of } H_2SO_4 \text{ per liter} = 1{,}802 \, \frac{\text{g solution}}{\text{liter}} \times \frac{88.0 \text{ g } H_2SO_4}{100 \text{ g solution}}$$

$$= 1{,}586 \text{ g } H_2SO_4/\text{liter}$$

PROBLEM:
Compute the molarity of the H_2SO_4 solution of the preceding problem.

SOLUTION:
A mole of H_2SO_4 is 98.1 g. Since the solution contains 1,586 g of H_2SO_4 per liter,

$$\text{molarity} = \frac{1{,}586 \, \dfrac{\text{g } H_2SO_4}{\text{liter}}}{98.1 \, \dfrac{\text{g } H_2SO_4}{\text{mole}}} = 16.2 \text{ mole/liter}$$

PROBLEM:
What volume of the $16.2M$ H_2SO_4 solution of the preceding problem is needed to prepare 3.0 liters of $6.0M$ solution?

SOLUTION:

This problem illustrates a method we shall often have occasion to use, the preparation of a solution by dilution of a more concentrated solution of the same reagent. Since no solute is added or removed in the process, the number of moles of solute in the given volume of original solution is the same as the number in the final volume of diluted solution.

$$\text{moles of } H_2SO_4 \text{ needed} = V \times M = 3.0 \text{ liters} \times 6.0 \frac{\text{moles } H_2SO_4}{\text{liter}}$$

$$= 18.0 \text{ moles of } H_2SO_4$$

$$\text{volume of } 16.2M \text{ solution} = \frac{18.0 \text{ moles } H_2SO_4}{16.2 \frac{\text{moles } H_2SO_4}{\text{liter}}}$$

$$= 1.11 \text{ liter}$$

Note that in a dilution, the product of molarity × volume remains constant, or $M_1V_1 = M_2V_2 =$ etc.

PROBLEM:

What is the molality of the $16.2M$ H_2SO_4 solution of the preceding examples?

SOLUTION:

As previously computed, a liter of the solution weighs 1,802 g and contains 1,586 g of H_2SO_4. The difference, 216 g, is the weight of water per liter of solution. The ratio, moles of H_2SO_4/weight of water in kg, gives the moles of H_2SO_4 per kilogram of water or the molality of the solution.

$$\text{molality} = \frac{16.2 \text{ moles } H_2SO_4}{0.216 \text{ kg } H_2O} = 75 \frac{\text{moles } H_2SO_4}{\text{kg } H_2O}$$

PROBLEM:

What are the mol fractions of H_2O and H_2SO_4 in the solution of the preceding problems?

SOLUTION:

To find the mol fractions we need to know the number of moles of each substance in a given amount of solution. For convenience we select a liter of solution, containing 16.2 moles of H_2SO_4 and 216 g of H_2O.

$$\text{moles of } H_2O = \frac{216 \text{ g } H_2O}{18.0 \frac{\text{g } H_2O}{\text{mole } H_2O}} = 12.0 \text{ moles}$$

total moles in a liter of solution

$$= 16.2 \text{ moles } H_2SO_4 + 12.0 \text{ moles } H_2O = 28.2 \text{ moles}$$

$$\text{mol fraction of } H_2O = \frac{12.0 \text{ moles } H_2O}{28.2 \text{ total moles}} = 0.426$$

$$\text{mol fraction of } H_2SO_4 = \frac{16.2 \text{ moles } H_2SO_4}{28.2 \text{ total moles}} = 0.574$$

Note that the mole fractions total 1.000.

PROBLEM:
A test solution used in the laboratory is made up to contain 10 mg of the metal ion per milliliter of the solution. What weight of $CuSO_4 \cdot 5H_2O$ is used to prepare 2.000 liters of test solution for cupric ion? What is the molarity of this solution?

SOLUTION:
Since we require 2,000 ml of solution,

$$\text{total weight of } Cu^{++} = 2,000 \text{ ml} \times 10 \frac{mg}{ml} = 20,000 \text{ mg} = 20.000 \text{ g}$$

$$\text{moles of } Cu^{++} = \frac{20.000 \text{ g } Cu^{++}}{63.5 \frac{\text{g } Cu^{++}}{\text{mole } Cu^{++}}} = 0.315 \text{ moles}$$

Moles of $CuSO_4 \cdot 5H_2O = 0.315$ (since each mole of the salt produces a mole of cupric ion.)

$$\text{weight of } CuSO_4 \cdot 5H_2O = 0.315 \text{ moles} \times 249.6 \frac{g}{\text{mole}}$$

$$= 78.6 \text{ g}$$

$$\text{molarity of solution} = \frac{0.315 \text{ moles}}{2.000 \text{ liters}} = 0.1575 \text{ moles/liter}$$

VOLUMES AND MOLARITIES
OF SOLUTIONS IN REACTIONS

When a solution is involved in a reaction the stoichiometric calculations usually deal with the volume and the concentration of the solution rather than with the weight of the dissolved substance.

PROBLEM:
Cupric nitrate is prepared by dissolving a weighed amount of copper metal in nitric acid solution:

$$3Cu + 8HNO_3 \rightarrow 3Cu(NO_3)_2 + 2NO + 4H_2O$$

What volume of $6M$ HNO_3 should be used to prepare 10.00 g of $Cu(NO_3)_2$?

SOLUTION:

$$\text{moles of } Cu(NO_3)_2 = \frac{10.00 \text{ g } Cu(NO_3)_2}{187.5 \dfrac{\text{g } Cu(NO_3)_2}{\text{mole } Cu(NO_3)_2}}$$

$$= 0.0533 \text{ mole}$$

$$\text{moles of } HNO_3 = \frac{8 \text{ moles } HNO_3}{3 \text{ moles } Cu(NO_3)_2} \times 0.0533 \text{ moles } Cu(NO_3)_2$$

$$= 1.425 \text{ moles } HNO_3$$

At this stage we could compute the weight of HNO_3 as in the calculations of the preceding chapter, then convert to the volume of $6M$ solution. But, since we know that a liter of solution contains 6 moles of HNO_3, we can compute the volume of solution directly from the number of moles required:

$$\text{volume of } 6M \ HNO_3 = \frac{1.425 \text{ moles } HNO_3}{6 \dfrac{\text{moles } HNO_3}{\text{liter}}}$$

$$= 0.237 \text{ liter}$$

Titration

The amount of a solution needed to react with a given quantity of another substance is often determined by a process known as *titration*. The reagent solution, the titrant, is added from a buret to the sample solution until chemically equivalent amounts of the two are present, as shown by a color change in the solution or by some other type of end-point indicator.

Titrations are frequently used to analyze samples. The process is rapid and the measurement, that of the volume of solution used, can be made with high accuracy. If this method is to be used for an analysis, the titrant solution must first be standardized (its strength accurately determined) by comparison with a known sample of a substance of known purity. Standardization of solutions may be done by any one of three ways.

1. If the reagent chemical is pure and dry, a weighed amount is dissolved and diluted to an exactly known volume in a volumetric flask. The molarity is computed from the weight and volume.

2. A weighed sample of a pure, dry chemical that will react with the solution to be standardized is dissolved and titrated by the solution. From the weight of the pure chemical, known as a primary standard, and the volume of the titrant solution used, we compute the molarity of the solution.

3. A measured volume of the solution to be standardized is titrated with a standard solution of a reagent that will react with it.

These three methods of standardizing solutions are illustrated in the following problems.

PROBLEM:
A 8.650-g sample of pure dry $AgNO_3$ is dissolved and diluted to exactly 500 ml in a volumetric flask, then mixed thoroughly to insure uniformity. What is the molarity of the solution?

SOLUTION:

$$\text{moles of } AgNO_3 = \frac{8.650 \text{ g}}{169.9 \frac{\text{g}}{\text{mole}}} = 0.05090 \text{ moles}$$

$$\text{molarity of solution} = \frac{0.05090 \text{ moles}}{0.5000 \text{ liter}} = 0.1018 \frac{\text{moles}}{\text{liter}}$$

PROBLEM:
A liter of approximately $0.1M$ HCl is prepared by diluting 8.5 ml of $12M$ solution to about 1 liter. The solution is standardized by titrating a weighed sample of pure dry Na_2CO_3:

$$Na_2CO_3 + 2HCl \rightarrow 2NaCl + H_2O + CO_2$$

$$\text{weight of } Na_2CO_3 = 0.275 \text{ g}$$

$$\text{volume of HCl} = 48.5 \text{ ml}$$

Compute the molarity of the HCl solution.

SOLUTION:
Since the titration volume is only 48.5 ml, we use millimoles instead of moles:

$$\text{m mol of } Na_2CO_3 = \frac{275 \text{ mg } Na_2CO_3}{106.0 \frac{\text{mg}}{\text{m mol}}} = 2.595 \text{ m mol}$$

$$\text{m mol of HCl} = 2.595 \text{ m mol } Na_2CO_3 \times \frac{2 \text{ m mol HCl}}{1 \text{ m mol } Na_2CO_3}$$

$$= 5.180 \text{ m mol}$$

$$\text{molarity of HCl} = \frac{5.180 \text{ m mol}}{48.5 \text{ ml}} = 0.107 \text{ m mol/ml}$$

PROBLEM:
The $0.107M$ HCl solution of the preceding problem is used to standardize a solution of $Ba(OH)_2$:

$$Ba(OH)_2 + 2HCl \rightarrow BaCl_2 + 2H_2O$$

It was found that 37.6 ml of $Ba(OH)_2$ was titrated by 43.8 ml HCl. What is the molarity of the $Ba(OH)_2$ solution?

SOLUTION:

$$\text{m mol of HCl} = 0.107 \frac{\text{m mol}}{\text{ml}} \times 43.8 \text{ ml} = 4.68 \text{ m mol}$$

$$\text{m mol of Ba(OH)}_2 = 4.68 \text{ m mol HCl} \times \frac{1 \text{ m mol Ba(OH)}_2}{2 \text{ m mol HCl}} = 2.34 \text{ m mol}$$

$$\text{molarity of Ba(OH)}_2 = \frac{2.34 \text{ m mol}}{37.6 \text{ ml}} = 0.0623 \text{ m mol/ml} = 0.0623M$$

Typical Titration Analyses

The following problems illustrate analyses using standard solutions.

PROBLEM:

A 0.500-g sample of impure $CaCO_3$ is dissolved in exactly 50 ml of $0.0985M$ HCl solution. After the sample is all dissolved, the excess HCl is titrated by 6.00 ml of $0.1050M$ NaOH solution. The reactions are:

$$CaCO_3 + 2HCl \rightarrow CaCl_2 + H_2O + CO_2$$
$$HCl + NaOH \rightarrow NaCl + H_2O$$

Compute the percentage of $CaCO_3$ in the sample, assuming that it contains no other substance that will react with HCl or NaOH.

SOLUTION:

We note that part of the HCl added reacts with the $CaCO_3$ sample and part with the NaOH solution used for back-titration after the sample is all dissolved.

$$\text{total m mol of HCl} = 50.00 \text{ ml} \times 0.0985 \frac{\text{m mol}}{\text{ml}}$$
$$= 4.925 \text{ m mol}$$

$$\text{m mol of NaOH} = 6.00 \text{ ml} \times 0.1050 \frac{\text{m mol}}{\text{ml}} = 0.630 \text{ m mol}$$

$$\text{m mol of HCl reacting with NaOH} = 0.630 \text{ (see equation)}$$
$$\text{m mol of HCl reacting with CaCO}_3 = 4.925 \text{ m mol} - 0.630 \text{ m mol}$$
$$= 4.295 \text{ m mol}$$

Note that this is the "net" HCl used to react with the sample. Then

$$\text{m mol of CaCO}_3 = 4.295 \text{ m mol HCl} \times \frac{1 \text{ m mol CaCO}_3}{2 \text{ m mol HCl}}$$
$$= 2.148 \text{ m mol}$$

$$\text{weight of } CaCO_3 = 2.148 \text{ m mol} \times 100.1 \frac{mg}{m \, mol}$$

$$= 215 \text{ mg}$$

$$\text{per cent of } CaCO_3 = \frac{215 \text{ mg}}{500 \text{ mg}} \times 100 = 43.0\%$$

PROBLEM:

An iron ore is a mixture of Fe_2O_3 and inert impurities. One method of analysis is to dissolve the ore sample in HCl, reduce all the iron to the ferrous condition, then titrate with a standard solution of $KMnO_4$ in acid solution. The permanganate oxidizes ferrous ion to ferric according to the equation

$$MnO_4^- + 8H^+ + 5Fe^{++} \rightarrow Mn^{++} + 4H_2O + 5Fe^{+++}$$

From the following data compute the percentage of iron, as Fe, in the ore sample. Also express the percentage of iron as Fe_2O_3.

   Data: volume of $KMnO_4$ = 38.60 ml
             molarity of $KMnO_4$ = 0.02100M
             weight of ore sample = 446.0 mg

SOLUTION:

$$\text{m mol of } KMnO_4 = 38.60 \text{ ml} \times 0.02100 \text{ m mol/ml} = 0.810 \text{ m mol}$$

$$\text{m mol of } Fe^{++} = 0.810 \text{ m mol } KMnO_4 \times \frac{5 \text{ m mol } Fe^{++}}{\text{m mol } MnO_4^-}$$

$$= 4.05 \text{ m mol}$$

$$\text{weight of } Fe^{++} = 4.05 \text{ m mol} \times 55.8 \frac{mg}{m \, mol} = 226 \text{ mg}$$

$$\text{per cent of } Fe = \frac{226 \text{ mg}}{446 \text{ mg}} \times 100 = 50.7\%$$

To express the percentage in terms of Fe_2O_3, we note that 1 m mol of Fe_2O_3 will produce 2 m mol of Fe^{++} when the sample is dissolved and the iron reduced:

$$\text{m mol of } Fe_2O_3 = 4.05 \text{ m mol } Fe^{++} \times \frac{1 \text{ m mol } Fe_2O_3}{2 \text{ m mol } Fe^{++}}$$

$$= 2.025 \text{ m mol}$$

$$\text{weight of } Fe_2O_3 = 2.025 \text{ m mol} \times 159.6 \frac{mg}{m \, mol}$$

$$= 323 \text{ mg}$$

$$\text{per cent of } Fe_2O_3 = \frac{323 \text{ mg}}{446 \text{ mg}} \times 100 = 72.4\%$$

PROBLEMS A   Answers on page 344

1. Tell how you would prepare each of the following solutions:
   (a) 3 liters of $0.75M$ $AgNO_3$ from solid $AgNO_3$
   (b) 55 ml of $2M$ $CuSO_4$ from solid $CuSO_4 \cdot 5H_2O$
   (c) 180 ml of $0.1M$ $Ba(NO_3)_2$ from solid $Ba(NO_3)_2$
   (d) 12 liters of $6.0M$ KOH from solid KOH
   (e) 730 ml of $0.07M$ $Fe(NO_3)_3$ from solid $Fe(NO_3)_3 \cdot 9H_2O$

2. Tell how you would prepare each of the following solutions:
   (a) 15 ml of $0.2M$ H_2SO_4 from $6.0M$ H_2SO_4
   (b) 280 ml of $0.6M$ $CoCl_2$ from $3M$ $CoCl_2$
   (c) 5.7 liters of $0.03M$ $ZnSO_4$ from $2.5M$ $ZnSO_4$
   (d) 60 ml of $0.0035M$ $K_3Fe(CN)_6$ from $0.8M$ $K_3Fe(CN)_6$
   (e) 25 ml of $2.7M$ $UO_2(NO_3)_2$ from $8.3M$ $UO_2(NO_3)_2$

3. Find molarities and molalities of each of the following solutions.

Solution	Density (g/ml)	Weight Percent	Molarity	Molality
(a) KOH	1.344	35.0		
(b) HNO_3	1.334	54.0		
(c) H_2SO_4	1.834	95.0		
(d) $MgCl_2$	1.119	29.0		
(e) $Na_2Cr_2O_7$	1.140	20.0		
(f) $Na_2S_2O_3$	1.100	12.0		
(g) Na_3AsO_4	1.113	10.0		
(h) $Al_2(SO_4)_3$	1.253	22.0		

4. Give the weight of metal ion in each of the following solutions:
   (a) 250 ml of $0.10M$ $CuSO_4$ (e) 75 ml of $0.10M$ $AlCl_3$
   (b) 125 ml of $0.05M$ $CdCl_2$ (f) 1.5 liters of $3.0M$ $AgNO_3$
   (c) 50 ml of $0.15M$ $MgSO_4$ (g) 2.0 liters of $0.333M$ $FeSO_4$
   (d) 50 ml of $0.075M$ $MgSO_4$

5. Tell how to prepare each of the following test solutions with concentrations of 1.0 mg of metal ion per milliliter of solution:
   (a) 500 ml of $AgNO_3$ from $0.5M$ $AgNO_3$ solution
   (b) 500 ml of $AgNO_3$ from solid $AgNO_3$
   (c) 1 liter of $CuSO_4$ from solid $CuSO_4 \cdot 5H_2O$
   (d) 1 liter of $CuSO_4$ from $0.05M$ $CuSO_4$ solution
   (e) 250 ml of $AlCl_3$ by dissolving solid Al in HCl
   (f) 2 liters of Na_2CO_3 from solid Na_2CO_3

6. The density of a $7M$ HCl solution is 1.113 g/ml. Find the percentage of HCl by weight.

7. If 3 liters of $6M$ HCl are added to 2 liters of $1.5M$ HCl, what is the resulting concentration? (Assume the final volume to be exactly 5 liters.)

8. What volume of $15M$ HNO_3 should be added to 1,250 ml of $2M$ HNO_3 to prepare 14 liters of $1M$ HNO_3? (Water is added to make the final volume exactly 14 liters.)

9. If 40.00 ml of an HCl solution is titrated by 45.00 ml of $0.15M$ NaOH, (a) what is the molarity of the HCl? (b) How many milligrams of HCl are in 1 ml of solution?

10. We need 35.45 ml of an NaOH solution to titrate a 2.0813-g sample of pure benzoic acid, $HC_7H_5O_2$. What is the molarity of the NaOH solution?

11. What is the molarity of a $K_2C_2O_4$ solution if 35.00 ml of it is needed for the titration of 47.65 ml of $0.0632M$ $KMnO_4$ solution? The reaction is

$$2MnO_4^- + 5C_2O_4^= + 16H^+ \rightarrow 2Mn^{++} + 10CO_2 + 8H_2O$$

12. How many milliliters of a solution containing 31.52 g $KMnO_4$ per liter will react with 3.814 g $FeSO_4 \cdot 7H_2O$? The reaction is

$$MnO_4^- + 5Fe^{++} + 8H^+ \rightarrow Mn^{++} + 5Fe^{+++} + 4H_2O$$

13. It takes 35.00 ml of $0.1500M$ KOH to react with 40.00 ml of H_3PO_4 solution. The titration reaction is

$$H_3PO_4 + 2OH^- \rightarrow HPO_4^= + 2H_2O$$

What is the molarity of the H_3PO_4 solution?

14. If we add 39.20 ml of $0.1333M$ H_2SO_4 to 0.4550 g of a sample of soda ash that is 59.95% Na_2CO_3, what volume of $0.1053M$ NaOH is required for back-titration?

15. A 19.75-ml sample of vinegar of density 1.061 g/ml requires 43.24 ml of $0.3982M$ NaOH for titration. What is the percentage by weight of acetic acid in the vinegar?

16. A 0.500-g sample of impure $CaCO_3$ is dissolved in 50.0 ml of $0.100M$ HCl, and the residual acid titrated by 5.00 ml of $0.120M$ NaOH. Find the percentage of $CaCO_3$ in the sample. The reaction is

$$CaCO_3 + 2HCl \rightarrow CaCl_2 + H_2O + CO_2$$

17. It takes 45.00 ml of a given HCl solution to react with 0.2435 g of calcite, $CaCO_3$. This acid is used to determine the percentage purity of a $Ba(OH)_2$ sample, as follows:

         wt of impure $Ba(OH)_2$ sample $= 0.4367$ g
         vol of HCl used $= 35.27$ ml
         vol of NaOH used for back-titration $= 1.78$ ml
         1.200 ml of HCl titrates 1.312 ml of NaOH

What is the percentage of $Ba(OH)_2$ in the sample?

18. Adding 25.00 ml of an HCl solution to an excess of $AgNO_3$ solution gave a precipitate of AgCl that weighed 0.5600 g. (a) How many milliliters of the HCl solution will be needed to prepare 250 ml of a 0.12M solution? (b) To what volume should 250 ml of the HCl solution be diluted to prepare a 0.10M solution?

19. (a) What weight of AgCl can be obtained by precipitating all the Ag^+ from 50 ml of 0.12M $AgNO_3$? (b) What weight of NaCl is required to precipitate the AgCl? (c) What volume of 0.24M HCl would be needed to precipitate the AgCl?

20. What volume of 10M HCl is needed to prepare 6.4 liters H_2S at 750 torr and 27°C? The reaction is

$$FeS + 2HCl \rightarrow FeCl_2 + H_2S$$

21. What volume of 12.5M NaOH is needed to prepare 25 liters of H_2 at 735 torr and 18°C by the reaction

$$2Al + 2NaOH + 2H_2O \rightarrow 2NaAlO_2 + 3H_2$$

22. What weight of silver and what volume of 6M HNO_3 are needed for the preparation of 500 ml of 3M $AgNO_3$? What volume of NO, collected over water at 725 torr and 27°C, will be formed? The reaction is

$$3Ag + 4HNO_3 \rightarrow 3AgNO_3 + NO + 2H_2O$$

23. The "saponification number" of a fat or oil is defined as the number of milligrams of KOH required to saponify 1 g of it. To a sample of peanut oil weighing 1.5763 g is added 25.00 ml of 0.4210M KOH solution. After saponification is complete, 8.46 ml of 0.2732M H_2SO_4 is needed to neutralize the excess KOH. What is the saponification number of this peanut oil?

24. Fuming sulfuric acid is a mixture of H_2SO_4 and SO_3. A 2.500-g sample of fuming sulfuric acid needed 47.53 ml of 1.1513M NaOH for titration. What is the percentage of SO_3 in the sample?

25. You have a 0.5000-g mixture of oxalic acid, $H_2C_2O_4 \cdot 2H_2O$, and benzoic acid, $HC_7H_5O_2$. This sample requires 47.53 ml of 0.1151M KOH for titration. What is the percentage composition of this mixture?

PROBLEMS B   No answers given

26. Tell how you would prepare each of the following solutions:
    (a) 125 ml of 0.62M NH_4Cl from solid NH_4Cl
    (b) 2.75 liters of 1.72M $Ni(NO_3)_2$ from solid $Ni(NO_3)_2 \cdot 6H_2O$
    (c) 65 ml of 0.25M $Al(NO_3)_3$ from solid $Al(NO_3)_3 \cdot 9H_2O$

(d) 230 ml of $0.46M$ LiOH from solid LiOH

(e) 7.57 liters of $1.1M$ KCr$(SO_4)_2$ from solid KCr$(SO_4)_2 \cdot 12H_2O$

27. Tell how you would prepare each of the following solutions:

(a) 750 ml of $0.55M$ H_3PO_4 from $3.6M$ H_3PO_4

(b) 12 liters of $3M$ NH_3 from $15M$ NH_3

(c) 25 ml of $0.02M$ $Pr_2(SO_4)_3$ from $0.5M$ $Pr_2(SO_4)_3$

(d) 365 ml of $0.075M$ $K_4Fe(CN)_6$ from $0.95M$ $K_4Fe(CN)_6$

28. Tell how you would prepare each of the following solutions (weight percentage given):

(a) 650 ml of $0.35M$ $AlCl_3$ from a 16% solution whose density is 1.149 g/ml

(b) 1.35 liters of $4.35M$ NH_4NO_3 from a 62% solution whose density is 1.294 g/ml

(c) 465 ml of $3.7M$ H_3PO_4 from an 85% solution whose density is 1.689 g/ml

(d) 75 ml of $1.25M$ $CuCl_2$ from a 36% solution whose density is 1.462 g/ml

(e) 8.32 liters of $1.50M$ $ZnCl_2$ from a 60.3% solution whose density is 1.747 g/ml

29. Find the molality of the original solutions used in Problem 28.

30. The density of a $3.68M$ sodium thiosulfate solution is 1.269 g/ml. (a) Find the percentage of $Na_2S_2O_3$ by weight. (b) What is the molality of this solution?

31. The solubility of Hg_2Cl_2 is 7×10^{-4} g per 100 g of water at 30°C. (a) Assuming that the density of water (1 g/ml) is not appreciably affected by the presence of this amount of Hg_2Cl_2, determine the molarity of this saturated solution of Hg_2Cl_2. (b) What is the molality of this solution?

32. A colorimetric test for cobalt that will detect as little as 1 part of cobalt ion in 250,000 parts of solution requires the following reagents: (a) 10% KI solution, (b) 1% solution of α-nitroso-β-naphthol in acetone (instead of water), (c) saturated sodium acetate, (d) $2M$ HCl solution. The density of acetone is 0.792 g/ml. How would you prepare these solutions?

33. A pharmaceutical house wishes to prepare a nonirritating nose-drop preparation. To do this, it will put the active agent in a "normal saline" solution, which is merely 0.90% NaCl. What quantities of material will be needed if it is desired to make 3,000 gal of nose-drop solution that is normal saline and contains 0.1% active agent? (The density of 0.90% NaCl solution is 1.005 g/ml.)

34. If 500 ml of $3M$ H_2SO_4 is added to 1.5 liters of $0.5M$ H_2SO_4, what is the resulting concentration?

35. What volume of $15M$ NH_3 should be added to 3.5 liters of $3M$ NH_3 in order to give 6 liters of $5M$ NH_3 on dilution with water?

36. The density of a 1.66M $Na_2Cr_2O_7$ solution is 1.244 g/ml. (a) Find the percentage of $Na_2Cr_2O_7$ by weight. (b) If 1.5 liters of water are added to 1 liter of this solution, what is the percentage by weight of $Na_2Cr_2O_7$ in the new solution?

37. If 35 ml of 0.75M $Al(NO_3)_3$ are added to 100 ml of 0.15M $Al(NO_3)_3$, what will be the resulting concentration?

38. (a) What is the percentage of HNO_3 by weight in a 21.2M solution of HNO_3 whose density is 1.483 g/ml? (b) What is the molality of this solution?

39. What volume of 4M NaOH should be added to 5 liters of 0.5M NaOH in order to get 15 liters of 1.0M NaOH on dilution with water?

40. The density of a 2.04M $Cd(NO_3)_2$ solution is 1.382 g/ml. If 500 ml of water is added to 750 ml of this solution, (a) what will be the percentage by weight of $Cd(NO_3)_2$ in the new solution? (b) What will be its molality?

41. How would you make a standard solution that contains 10.0 γ of Hg^{++} per liter of solution? (Start with $HgCl_2$ and use volumetric flasks, pipets, and a balance sensitive to 0.2 mg. A γ is a millionth of a gram.)

42. It takes 31.00 ml of 0.2500M H_2SO_4 to react with 48.00 ml NH_3 solution. What is the molarity of the NH_3?

43. How many milligrams of Na_2CO_3 will react with 45.00 ml of 0.2500M HCl?

44. What is the molarity of a $KMnO_4$ solution if 30.00 ml of it is needed for the titration of 45.00 ml of 0.1550M $Na_2C_2O_4$ solution? (See Problem 11 for the equation.)

45. What is the molarity of a ceric sulfate solution if 46.35 ml is required for the titration of a 0.2351-g sample of $Na_2C_2O_4$ that is 99.60% pure? The reaction is

$$2Ce^{++++} + C_2O_4^= \rightarrow 2Ce^{+++} + 2CO_2$$

46. A 0.212-g sample of pure iron wire is dissolved, reduced to Fe^{++}, and titrated by 40.0 ml $KMnO_4$ solution. The reaction is

$$MnO_4^- + 5Fe^{++} + 8H^+ \rightarrow Mn^{++} + 5Fe^{+++} + 4H_2O$$

Find the molarity of the $KMnO_4$ solution.

47. From the following data compute the molarity of (a) an H_2SO_4 solution and (b) a KOH solution:

    wt of sulfamic acid, $H(NH_2)SO_3 = 0.2966$ g
    vol of KOH to neutralize the $H(NH_2)SO_3 = 34.85$ ml
    31.08 ml of H_2SO_4 titrates 33.64 ml of KOH

48. A crystal of calcite, $CaCO_3$, is dissolved in excess HCl, and the solution boiled to remove the CO_2. The unneutralized acid is then titrated by a base solution that has previously been compared with the acid solution.

Calculate the molarity of (a) the acid and (b) the base solutions. The data are

wt of calcite $= 1.9802$ g
vol of HCl added to calcite $= 45.00$ ml
vol of NaOH used in back-titration $= 14.43$ ml
30.26 ml of acid titrates 21.56 ml of base

49. A 0.220-g sample of H_2SO_4 is diluted with water and titrated by 40.0 ml of $0.100M$ NaOH. Find the percentage by weight of H_2SO_4 in the sample.

50. A sample of vinegar weighs 14.36 g and requires 42.45 ml of $0.2080M$ NaOH for titration. Find the percentage of acetic acid in the vinegar.

51. A 2.500-g sample of an ammonium salt of technical grade is treated with concentrated NaOH, and the NH_3 that is liberated is distilled and collected in 50.00 ml of $1.2000M$ HCl; 3.65 ml of $0.5316M$ NaOH is required for back-titration. Calculate the percentage of NH_3 in the sample.

52. A 0.500-g sample of impure CaO is added to 50.0 ml of $0.100M$ HCl. The excess HCl is titrated by 5.00 ml of $0.125M$ NaOH. Find the percentage of CaO in the sample. The reaction is

$$CaO + 2HCl \rightarrow CaCl_2 + H_2O$$

53. Silver nitrate solution is prepared by dissolving 85.2 g of pure $AgNO_3$ and diluting to 500 ml. (a) What is its molarity? (b) A $CaCl_2$ sample is titrated by 40.00 ml of this $AgNO_3$ solution. What is the weight of $CaCl_2$ in the sample?

54. (a) What weight of MnO_2 and (b) what volume of $12M$ HCl are needed for the preparation of 750 ml of $2M$ $MnCl_2$? (c) What volume of Cl_2 at 745 torr and 23°C will be formed? The reaction is

$$MnO_2 + 4HCl \rightarrow MnCl_2 + Cl_2 + 2H_2O$$

55. The "saponification number" of a fat or oil is defined as the number of milligrams of KOH required to saponify 1 g of it. To a sample of vegetable oil weighing 1.8531 g is added 25.00 ml of $0.3500M$ KOH solution. After saponification is complete, 3.57 ml of $0.5505M$ HCl is needed to neutralize the excess KOH. What is the saponification number of this vegetable oil?

56. What volume of $12M$ HCl is needed to prepare 3 liters of Cl_2 at 730 torr and 25°C by the reaction

$$2KMnO_4 + 16HCl \rightarrow 2MnCl_2 + 5Cl_2 + 8H_2O + 2KCl$$

57. (a) What volume of $6M$ HNO_3 and what weight of copper are needed for the production of 1.5 liters of a $0.50M$ $Cu(NO_3)_2$ solution? (b) What volume of NO, collected over water at 745 torr and 18°C, will be produced at the same time? The reaction is

$$3Cu + 8HNO_3 \rightarrow 3Cu(NO_3)_2 + 2NO + 4H_2O$$

58. What volume of $10M$ HCl is needed to prepare 12.7 liters of CO_2 at 735 torr and 35°C? The reaction is

$$CaCO_3 + 2HCl \rightarrow CaCl_2 + CO_2 + H_2O$$

59. A 20.00-ml sample of a solution containing $NaNO_2$ and $NaNO_3$ is acidified with H_2SO_4 and then treated with an excess of NaN_3 (sodium azide). The hydrazoic acid so formed reacts with and completely removes the HNO_2. It does not react with the nitrate. The reaction is

$$HN_3 + HNO_2 \rightarrow H_2O + N_2 + N_2O$$

The volume of N_2 and N_2O is measured over water and found to be 36.5 ml at 740 torr and 27°C. What is the molar concentration of $NaNO_2$ in this solution?

Chemical Equilibrium in Gases

When reacting chemical systems come to a dynamic equilibrium, a quantitative relationship exists between the concentrations of all reactants and products in the system. For the general reaction

$$aA + bB + cC \ldots \rightleftarrows mM + nN + \ldots,$$

this relation (which is derived in your textbook) is

$$K_e = \frac{[M]^m[N]^n}{[A]^a[B]^b[C]^c} \tag{14-1}$$

K_e is known as the *equilibrium constant,* and each of the quantities in brackets is the *equilibrium* concentration of the substance shown. The size of the K_e value depends on the concentration units used. For gases, as discussed in this chapter, it is customary to give these concentrations as pressures (P) expressed in atmospheres. Thus, equation 14-1 could be rewritten as

$$K_e = \frac{P_M^m \cdot P_N^n}{P_A^a \cdot P_B^b \cdot P_C^c} \tag{14-2}$$

K_e varies with temperature, but at a given temperature each reversible reaction has a value of K_e that remains constant whether equilibrium is approached by mixing A, B, and C, or M and N, and regardless of the proportions in which the chemicals are mixed.

Determination of K_e

Values of the equilibrium constant may be obtained by allowing the reactants to come to equilibrium at a given temperature, analyzing the equilibrium mixture, and then substituting the equilibrium concentrations into equation 14-2.

PROBLEM:

Starting with a 3:1 mixture of H_2 and N_2 at 450°C, it is found that the equilibrium mixture is 9.6% NH_3, 22.6% N_2, and 67.8% H_2 by volume. The total pressure is 50 atm. Calculate K_e. The reaction is $N_2 + 3H_2 \rightleftarrows 2NH_3$.

SOLUTION:

According to Dalton's Law of Partial Pressures the partial pressure of a gas in a mixture is proportional to the volume fraction of that gas in the mixture. The equilibrium pressure of each gas is thus:

$$P_{NH_3} = (.096)(50 \text{ atm}) = 4.80 \text{ atm}$$
$$P_{N_2} = (.226)(50 \text{ atm}) = 11.30 \text{ atm}$$
$$P_{H_2} = (.678)(50 \text{ atm}) = \underline{33.90 \text{ atm}}$$
$$\text{the total is } 50.00 \text{ atm}$$

by substitution

$$K_e = \frac{P_{NH_3}^2}{P_{N_2} \times P_{H_2}^3} = \frac{(4.8)^2}{(11.3)(33.9)^3}$$
$$K_e = 5.22 \times 10^{-5}$$

Equilibrium constants for a number of gaseous reactions are given in Table 14-1.

The Effect of Total Pressure on K_e and Equilibrium Position

When the preceding experiment with H_2 and N_2 is conducted at 450°C and 100 atm total pressure, an equilibrium mixture is obtained that, on analysis, proves to be 16.36% NH_3, 20.96% N_2, and 62.68% H_2. Following the same line of reasoning as in the last problem, we can determine that the equilibrium pressures are $P_{NH_3} = 16.36$ atm, $P_{N_2} = 20.96$ atm, and $P_{H_2} = 62.68$ atm, and the calculated value of $K_e = 5.22 \times 10^{-5}$. Note that doubling the total pressure does *not* double the pressure of each gas as it would have done in a mixture of nonreacting gases. Instead, the chemical equilibrium shifts to

TABLE 14-1.
Equilibrium Constants
for Selected Gaseous Reactions

Equilibrium	Temp, °C	K_e
$H_2 \rightleftarrows 2H$	1,000	7.0×10^{-18}
	2,000	3.1×10^{-6}
$Cl_2 \rightleftarrows 2Cl$	1,000	2.45×10^{-7}
	2,000	0.570
$N_2O_4 \rightleftarrows 2NO_2$	25	0.143
	45	0.671
$2H_2O \rightleftarrows 2H_2 + O_2$	1,000	6.9×10^{-15}
	1,700	6.4×10^{-8}
$2H_2S \rightleftarrows 2H_2 + S_2$	1,130	0.0260
	1,200	0.0507
$H_2 + Cl_2 \rightleftarrows 2HCl$	1,200	2.51×10^4
	1,800	1.12×10^3
$SO_2 + \frac{1}{2}O_2 \rightleftarrows SO_3$	900	6.55
	1,000	1.86
$CO_2 + H_2 \rightleftarrows CO + H_2O$	700	0.534
	1,000	0.719

make more NH_3 at the expense of N_2 and H_2 in such a way that the new equilibrium pressures give exactly the same value of K_e.

We could generalize from this problem and say that, for any chemical equilibrium involving a different number of moles of gas on each side of the balanced equation, the equilibrium position will always shift with an increase in total pressure toward the side with the smaller number of gaseous moles; the value of K_e will remain unchanged.

The Effect on K_e and Equilibrium Position
of Changing the Partial Pressure of One Gas

It is often important to know the yield of a chemical reaction, that is the percentage of reactants converted to products. The following example shows how this may be calculated, and how conditions may be altered to increase the yield.

PROBLEM:
$K_e = 54.4$ at 355°C for the reaction $H_2 + I_2 \rightleftarrows 2HI$. What percentage of I_2 will be converted to HI if 0.2 mole each of H_2 and I_2 are mixed and allowed to come to equilibrium at 355°C and a total pressure of 0.5 atm?

SOLUTION:

Assume that X moles each of H_2 and I_2 are used up in reaching equilibrium to give $2X$ moles of HI, in accordance with the chemical equation, leaving $0.2 - X$ moles each of H_2 and I_2. The partial pressure of each gas will be in proportion to its fraction of the total moles present.

$$\text{moles of } H_2 \text{ at equilibrium} = 0.2 - X$$
$$\text{moles of } I_2 \text{ at equilibrium} = 0.2 - X$$
$$\text{moles of HI at equilibrium} = \underline{\quad 2X \quad}$$
$$\text{total moles at equilibrium} = 0.4 - 2X + 2X = 0.4$$

$$P_{\text{HI}} = \left(\frac{2X}{.4}\right)\left(.5 \text{ atm}\right)$$

$$P_{\text{H}_2} = P_{\text{I}_2} = \left(\frac{0.2 - X}{.4}\right)\left(.5 \text{ atm}\right)$$

$$K_e = \frac{P_{\text{HI}}^2}{P_{\text{H}_2} \times P_{\text{I}_2}} = \frac{\left[\left(\frac{2X}{.4}\right)\left(.5 \text{ atm}\right)\right]^2}{\left[\left(\frac{0.2 - X}{.4}\right)\left(.5 \text{ atm}\right)\right]^2}$$

$$54.4 = \left(\frac{2X}{0.2 - X}\right)^2$$

Taking the square root of each side gives

$$7.4 = \frac{2X}{0.2 - X}$$

$$X = \frac{1.48}{9.4} = .157 = \text{moles of } H_2 \text{ and } I_2 \text{ used up}$$

$$\text{Per cent conversion (yield)} = \frac{.157}{.200} \times 100 = 78.5\%$$

PROBLEM:

What percentage of I_2 will be converted to HI at equilibrium at 355°C if 0.2 mole of I_2 is mixed with 2.0 moles of H_2 at a total pressure of 0.5 atm?

SOLUTION:

In this problem it is advantageous to first assume that the large excess of H_2 will use almost the entire amount of I_2, leaving only X moles of it unused. If X is very small, then it is practical to assume that 0.4 mole of HI will be formed instead of $0.4 - 2X$, and that $2 - 0.2 = 1.8$ moles of H_2 will remain unused instead of $2 - (0.2 - X) = 1.8 + X$. At equilibrium, then:

$$\text{moles of } H_2 = 1.8 + X \cong 1.8$$
$$\text{moles of } I_2 = \qquad X$$
$$\text{moles of HI} = \underline{.4 - 2X} \cong .4$$
$$\text{total moles} = 2.2$$

$$P_{H_2} = \left(\frac{1.8}{2.2}\right)(0.5 \text{ atm})$$

$$P_{I_2} = \left(\frac{X}{2.2}\right)(0.5 \text{ atm})$$

$$P_{HI} = \left(\frac{0.4}{2.2}\right)(0.5 \text{ atm})$$

$$54.4 = \frac{\left[\left(\frac{0.4}{2.2}\right)(0.5)\right]^2}{\left(\frac{1.8}{2.2}\right)\left(\frac{X}{2.2}\right)(0.5)^2} = \frac{0.4^2}{1.8X}$$

$$X = \frac{0.4^2}{(54.4)(1.8)} = 0.0016 = \text{moles of } I_2 \text{ not used}$$

$$0.2 - 0.0016 = 0.1984 \text{ moles of } I_2 \text{ used up}$$

$$\text{percentage of } I_2 \text{ used up} = \frac{0.1984}{0.200} \times 100 = 99.2\%$$

This problem illustrates the fact that, although the value of K_e does not change with changes in concentration, the *equilibrium position* will change to use up part of the excess of any one reagent that has been added. In this problem the large excess of H_2 shifted the equilibrium position to the right, causing more of the I_2 to be used up (99.2% compared to 78.5%) than when H_2 and I_2 were mixed in equal proportions. Advantage may be taken of this principle by using a large excess of a cheap chemical to convert the maximum amount of an expensive chemical to a desired product. In this case I_2, the more expensive chemical, is made to yield more HI by using more of the cheaper H_2.

The Decomposition of Gases

Many gases decompose into simpler ones at elevated temperatures, and it is often important to know the extent to which decomposition takes place.

PROBLEM:
$K_e = 1.78$ at 250°C for the decomposition reaction $PCl_5 \rightleftarrows PCl_3 + Cl_2$. Calculate the percentage of PCl_5 that dissociates if 0.05 mole of PCl_5 is placed in a closed vessel at 250°C and 2 atm pressure.

SOLUTION:
Assume that X moles of PCl_5 dissociate giving X moles each of PCl_3 and Cl_2 and leaving $0.05 - X$ moles of PCl_5 at equilibrium:

moles of PCl_5 at equilibrium $= 0.05 - X$
moles of PCl_3 at equilibrium $= \qquad X$
moles of Cl_2 at equilibrium $= \underline{\qquad X}$
total moles at equilibrium $= 0.05 + X$

$$K_e = 1.78 = \frac{P_{PCl_3} \times P_{Cl_2}}{P_{PCl_5}} = \frac{\left[\left(\dfrac{X}{.05 + X}\right)(2)\right]^2}{\left(\dfrac{.05 - X}{.05 + X}\right)(2)}$$

$$1.78 = \frac{2X^2}{(.05 + X)(.05 - X)} = \frac{2X^2}{.0025 - X^2}$$

$$.00445 - 1.78X^2 = 2X^2$$

$$X^2 = \frac{.00445}{3.78} = 0.00118$$

$$X = .0342 \text{ moles } PCl_5 \text{ dissociate}$$

$$\text{percentage of } PCl_5 \text{ dissociated} = \frac{0.0342}{0.05} \times 100 = 68.4\%$$

Effect of Temperature on K_e and Equilibrium Position

The important practical applications of the NH_3 synthesis reaction mentioned previously have led to exhaustive studies of this reaction over a wide range of temperatures. It has been found that at a total pressure of 50 atm the values of K_e are 7.73×10^{-4} at 350°C, 1.69×10^{-4} at 400°C, 5.22×10^{-5} at 450°C, 1.5×10^{-5} at 500°C. This steady decrease in the value of K_e with increasing temperature is typical of all reactions that are exothermic. Since a small value of K_e is interpreted as meaning that a relatively small amount of products will exist at equilibrium compared to reactants, you can say that the yield of an exothermic chemical reaction always decreases with increasing temperature. You can also say that an increase in temperature always shifts the equilibrium position of an exothermic reaction to the left. The reverse is true for endothermic reactions.

Catalysts

A catalyst affects only the *rate* of a chemical reaction; it has no effect on K_e or on the position of equilibrium at a given temperature. You cannot, therefore, increase the yield of a chemical reaction at a given temperature by adding a catalyst to the reaction mixture. Catalysts are, however, of great practical value because they make an impractically slow reaction reach equilibrium at a practical rate, or may permit such a reaction to go at a practical rate at a lower temperature, where a more favorable equilibrium position exists.

Summary

The Principle of Le Chatelier summarizes the conclusions that may be drawn from the illustrative examples in this chapter: "Whenever a stress is placed on a system at equilibrium, the equilibrium position shifts in such a way as to relieve that stress."

PROBLEMS A   Answers on page 345

1. In the following table columns, A, B, and C refer to the gaseous equilibria listed below:

   A. $N_2O_4 \rightleftharpoons 2NO_2 - E$
   B. $CO_2 + H_2 \rightleftharpoons CO + H_2O - E$
   C. $SO_2 + \frac{1}{2}O_2 \rightleftharpoons SO_3 + E$

   The symbol $+E$ means that the reaction is exothermic; $-E$ means that the reaction is endothermic. For each reaction tell the effect on K_e and on the equilibrium position of each change listed in the table. Assume that each change affects the reaction mixture only after it has already once reached equilibrium. Use the following symbols in completing the table: $+$, $-$, 0, R, and L to mean increase, decrease, no effect, shift right, and shift left, respectively.

Change	A. Effect on position	K_e	B. Effect on position	K_e	C. Effect on position	K_e
Decrease in total pressure						
Increase in temperature						
Decrease in partial pressure of last-named gas						
A catalyst added						
Argon gas added, keeping total pressure constant						

2. A 0.5-mole sample of $SbCl_5$ is put into a closed container and heated to 248°C at 1 atm. At equilibrium, analysis shows 42.8% by volume of Cl_2 in the mixture. Calculate K_e at this temperature for the dissociation reaction $SbCl_5 \rightleftharpoons SbCl_3 + Cl_2$.

3. Analysis of the equilibrium mixture that resulted from heating 1 mole of CO_2 to a temperature of 1100°C at a pressure of 10 atm shows the presence of $1.4 \times 10^{-3}\%$ O_2 by volume. Calculate the value of K_e for the dissociation reaction $2CO_2 \rightleftharpoons 2CO + O_2$ at this temperature.

4. Calculate the percentage of H_2 that dissociates to atoms at a total pressure of 0.1 atm H_2 and (a) 1,000°C and (b) 2,000°C.

* 5. Calculate the percentage of H_2S that dissociates at a pressure of 1 atm and (a) 1,130°C and (b) 1,200°C.

* 6. What per cent of SO_2 will be converted to SO_3 at 900°C using (a) a 2:1 mixture of SO_2 to O_2 and a total pressure of 0.6 atm and (b) a 1:9 mixture of SO_2 to O_2 at a total pressure of 0.6 atm?

7. Calculate the composition of the equilibrium gas mixture that results when (a) 0.5 mole each of CO_2 and H_2 are mixed at 1,000°C and a total pressure of 2 atm, and (b) 1 mole of CO and 5 moles of H_2O are mixed at 1,000°C and a total pressure of 2 atm.

PROBLEMS B   No answers given

8. In the following table columns, A, B, and C refer to the gaseous equilibria listed below:

A. $H_2 + Cl_2 \rightleftarrows 2HCl + E$
B. $2H_2O \rightleftarrows 2H_2 + O_2 - E$
C. $H_2 + C_2H_4 \rightleftarrows C_2H_6 + E$

Read the explanation given in Problem 1.

Change	A. Effect on		B. Effect on		C. Effect on	
	position	K_e	position	K_e	position	K_e
Decrease in temperature						
Increase in total pressure						
A catalyst added						
Decrease in partial pressure of first-named gas						
Helium gas added, keeping total pressure constant						

9. Four moles of $COCl_2$ are put into a sealed vessel and heated to 395°C at a pressure of 0.2 atm. At equilibrium, analysis shows 30.0% by volume of CO in the mixture. Calculate K_e for the dissociation reaction $COCl_2 \rightleftarrows CO + Cl_2$ at this temperature.

---

* Solve by successive approximation.

10. Analysis of the equilibrium mixture that resulted from heating 2 moles of NOCl to a temperature of 225°C at a pressure of 0.2 atm shows the presence of 34.0% NO by volume. Calculate the value of K_e for the dissociation reaction $2NOCl \rightleftarrows 2NO + Cl_2$ at this temperature.

11. Calculate the percentage of Cl_2 that dissociates to atoms at a total pressure of 1 atm and (a) 1,000°C and (b) 2,000°C.

* 12. Calculate the percentage of H_2O that dissociates at a pressure of 0.5 atm and (a) 1,000°C and (b) 1,700°C.

13. What per cent of H_2 will be converted to HCl at 1,800°C using (a) an equimolar mixture of H_2 and Cl_2 at a total pressure of 0.9 atm, and (b) an 8:1 mixture of Cl_2 to H_2 at a total pressure of 0.9 atm?

* 14. Calculate the composition of the equilibrium gas mixture that results when (a) 2 moles of SO_2 and 6 moles of O_2 are mixed at 1,000°C and a total pressure of 0.8 atm, and (b) 1 mole of SO_2 and 2 moles of SO_3 are mixed at 1,000°C and a total pressure of 0.8 atm.

15. Derive a general mathematical equation that relates K_e, equilibrium pressure (P), and fraction (α) of A dissociated in the gaseous equilibrium $A \rightleftarrows B + C$. Assume that B and C come only from the dissociation of A, and that the temperature remains constant. What would be the form of this equation if the density (D) of the gas mixture was used instead of the pressure?

---

* Solve by successive approximation.

Electrochemistry
Electron Transfer I.

Many of the simplest chemical reactions involve only an interchange of atoms or ions between reactants, or perhaps only the dissociation of one reactant into two parts. In such reactions there is no change in the electrical charge of any of the atoms involved. This chapter deals with another type of reaction, in which one or more electrons are transferred between atoms, with the result that some of the atoms involved do have their electrical charges changed. They are known as electron-transfer reactions.

Oxidation and Reduction

In the transfer of electrons between atoms, an electron donor is called a *reducing agent;* an electron acceptor an *oxidizing agent.* Whenever a reducing agent donates electrons we say that it has been oxidized, and that an oxidizing agent on accepting electrons has been reduced. Oxidation and reduction always occur simultaneously; if one atom or ion donates an electron, another atom or ion must accept it.

The process of donating and accepting electrons is reversible. For example, under one set of conditions metallic cadmium may donate electrons and become Cd^{++} ions, as it does when immersed in HCl,

$$Cd + 2H^+ \rightarrow Cd^{++} + H_2 \uparrow$$

or, under another set of conditions the Cd^{++} ion may accept electrons and be reduced to Cd metal, as it is when it comes in contact with metallic Zn:

$$Cd^{++} + Zn \rightarrow Zn^{++} + Cd$$

These two reactions illustrate the fact that reducing agents differ in their ability to donate electrons; metallic Zn can donate electrons to Cd^{++} to produce metallic Cd, but metallic Cd can donate electrons to H^+ to produce H_2 gas.

A reversible electron-transfer reaction written in the form

$$n\ e^- + oxidizing\ agent \rightleftarrows reducing\ agent$$

is referred to as a *half-reaction*, since it cannot occur unless it is coupled with another half-reaction going in the opposite direction. If we use half-reactions in the manner described in the next few paragraphs, we can assign a number to each reducing agent to describe its strength, or ability to donate electrons.

Galvanic Cells

An electric current is a flow of electrons through a conductor. Many electron-transfer reactions can be arranged so that the electrons donated by the reducing agent are forced to flow through a conducting wire to reach the oxidizing agent. Such an arrangement is called a *battery,* or *electrochemical* (*galvanic*) *cell;* a simple form is shown in Figure 15-1. The electron-transfer reaction that produces the current is

$$Zn + Cu^{++} \rightarrow Zn^{++} + Cu$$

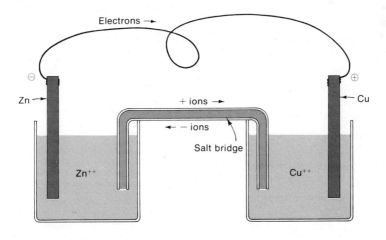

FIGURE 15-1.
Simple galvanic cell.

Note that *all* the components of one half-reaction are placed in one beaker, and *all* the components of the second half-reaction are placed in the other beaker. One must *not* put the materials of the lefthand side of the equation in one beaker, and those of the righthand side in the other! If one did, electron transfer could occur on contact of the two reactants, and there would be no flow of current from one electrode to the other. The *negative electrode* always involves the half-reaction with the greatest reducing strength; in this case it is a strip of zinc dipping into a solution of Zn^{++} (any soluble zinc salt). The *positive electrode,* the one to which the electrons flow in the connecting wire, is a strip of copper dipping into a solution of Cu^{++} (any soluble copper salt). The two solutions are connected by an inverted U-tube filled with a salt solution (held in place by a gel such as agar-agar), to permit ions to pass from one beaker to the other. Each beaker with its contents is called a half-cell; traditionally, the negative electrode is shown at the left.

Now, let us trace the reaction that occurs when one atom of Zn donates its electrons according to the half-reaction

$$Zn \rightarrow Zn^{++} + 2\ e^{-}$$

The Zn^{++} ion goes into solution while the electrons pass through the wire to the Cu electrode. Here, the electrons combine with a Cu^{++} ion from solution, according to the half-reaction

$$Cu^{++} + 2\ e^{-} \rightarrow Cu$$

and the atom of Cu is deposited at the metal surface where the electron transfer occurs. The net result of this is to put a zinc ion into solution in the left-hand beaker and to remove a copper ion from the right-hand beaker. This intolerable situation would quickly lead to the accumulation of an excess of positive ions in one beaker and of an excess of negative ions in the other. The reaction would stop immediately if it were not for the "salt bridge," which permits negative ions to migrate into the lefthand beaker and positive ions to migrate into the righthand beaker, in order to maintain electroneutrality in each beaker at all times. The completed circuit thus involves the *uni*-directional flow of electrons through the copper wire, but the *bi*directional flow of ions in solution.

If, in the construction of a galvanic cell, a combination of half-reactions is used that does not involve metals that can be used for electrodes, then a piece of platinum or other inert conducting substance is used to transfer the electrons to and from the solution. Figure 15-2 shows a galvanic cell in which one half-reaction involves a gas (H_2) that must make electrical contact with both the H^+ ions in solution and the platinum electrode. The other half-reaction involves only *ions* (Fe^{++} and Fe^{+++}) in solution that make contact with a platinum-wire electrode. The liquid mercury provides a simple electrical connection to the wires.

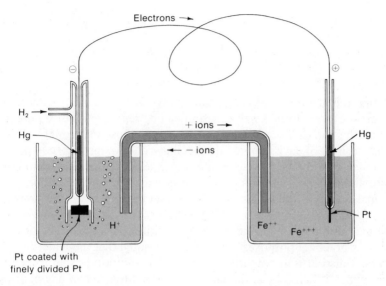

FIGURE 15-2.
Galvanic cell that uses the hydrogen electrode and the Fe^{++}, Fe^{+++} electrode.

A standard shorthand notation is used to describe the construction of galvanic cells and avoid the necessity of drawing pictures. It is based on the convention that the negative electrode is shown at the left. The notation for the galvanic cells shown in Figures 15-1 and 15-2 is

$$Zn/Zn^{++}(M_1)//Cu^{++}(M_2)/Cu$$

and

$$Pt/H_2(P), H^+(M_1)//Fe^{++}(M_2), Fe^{+++}(M_3)/Pt$$

where M_1, M_2, and M_3 are the numerical values of the concentrations of the ions, and P is the pressure of the H_2 gas. A single / indicates the interface between the solid electrode and the solution, and a double // indicates the salt bridge or some other junction between the two half-cells.

Standard Electrode Potentials

If, in Figures 15-1 and 15-2, we cut the wire connecting the electrodes and connect the two ends to a voltmeter that draws essentially no current (such as a vacuum-tube voltmeter) or to a potentiometer, the observed voltage (E_{cell}) reading will be an accurate measure of the *difference* in the reducing strengths of the two half-reactions involved. If the voltage were zero, we would know that they were of equal reducing strength. By definition,

$$E_{cell} = E_{\text{half-cell accepting electrons}} \quad - E_{\text{half-cell donating electrons}}$$

$$= E_{\text{oxidizing half-reaction}} \quad - E_{\text{reducing half-reaction}}$$

$$= E_{ox} - E_{red} \tag{15-1}$$

Note that this definition of cell potential involves the basic assumption that the reducing strength of a half-reaction can be represented by a potential (a voltage), the *half-cell* potential. In order to use these cell voltages to measure the strength of each individual reducing agent, we must (a) compare voltages under conditions that eliminate the effect of temperature and concentration, and (b) know the value of E for at least one half-reaction.

To make a fair comparison we choose a set of reference conditions called the *standard state*. For a *pure* substance, this is taken as the physical form stable at 1 atm and 25°C; under these conditions it is said to be at unit activity. For practical purposes we shall also assume that the water in a dilute solution is also at unit activity. The *solute* in solution is said to be at unit activity when it behaves as though it were a fictitious *ideal* one molar solution in which there are no electrical interactions between ions or molecules. The *actual* solution concentration required to produce unit activity varies considerably from solute to solute: for HCl it is 1.20 M; for LiCl it is 1.26 M. We shall not dwell at this time on the problems connected with finding the actual concentrations of solutions associated with unit activity. If we restrict our solution concentrations to 0.1M or less, our computational errors will generally be less than 5% if we use molar concentrations instead of activities; the more dilute the solution, the less the error.

For a variety of reasons it is impossible to find the absolute value of E for the strength of any reducing agent, even though the difference between any two of them can be measured very accurately by equation 15-1. Instead, we *arbitrarily* select a voltage of zero for the half-reaction

$$2\ e^- + 2\ H^+ \rightleftharpoons H_{2(g)}$$

when all of the components are at unit activity. This special electrode is called the "Standard Hydrogen Electrode." Letting a superscript ° indicate the standard state of unit activity, our fundamental convention is that for the standard hydrogen electrode

$$E^\circ_{H_2-H^+} = 0.00$$

If we now make a whole series of galvanic cells with solutions at unit activity and use the standard hydrogen electrode as one half of every cell, then the measured cell voltage (E°_{cell}) for each cell will be given, according to equation 15-1, by either

TABLE 15-1.
Standard Electrode Potentials ($E°$) at 25°C.

Half-reaction	$E°$ in volts	Half-reaction	$E°$ in volts
$e^- + Li^+ \rightleftarrows Li$	−3.04	$2\,e^- + Hg_2Cl_2 \rightleftarrows 2\,Cl^- + 2\,Hg$	0.27
$e^- + K^+ \rightleftarrows K$	−2.92	$2\,e^- + Cu^{++} \rightleftarrows Cu$	0.34
$2\,e^- + Ca^{++} \rightleftarrows Ca$	−2.87	$e^- + Cu^+ \rightleftarrows Cu$	0.52
$e^- + Na^+ \rightleftarrows Na$	−2.71	$2\,e^- + I_2 \rightleftarrows 2\,I^-$	0.54
$3\,e^- + La^{+++} \rightleftarrows La$	−2.52	$2\,e^- + 2\,H^+ + O_2 \rightleftarrows H_2O_2$	0.68
$3\,e^- + Ce^{+++} \rightleftarrows Ce$	−2.48	$e^- + Fe^{+++} \rightleftarrows Fe^{++}$	0.77
$2\,e^- + Mg^{++} \rightleftarrows Mg$	−2.36	$2\,e^- + Hg_2^{++} \rightleftarrows 2\,Hg$	0.79
$3\,e^- + Lu^{+++} \rightleftarrows Lu$	−2.26	$e^- + Ag^+ \rightleftarrows Ag$	0.80
$3\,e^- + Al^{+++} \rightleftarrows Al$	−1.66	$2\,e^- + 3\,H^+ + NO_3^- \rightleftarrows H_2O + HNO_2$	0.94
$2\,e^- + Zn^{++} \rightleftarrows Zn$	−0.76	$3\,e^- + 4\,H^+ + NO_3^- \rightleftarrows 2\,H_2O + NO$	0.96
$2\,e^- + 2\,H^+ + 2\,CO_2 \rightleftarrows H_2C_2O_4$	−0.49	$e^- + H^+ + HNO_2 \rightleftarrows H_2O + NO$	1.00
$2\,e^- + Fe^{++} \rightleftarrows Fe$	−0.44	$2\,e^- + Br_{2(l)} \rightleftarrows 2\,Br^-$	1.07
$2\,e^- + Cd^{++} \rightleftarrows Cd$	−0.40	$4\,e^- + 4\,H^+ + O_2 \rightleftarrows 2\,H_2O$	1.23
$2\,e^- + Co^{++} \rightleftarrows Co$	−0.28	$6\,e^- + 14\,H^+ + Cr_2O_7^= \rightleftarrows 7\,H_2O + 2\,Cr^{+++}$	1.33
$2\,e^- + Ni^{++} \rightleftarrows Ni$	−0.25	$2\,e^- + Cl_2 \rightleftarrows 2\,Cl^-$	1.36
$e^- + AgI \rightleftarrows I^- + Ag$	−0.15	$3\,e^- + Au^{+++} \rightleftarrows Au$	1.50
$2\,e^- + Sn^{++} \rightleftarrows Sn$	−0.14	$5\,e^- + 8\,H^+ + MnO_4^- \rightleftarrows 4\,H_2O + Mn^{++}$	1.51
$2\,e^- + Pb^{++} \rightleftarrows Pb$	−0.13	$e^- + Ce^{++++} \rightleftarrows Ce^{+++}$	1.61
$2\,e^- + 2\,H^+ \rightleftarrows H_2$	0.00	$2\,e^- + 2\,H^+ + 2\,HClO \rightleftarrows 2\,H_2O + Cl_2$	1.63
$e^- + AgBr \rightleftarrows Br^- + Ag$	0.07	$e^- + Au^+ \rightleftarrows Au$	1.69
$2\,e^- + S_4O_6^= \rightleftarrows 2\,S_2O_3^=$	0.08	$2\,e^- + 2\,H^+ + H_2O_2 \rightleftarrows 2\,H_2O$	1.78
$2\,e^- + 2\,H^+ + S \rightleftarrows H_2S$	0.14	$6\,e^- + 6\,H^+ + XeO_3 \rightleftarrows 3\,H_2O + Xe$	1.80
$2\,e^- + Sn^{++++} \rightleftarrows Sn^{++}$	0.15	$2\,e^- + S_2O_8^= \rightleftarrows 2\,SO_4^=$	2.01
$e^- + Cu^{++} \rightleftarrows Cu^+$	0.15	$2\,e^- + 2\,H^+ + O_3 \rightleftarrows H_2O + O_2$	2.07
$e^- + AgCl \rightleftarrows Cl^- + Ag$	0.22	$2\,e^- + F_2 \rightleftarrows 2\,F^-$	2.87

$$E°_{cell} = E°_{ox} - E°_{H_2-H^+} = E°_{ox}$$

or

$$E°_{cell} = E°_{H_2-H^+} - E°_{red} = -E°_{red}$$

depending on whether the standard hydrogen electrode is a stronger or weaker reducing agent than the other half-reaction. If we measure the voltage of the cell shown in Figure 15-2, with all components at unit activity, we find $E°_{cell} = 0.77$ volts. From our fundamental equation and the assumption for the standard hydrogen electrode,

$$E^\circ_{cell} = E^\circ_{Fe^{++}-Fe^{+++}} \quad - E^\circ_{H_2-H^+} = E^\circ_{Fe^{++}-Fe^{+++}}$$

$$E^\circ_{Fe^{++}-Fe^{+++}} \quad = +0.77 \text{ volts}$$

This direct determination of the E° of every reducing agent by comparison with another to which an arbitrary value of 0 has been assigned is the way, in principle, in which all the E° values (or *standard electrode potentials*) in Table 15-1 were found. Reducing agents that are stronger than H_2 have a $-$ value of E°, and those that are weaker have a $+$ value; the more negative the E° value, the stronger the reducing agent. Table 15-1 can always be used to calculate the voltage of any galvanic cell whose concentrations are at unit activity.

PROBLEM:

Calculate the voltage of the cell shown in Figure 15-1 if the solutions are at unit activity.

SOLUTION:

We get the appropriate values for $E^\circ_{Zn-Zn^{++}}$ and $E^\circ_{Cu-Cu^{++}}$ from Table 15-1 and substitute them into equation (15-1):

$$E^\circ_{cell} = E^\circ_{ox} - E^\circ_{red} = E^\circ_{Cu-Cu^{++}} - E^\circ_{Zn-Zn^{++}}$$

$$= +0.34 - (-0.76) = +1.10 \text{ volts}$$

Variation of Cell Voltage with Concentration

When concentrations of reactants are not at unit activity, the cell potential may be calculated by the relation

$$E = E^\circ - \frac{RT}{nF} \ln Q_{1/2}$$

which is known as the Nernst equation. This equation will be derived and discussed in more advanced courses. Here n is the number of moles of electrons as shown in the half-reaction equation, and $Q_{1/2}$ is the "ion product" that has the same form as the usual equilibrium-constant expression but that uses the actual ionic concentrations, *not* the equilibrium values. It also differs because half-reactions by themselves are fictitious, and isolated electrons do not exist in aqueous solution; e.g., there is no factor in the $Q_{1/2}$ expression for $[e^-]^n$, the "electron concentration." If we express the gas constant R in fundamental units (using the value of 1.013×10^6 dynes/cm^2 for one atm that we determined on p. 126, along with the volume of 22,400 cm^3/ mole at 1 atm and 273°K), we get $R = 8.314 \times 10^7 \dfrac{\text{ergs}}{\text{mole deg}}$. Reexpressed in still different units, $R = 8.314 \dfrac{\text{volt coulombs}}{\text{mole deg}}$. F, the total electrical charge on one mole of electrons, is 96,487 coulombs/mole of electrons. If we use these values of R and F, limit outselves to 25°C for convenient use of Table 15-1, and convert the logarithm to base 10, then the general expression for the *voltage of a half-cell* is

$$E = E° - \frac{\left(8.314 \dfrac{\text{volt coulomb}}{\text{mole deg}}\right)(298 \text{ deg})(2.303)}{\left(n\dfrac{\text{mole of electrons}}{\text{mole}}\right)\left(96,487 \dfrac{\text{coulombs}}{\text{mole of electrons}}\right)} \log Q_{1/2}$$

$$= E° - \frac{0.0591}{n} \log Q_{1/2} \tag{15-2}$$

PROBLEM:
What is the voltage of a cell constructed as in Figure 15-1, but with a $0.1M$ $ZnSO_4$ solution in the lefthand beaker and a $10^{-4}M$ $CuSO_4$ solution in the righthand beaker?

SOLUTION:
The voltage of the cell is

$$E_{\text{cell}} = E_{\text{ox}} - E_{\text{red}} = E_{Cu-Cu^{++}} - E_{Zn-Zn^{++}}$$

Each of the half-cell potentials is given by fundamental equation 15-2, with $E°$ values taken from Table 15-1:

$$E_{Cu-Cu^{++}} = E°_{Cu-Cu^{++}} - \frac{0.0591}{2} \log \frac{[Cu]}{[Cu^{++}]}$$

$$= +0.34 - \frac{0.0591}{2} \log \frac{1}{10^{-4}} = +0.22 \text{ volt}$$

$$E_{Zn-Zn^{++}} = E°_{Zn-Zn^{++}} - \frac{0.0591}{2} \log \frac{[Zn]}{[Zn^{++}]}$$

$$= -0.76 - \frac{0.0591}{2} \log \frac{1}{10^{-1}} = -0.79 \text{ volt}$$

$$E_{\text{cell}} = +0.22 - (-0.79) \doteq +1.01 \text{ volt}$$

We could have combined the equations for E_{ox} and E_{red} into one equation before solving for E_{cell}, as follows.

$$E_{\text{cell}} = \left[E°_{\text{ox}} - \frac{0.0591}{n} \log (Q_{1/2})_{\text{ox}} \right] - \left[E°_{\text{red}} - \frac{0.0591}{n} \log (Q_{1/2})_{\text{red}} \right]$$

$$= [E°_{\text{ox}} - E°_{\text{red}}] - \frac{0.0591}{n} \log \frac{(Q_{1/2})_{\text{ox}}}{(Q_{1/2})_{\text{red}}}$$

$$E_{cell} = E_{cell}^{\circ} - \frac{0.0591}{n} \log \frac{(Q_{1/2})_{ox}}{(Q_{1/2})_{red}}$$

Of particular importance here is the fact that the *ratio* $(Q_{1/2})_{ox}/(Q_{1/2})_{red}$ is identical to the ion product for the over-all cell reaction, which we shall simply call Q. In other words, we can write our expression for *the voltage of the whole cell* as

$$E_{cell} = E_{cell}^{\circ} - \frac{0.0591}{n} \log Q \tag{15-3}$$

For the cell reaction

$$Zn + Cu^{++} \rightleftarrows Zn^{++} + Cu$$

the corresponding calculation is

$$E_{cell} = [(+0.34) - (-0.76)] - \frac{0.0591}{2} \log \frac{[Zn^{++}]}{[Cu^{++}]}$$

$$= +1.10 - \frac{0.0591}{2} \log \frac{10^{-1}}{10^{-4}}$$

$$= +1.10 - \frac{(0.0591)(3)}{2} = +1.01 \text{ volts}$$

You should note that equation 15-3 applies only to chemical equations that are *complete* and *balanced*, and that for a given electron-transfer reaction n is the number of electrons lost by the reducing agent (or gained by the oxidizing agent) in the balanced chemical equation. Equation 15-2 on the other hand applies only to a *balanced, half*-reaction. When equation 15-2 is applied to two different half-reactions that are to be combined to make one complete reaction, it is essential that the half-reactions be balanced to give the same value of n, the value that will be used in equation 15-3 and that will refer to the complete reaction.

Oxidation-Reduction Equilibrium

If you permit a battery to completely "run down" or become "dead," its E_{cell} becomes zero and the cell reaction reaches an equilibrium condition. Now, making a battery and letting it go dead is not the only way to let an electron-transfer reaction reach equilibrium; the components could just as well be mixed together in a single beaker. The important thing to realize is that no matter how you reach equilibrium, the ion product Q at equilibrium is now equal to the equilibrium constant, so that *for any electron-transfer reaction at equilibrium* we could write

$$E_{cell} = 0 = E^{\circ}_{cell} - \frac{0.0591}{n} \log K_e$$

$$E^{\circ}_{cell} = \frac{0.0591}{n} \log K_e \tag{15-4}$$

The most important thing about equation 15-4 is that the equilibrium constants for electron-transfer reactions can be calculated from standard electrode potentials without ever having to make experimental measurements.

PROBLEM:
Calculate the equilibrium constant at 25°C for the reaction

$$Fe^{++} + Ag^+ \rightleftarrows Fe^{+++} + Ag$$

SOLUTION:
From Table 15-1 we get

$$E^{\circ}_{cell} = E^{\circ}_{Ag-Ag^+} - E^{\circ}_{Fe^{++}-Fe^{+++}}$$
$$= (+0.80) - (+0.77) = +0.03 \text{ volt}$$

Only 1 mole of electrons is transferred in the equation as written, so

$$\log K_e = \frac{nE^{\circ}_{cell}}{0.0591} = \frac{(1)(0.03)}{0.0591} = 0.51$$

$$K_e = 3.2$$

Electrical Energy from Chemical Energy

Up to this point we have emphasized the *voltage* of a galvanic cell. We are also in a position to consider the conversion of chemical energy to electrical energy in a galvanic cell. Electrical energy is calculated as the product of the voltage of a cell and the total electrical charge, in coulombs, that passes from the battery; i.e.,

$$\text{electrical energy} = \text{volts} \times \text{coulombs} = \text{joules}$$

The total charge that passes from a battery will be determined by the number of moles of electrons (n) that pass through the circuit. By definition,

$$1 \text{ faraday } (F) = 96,487 \, \frac{\text{coulombs}}{\text{mole of electrons}}$$

and

$$1 \text{ cal} = 4.184 \text{ joules} = 4.184 \text{ volt coulombs}$$

Therefore,

electrical energy

$$= (\text{volts})(\text{moles of electrons})\left(\frac{\text{coulombs}}{\text{mole of electrons}}\right)\left(\frac{\text{calories}}{\text{volt coulomb}}\right)$$

$$= (E)(n)(F)\left(\frac{1}{4.184}\right) = nE \times \frac{96,487}{4.184}$$

$$= 23,060 \times nE \text{ cal}$$

The cell voltages we have been talking about in this chapter are all maximum voltages, voltages measured with a potentiometer that just matches the voltage of the cell without actually draining any current from the cell, or with a vacuum-tube voltmeter whose resistance is so high that the result is essentially the same. If a battery is used to do work, the voltage will be less; not all the energy that is produced can be employed usefully because of partial dissipation as heat. In fact, *all* the electrical energy could be wasted if you wanted to. Nevertheless, our potentiometrically measured (*maximum*) voltages can be used for the calculation of the *maximum* amount of available electrical energy that can be obtained from a chemical reaction, *regardless of whether it is actually used for work*. For a given chemical reaction, we can equate maximum available electrical energy with maximum available work, and write

$$\text{maximum available work} = \text{maximum available electrical energy}$$

$$= 23,060 \times nE \text{ cal}$$

The Concept of Free Energy

The preceding statements imply that every substance has an amount of energy, called its *free energy* (G), which could be used for useful work. If, as a result of a chemical reaction that occurs at constant temperature and pressure, the sum of the products ("state 2") possesses an amount of free energy G_2 while the sum of the reactants ("state 1") possesses an amount of free energy G_1, then the change in free energy (ΔG) for the reaction will be

$$\Delta G = G_2 - G_1 = G_{\text{products}} - G_{\text{reactants}}$$

If the *sign* of ΔG is negative, it means that the products have less free energy than the reactants. The *magnitude* of the change is the maximum amount of work that might have been obtained from the reaction. The maximum obtainable work is associated with a *decrease* in free energy ($-\Delta G$); i.e., the

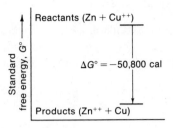

FIGURE 15-3.
The decrease in standard free energy ($\Delta G°$)
that occurs when Zn reacts with Cu^{++}.

work is obtained at the expense of the chemical system. This is shown in Figure 15-3. In equation form we would write

$$\text{Maximum available work} = -\Delta G$$

and, for maximum available electrical work,

$$\Delta G = -nFE$$
$$= -23,060 \times nE \text{ cal}$$

Changes in Standard Free Energy

If the reactants and products were in their standard states of unit activity, then of course the voltage ($E°$) is the "standard cell potential," and the change in free energy is the change in "standard free energy" ($\Delta G°$):

$$\Delta G° = -nFE°$$

For the reaction

$$Zn + Cu^{++} \rightleftarrows Zn^{++} + Cu$$

the change in standard free energy that accompanies the reaction is

$$\Delta G° = -23,060 \times nE° \text{ cal}$$
$$= -(23,060)(2)(1.10) = -50,800 \text{ cal/mole of Zn or Cu}$$

The change in standard free energy is $-50,800$ cal/mole whether metallic Zn is wastefully put into a beaker of $CuSO_4$ or whether the reaction is usefully employed as a battery as in Figure 15-1. Either way, the products end up being less capable of doing work than they were before starting the reaction.

The Relationship Between $\Delta G°$ and K_e

There is a very important extension of the concepts associated with ΔG, which up to this point have been closely linked with electron-transfer reactions and the production of electrical energy. Equations 15-4 and 15-5 state that

$$\Delta G° = -nFE°$$

and

$$E° = \frac{2.3\,RT}{nF}\log K_e$$

If we combine these two expressions, we get

$$\Delta G° = -2.3\,RT\log K_e$$

$$= -(2.3)\left(1.987\,\frac{\text{cal}}{\text{mol deg}}\right)(T\ \text{deg})\log K_e$$

$$= -4.57\,T\log K_e\ \text{cal/mole} \tag{15-6}$$

where T is the temperature at which the equilibrium constant is known. With this last expression you can see that the concept of electron-transfer has been eliminated ("n" is no longer involved), and you see that *the change in standard free energy for a reaction is associated with its equilibrium constant.* In other words, for *any* chemical reaction, regardless of whether it involves electron-transfer, it is possible to calculate the change in standard free energy at a given temperature if its equilibrium constant is known for that temperature. In eliminating "n" to obtain equation 15-6, we must not overlook the fact that the value of $\Delta G°$ still depends on the amount of material used and the way the chemical equation is written. For the reaction written as $20\ H_2 + 10\ O_2 \rightleftarrows 20\ H_2O$, the value of the equilibrium constant is $(K_e)^{10}$ compared with K_e for the same reaction written as $2\ H_2 + O_2 \rightleftarrows 2\ H_2O$; the value of $\Delta G°$ for the first is ten times that for the second, as expected from equation 15-6.

It is not always easy to harness nonelectron-transfer reactions to do useful work, but the *potential* to do the work is still present. To see how this wider concept ties in with other equilibrium reactions studied in this book, consider the following examples:

1. At 25°C for the reaction (p. 245)

$$HNO_2 \rightleftarrows H^+ + NO_2^-$$

$$\Delta G° = -(4.57)(298)\log K_e = -(4.57)(298)\log(4.5 \times 10^{-4})$$

$$= +4{,}550\ \text{cal/mole}$$

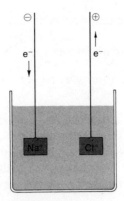

FIGURE 15-4.
Electrolysis of molten NaCl.

2. At 25°C for the reaction (p. 263)

$$AgCl \rightleftharpoons Ag^+ + Cl^-$$

$$\Delta G^\circ = -(4.57)(298) \log K_e = -(4.57)(298) \log (1.6 \times 10^{-10})$$

$$= +13,350 \text{ cal/mole}$$

3. At 355°C for the reaction (p. 173)

$$H_{2(g)} + I_{2(g)} \rightleftharpoons 2 HI_{(g)}$$

$$\Delta G^\circ = -(4.57)(628) \log K_e = -(4.57)(628) \log (54.4)$$

$$= -4994 \text{ cal per 2 moles of HI}$$

$$= -2497 \text{ cal/mole}$$

ELECTROLYSIS

So far in this chapter we have emphasized the principles associated with *obtaining* electrical energy from electron-transfer reactions in solution. Now, we want to examine what happens when electrical energy is *applied* to solutions in the operation of *electrolytic* cells. The oxidation and reduction processes that take place in an electrolytic cell are referred to as *electrolysis*.

Unlike a galvanic cell, an electrolytic cell needs only a single beaker; both electrodes are immersed in it. The electrode to which the electrons are furnished by the power supply is the negative electrode; the other electrode is positive. In solution the positive ions are attracted to the negative electrode and the negative ions to the positive electrode; i.e., conduction of current in solution is a *bi*directional flow of ions, just as in the salt bridge of a galvanic cell.

The electron-transfer process that occurs at each electrode (removal of electrons *from* the negative electrode and supply of electrons *to* the positive electrode) depends on the material between the electrodes, as discussed below.

Electrolysis of Molten Salts

The simplest processes are those that occur when an electrolytic cell contains a molten salt, such as NaCl (see Figure 15-4). In this situation the Na^+ ions migrate to the negative electrode, and, if the applied voltage is sufficiently high, accept electrons from the electrode according to the half-reaction

$$Na^+ + e^- \rightarrow Na$$

At the positive electrode to which the Cl^- ions have migrated, Cl^- ions give up electrons to the electrode according to the half-reaction

$$Cl^- \rightarrow e^- + Cl$$

and the Cl atoms combine with others to form Cl_2 gas,

$$Cl + Cl \rightarrow Cl_2 \uparrow$$

If proper mechanical arrangements are provided, the Na is collected (as a vapor at the temperature of molten NaCl) in the absence of air at the negative electrode, and Cl_2 gas is collected at the positive electrode. Different mechanical provisions must be made if the metal is produced as a liquid or a solid, but the principle is the same in every case. If we look at the completed circuit, we see that electrons have come *from* the power supply to the negative electrode and have gone *to* the power supply from the positive electrode, with a *bi*directional flow of ions within the cell.

Electrolysis of Aqueous Solutions (the Negative Electrode)

The situation is more complicated when the electrolytic cell contains aqueous solutions, because in addition to the ions of the solute, there are also present the H^+ and OH^- ions of water. The basic principle is still the same: positive ions go to the negative electrode, and negative ions go to the positive electrode. But now, at the negative electrode, we must decide whether it is the plus ions of the solute or the H^+ ions of water that accept the electrons. Our natural reaction would be to say that we will first form the atom that has the least tendency to give off electrons, and by reference to the table of standard electrode potentials (Table 15-1) we can see that all ions lying above H^+ will accept electrons *less* readily than H^+ *at the same concentration.* In water, of course, the H^+ concentration is $10^{-7}M$, a concentration we can use in the Nernst equation (p. 186) to show that the hydrogen electrode potential in water is -0.41 volts. With this knowledge in hand, we would be disposed to say that all those ions lying above -0.41 volts in Table 15-1 will accept electrons less readily than H^+ in aqueous solution. This would imply that metal ions lying below H^+ would plate out as metals from aqueous solution, but only H_2 gas would be obtained from the electrolysis of aqueous solutions of metal ions lying above H^+.

In actual fact this is not true, because a second problem arises. For reasons not yet clearly understood in spite of enormous man-years of research, H^+ ions accept electrons at metal surfaces with much greater difficulty than expected, and this difficulty varies with the kind of metal surface. Only on a platinum surface covered with finely divided platinum does this difficulty disappear. But even if you start with platinum and begin to plate out another metal on it, the electrode surface becomes the "other metal," and it is difficult to produce H_2. Extra voltage is required to produce H_2. This extra voltage is called "hydrogen overvoltage," and it varies with different metal surfaces. The net result of "hydrogen overvoltage" combined with a low H^+ ion concentration ($10^{-7}M$) is that most metal ions can be plated out from aqueous solution, but no H_2 gas is produced; the alkali and alkaline earth metals are the major exceptions. Two common metals that *cannot* be plated out from aqueous solution under any circumstances are iron and aluminum.

To summarize for the negative electrode we would say:

(a) Metal is plated out in the electrolysis of solutions of metal salts, other than those of the alkali and alkaline earth metals (and iron and aluminum). For example, in the electrolysis of a $ZnCl_2$ solution, we would obtain metallic zinc,

$$Zn^{++} + 2\,e^- \rightarrow Zn$$

because the H_2-overvoltage on metallic zinc changes the H-electrode potential to some value more negative than the Zn-electrode potential of -0.76 volt.

(b) H_2 gas is evolved in the electrolysis of solutions of alkali and alkaline earth metal salts. For example, in the electrolysis of an NaCl solution we would obtain H_2 gas,

$$2\,H^+ + 2\,e^- \rightarrow H_2\uparrow$$

because the electrode potential (-2.71 volts) for Na (an alkali metal) is more negative than the H-electrode potential, including the H_2 overvoltage.

Electrolysis of Aqueous Solutions (the Positive Electrode)

At the positive electrode, which is composed of a relatively inert conductor such as Pt metal or carbon, the reaction is simple if monatomic ions such as Cl^-, Br^-, or I^- are present in solution; they lose their electrons and form the corresponding halogen. But if the negative ions contain oxygen, such as $SO_4^=$, NO_3^-, ClO_4^-, then they compete with OH^- from water to give up electrons — and OH^- always gives up its electrons more easily than the others to form O_2 gas, according to the half-reaction

$$4\,OH^- \longrightarrow O_2 + 2\,H_2O + 4\,e^-$$

Thus, at a positive Pt or C electrode,
  (a) a solution of NaCl will yield Cl_2,

$$2\,Cl^- \longrightarrow Cl_2 + 2\,e^-$$

  (b) a solution of $NaNO_3$ will yield O_2,

$$4\,OH^- \longrightarrow O_2 + 2\,H_2O + 4\,e^-$$

If the positive electrode is of some common metal such as silver, copper, nickel, etc., then a still simpler source of electrons is available for the external circuit — it is the electrode itself. The electrode "dissolves" in solution as a result. Thus, at a positive Cu electrode, a solution of NaCl or $NaNO_3$ will yield neither Cl_2 nor O_2, but a solution of Cu^{++} ions,

$$Cu \longrightarrow Cu^{++} + 2\,e^-$$

As a matter of fact, this is the basis for the electrorefining of metals. The impure metal electrode (say Cu) is used as the positive electrode, and a piece of pure Cu metal is used as the negative electrode; the solution is $CuSO_4$. The composition of the solution is so designed, and the applied voltage so chosen, that the impurities stay in solution or precipitate out, and only copper plates out at the negative electrode:
  at the positive electrode,

$$Cu \longrightarrow Cu^{++} + 2\,e^-$$

  at the negative electrode,

$$Cu^{++} + 2\,e^- \longrightarrow Cu$$

Faraday's Law

Whenever one mole of electrons is involved in an electron-transfer reaction, in either a galvanic or an electrolytic cell, the total quantity of electricity involved is 96,487 coulombs. This quantity is known as one faraday (F), in honor of Michael Faraday, who first noted the relation. It is easy to find the quantity of material that is oxidized or reduced in an electron-transfer reaction when one F of electricity is passed; simply divide the coefficient of each component of the half-reaction in Table 15-1 by the number of moles of electrons (n) in the half-reaction. The following are examples:

$$e^- + \tfrac{1}{2} H_2O + \tfrac{1}{4} O_2 \rightleftarrows OH^-$$

$$e^- + \tfrac{1}{3} Al^{+++} \rightleftarrows \tfrac{1}{3} Al$$

$$e^- + \tfrac{1}{2} Zn^{++} \rightleftarrows \tfrac{1}{2} Zn$$

$$e^- + Fe^{+++} \rightleftarrows Fe^{++}$$

We see that for the half-reaction in question, $\tfrac{1}{4}$ mole of O_2 or 1 mole of OH^-, $\tfrac{1}{3}$ mole of Al or Al^{+++}, $\tfrac{1}{2}$ mole of Zn or Zn^{++}, or 1 mole of Fe^{++} or Fe^{+++} are oxidized (or reduced) by one F. The weight of material oxidized or reduced by 1 F is called an *electron-transfer equivalent weight* (as distinguished from an acid-base equivalent weight); i.e.,

$$\text{E-T equiv wt} = \frac{\text{grams}}{F} \quad \text{or} \quad \frac{\text{grams}}{\text{mole of electrons}}$$

For aluminum,

$$\text{E-T equiv wt} = \left(27.0 \ \frac{g}{\text{mole}}\right)\left(\frac{1}{3} \frac{\text{mole}}{F}\right)$$

$$= 9.0 \ \frac{g}{F} = 9.0 \ \frac{g}{\text{mole of electrons}}$$

Some substances have more than one electron-transfer equivalent weight as, for example, iron in the reactions

$$2 \ Fe^{+++} + Zn \longrightarrow 2 \ Fe^{++} + Zn^{++}$$

$$Fe + 2 \ H^+ \longrightarrow Fe^{++} + H_2 \uparrow$$

$$2 \ Fe + 3 \ Cl_2 \longrightarrow 2 \ FeCl_3$$

One mole of Fe is involved in the transfer of 1 mole of electrons in the first, 2 moles of electrons in the second, and 3 moles in the third reaction. This, in turn, means that the equivalent weight of Fe in the first is 55.85 g/F, 27.92 g/F in the second, and 18.62 g/F in the third.

The quantity of electricity involved in an electrolysis reaction is determined by the current (amperes) and the length of time the current is passed.

By definition,

$$ampere = \frac{coulomb}{sec}$$

or,

$$coulombs = (amperes)(sec)$$

A number of practical problems can be solved by applying these simple principles. The following problems are typical.

PROBLEM:
Calculate the weight of copper and the volume of O_2 (at standard conditions, dry) that would be produced by passing a current of 0.5 amp through a $CuSO_4$ solution between Pt electrodes for a period of 2 hr.

SOLUTION:
The number of faradays of electricity passed is

$$Faradays = \frac{(amp)(sec)}{96,487 \ coulombs/F} = \frac{(coulombs/sec)(sec)}{96,487 \ coulombs/F}$$

$$= \frac{(0.5)(2 \times 60 \times 60)}{96,487} = 0.373 \ F$$

Since 0.373 F will liberate 0.373 E-T equiv, for

$$\tfrac{1}{2} Cu^{++} + e^- \rightarrow \tfrac{1}{2} Cu$$

$$weight \ of \ Cu = (0.373 \ equiv) \left(63.5 \ \frac{g}{mole}\right) \left(\frac{1 \ mole}{2 \ equiv}\right)$$

$$= 11.8 \ g \ Cu$$

and, for the reaction

$$OH^- \rightarrow \tfrac{1}{4} O_2 + \tfrac{1}{2} H_2O + e^-$$

$$volume \ of \ O_2 = (0.373 \ equiv) \left(22.4 \ \frac{liter}{mole}\right) \left(\frac{1 \ mole}{4 \ equiv}\right)$$

$$= 2.09 \ liters \ O_2$$

PROBLEM:
What length of time is required to plate out 0.1000 g Ag from an $AgNO_3$ solution using a current of 0.200 amp?

SOLUTION:
The number of electron-transfer equivalents of Ag plated out is

$$\text{E-T equiv of Ag} = \frac{0.1000 \ g}{(107.9 \ g/mole)(1 \ mole/equiv)} = 9.28 \times 10^{-4}$$

9.28×10^{-4} equiv Ag requires 9.28×10^{-4} F of electricity, so

$$\text{seconds} = \frac{\text{coulombs}}{\text{amperes}} = \frac{(96{,}487 \text{ coulombs}/F)(9.28 \times 10^{-4} \, F)}{(0.200 \text{ coulomb/sec})}$$

$$= 447 \text{ sec}$$

$$= \frac{447 \text{ sec}}{60 \text{ sec/min}} = 7.45 \text{ min}$$

PROBLEM:
What current is required to plate out 0.02 mole of gold from a $AuCl_3$ solution in 3 hr?

SOLUTION:
From the half-cell reaction

$$\tfrac{1}{3} Au^{+++} + e^- \longrightarrow \tfrac{1}{3} Au$$

we can see that a mole of Au^{+++} contains three E-T equivalents. Thus

$$\text{equivalents of Au} = (0.02 \text{ mole Au}) \left(3 \, \frac{\text{equiv}}{\text{mole}} \right) = 0.06 \text{ equiv Au}$$

and, since 0.06 equiv Au requires 0.06 F of electricity,

$$\text{amperes} = \frac{\text{coulombs}}{\text{seconds}} = \frac{(0.06 \, F)(96{,}487 \text{ coulombs}/F)}{(3 \text{ hr})(3600 \text{ sec/hr})} = 0.536 \text{ amp required}$$

PROBLEMS A   Answers on page 346

1. Using the table of standard electrode potentials, calculate the voltage you could obtain from batteries that use the following reactions with (i) concentrations at unit activity and (ii) oxidizing half-cell ionic concentrations at 10^{-2} molar and reducing half-cell ionic concentrations at 10^{-4} molar.

   (a) $Mg + 2Ag^+ \rightleftarrows Mg^{++} + 2Ag$ (c) $Fe + 2Fe^{+++} \rightleftarrows 3Fe^{++}$

   (b) $Cu + Hg_2^{++} \rightleftarrows Cu^{++} + 2Hg$ (d) $Sn^{++} + Br_2 \rightleftarrows Sn^{++++} + 2Br^-$

2. Make a simple sketch to show how you would arrange the materials used to construct each of the batteries in Problem 1. Show also polarity and direction of flow of electrons and ions.

3. Calculate the equilibrium constant for each of the reactions listed in Problem 1.

4. Calculate the change in standard free energy that accompanies each of the reactions listed in Problem 1.

5. The Edison cell is a rugged storage battery that may receive hard treatment and yet give good service for years. It may even be left uncharged indefinitely and still be recharged. It gives 1.3 volts. Its electrolyte is a

21% KOH solution to which a small amount of LiOH is added. The chemical reaction that occurs on discharge is

$$Fe + 2Ni(OH)_3 \rightleftarrows Fe(OH)_2 + 2Ni(OH)_2$$

Write the half-cell reactions for the electrodes. Which pole must be the negative pole?

6. Tell what is liberated at the platinum anode and cathode during electrolysis of each of the following aqueous solutions: (a) Na_3PO_4; (b) CdI_2; (c) $ZnSO_4$; (d) $AuCl_3$; (e) KNO_3.

7. What volume of each of (a) O_2, (b) Cl_2, and (c) H_2, at standard conditions, will be liberated by the passage of 5 faradays of electricity?

8. What weight of each of the following will be liberated by the passage of 1 faraday of electricity? (a) O_2 from Na_2SO_4; (b) Cl_2 from $AlCl_3$; (c) Mg from $MgCl_2$; (d) Co from $CoSO_4$; (e) PbO_2 from $Pb(NO_3)_2$.

9. What current will be needed to deposit 6 g Ag in 30 min?

10. How long will it take to deposit the cadmium from 350 ml of a 0.3M $CdSO_4$ solution, using a current of 1.75 amp?

11. What will be the concentration of $Cd(NO_3)_2$ in solution after 2.5 amp has passed for 5 hr through 900 ml of a solution that was originally 0.30M? (Platinum electrodes are used.)

12. A current of 3.7 amp is passed for 6 hr between nickel electrodes in a 2.3M $NiCl_2$ solution. What will be the concentration of the $NiCl_2$ at the end of the 6 hr? Assume that no other reactions occur.

13. Using a current of 5.5 amp, how long will it take to produce 47 liters of O_2 measured over water at 735 torr and 35°C, by the electrolysis of a $CuSO_4$ solution?

14. How many ampere-hours of electricity will be needed to refine electrolytically half a ton of silver by removing it from an impure anode and depositing it in a pure form on the cathode?

15. During the discharge of a storage battery, the specific gravity of the sulfuric acid fell from 1.294 to 1.139. Calculate the number of ampere-hours the battery must have been used. (Sulfuric acid of specific gravity 1.294 is 39.0% H_2SO_4 by weight, and sulfuric acid of specific gravity 1.139 is 20.0% H_2SO_4 by weight. The battery holds 3.5 liters of acid. Assume that the volume remains unchanged during discharge.)

PROBLEMS B   No answers given

16. Using the table of standard electrode potentials, calculate the voltage you could obtain from batteries that use the following reactions with (i) con-

centrations at unit activity, and (ii) oxidizing half-cell ionic concentrations at 10^{-5} molar and reducing half-cell ionic concentrations at 10^{-1} molar.

(a) $2Al + 3Cu^{++} \rightleftarrows 2Al^{+++} + 3Cu$ (c) $Sn + Sn^{++++} \rightleftarrows 2Sn^{++}$

(b) $Zn + Hg_2^{++} \rightleftarrows Zn^{++} + 2Hg$ (d) $Fe + Cl_2 \rightleftarrows Fe^{++} + 2Cl^-$

17. Make a simple sketch to show how you would arrange the materials used to construct each of the batteries in Problem 16. Show also polarity and direction of flow of electrons and ions.

18. Calculate the equilibrium constant for each of the reactions listed in Problem 16.

19. Calculate the change in standard free energy that accompanies each of the reactions listed in Problem 16.

20. Tell what is liberated at the anode and cathode during electrolysis of each of the following aqueous solutions with Pt electrodes: (a) $AgNO_3$; (b) $CuBr_2$; (c) K_2CO_3; (d) $PtCl_4$; (e) Li_2SO_4.

21. What weight of each of the following will be liberated by the passage of 1 faraday of electricity? (a) Pt from $PtCl_4$; (b) O_2 from $Al(NO_3)_3$; (c) Ba from $BaCl_2$; (d) H_2 from H_3PO_4; (e) Zn from $ZnSO_4$.

22. What volume of (a) F_2, (b) O_2, and (c) H_2, at standard conditions, will be liberated by the passage of 0.25 faraday of electricity?

23. How long will it take a current of 2.5 amp to deposit the silver from 650 ml of a $0.2M$ $AgNO_3$ solution?

24. What current will be needed to deposit 2.5 Cu in 15 min?

25. Platinum electrodes are placed in 400 ml of a $0.35M$ $NiSO_4$ solution. If a current of 1.75 amp is passed for 3 hr, what will be the $NiSO_4$ concentration at the end of that time?

26. A current of 4.5 amp is passed between copper electrodes in a $1.85M$ $CuSO_4$ solution. What will be the concentration of the $CuSO_4$ at the end of 5 hr?

27. Using a current of 3.5 amp, how long will it take to produce 25 liters of H_2, measured over water at 730 torr and 18°C, by the electrolysis of an H_2SO_4 solution?

28. How many ampere-hours of electricity will be needed to refine electrolytically 1 ton of Cu by removing it from an impure anode and depositing it in a pure form on the cathode?

29. During the discharge of a storage battery, the specific gravity of the sulfuric acid falls from 1.277 to 1.155. Calculate the number of ampere-hours the battery must have been used. (Sulfuric acid of specific gravity 1.277 is 37.0% H_2SO_4 by weight, and sulfuric acid of specific gravity 1.155 is 22.0% H_2SO_4 by weight. The battery holds 4 liters of acid. Assume that the volume remains unchanged during discharge.)

Electron Transfer II.
Equations

The basic principles discussed at the beginning of the last chapter in connection with the construction of simple electrochemical cells are exactly the ones used to write and balance chemical equations for electron-transfer reactions. These principles also enable you to predict whether or not a given electron-transfer reaction will actually take place.

Let us review these principles.

1. Every electron-transfer reaction may be considered as composed of two half-reactions, each half-reaction being written in the form

$$ne^- + \text{oxidizing agent} \rightleftarrows \text{reducing agent}$$

2. Electron-transfer half-reactions may be listed in order of decreasing tendency for the reducing agents to give up electrons, as in Table 15-1.

3. An electron-transfer reaction will occur spontaneously only if the reducing half-reaction lies above the oxidizing half-reaction in such a table.

4. In all electron-transfer reactions, there must be the same number of electrons gained by the oxidizing agent as are lost by the reducing agent.

This chapter concerns the application of these four principles to the balancing of equations for electron-transfer reactions.

Balancing Equations Containing Inorganic Compounds

PROBLEM:
Write a balanced ionic equation for the reaction between MnO_4^- and H_2S in acid solution. Assume unit activity for all concentrations.

SOLUTION:
The reaction is spontaneous because in Table 15-1 the reducing half-reaction

$$2\,e^- + 2\,H^+ + S \rightleftharpoons H_2S$$

lies above the oxidizing half-reaction

$$5\,e^- + 8\,H^+ + MnO_4^- \rightleftharpoons 4\,H_2O + Mn^{++}$$

In order to combine these two half-reactions to give the complete reaction, we must multiply each one by a factor that will yield the same number of electrons lost as gained. The factor 5 is needed for the H_2S half-reaction and the factor 2 is needed for the MnO_4^- half-reaction, in order to provide a loss of 10 electrons by H_2S and a gain of 10 by MnO_4^-:

$$10\,e^- + 10\,H^+ + 5\,S \rightleftharpoons 5\,H_2S$$
$$10\,e^- + 16\,H^+ + 2\,MnO_4^- \rightleftharpoons 8\,H_2O + 2\,Mn^{++}$$

Now, if we subtract the first half-reaction from the second, the ten electrons cancel to give

$$5\,H_2S + 2\,MnO_4^- + 16\,H^+ \rightarrow 5\,S + 2\,Mn^{++} + 10\,H^+ + 8\,H_2O$$

This can be simplified by subtracting $10\,H^+$ from each side of the equation to give

$$5\,H_2S + 2\,MnO_4^- + 6\,H^+ \rightarrow 5\,S + 2\,Mn^{++} + 8\,H_2O$$

We can make the general statement that *to balance any electron-transfer equation, you must subtract the reducing half-reaction equation from the oxidizing half-reaction equation after the two have first been written to show the same number of electrons.* Simplify the final equation, if needed.

The general method of balancing electron-transfer equations requires that half-reaction equations be available. Short lists of common half-reactions, similar to Table 15-1, are given in most textbooks, and chemistry handbooks have extensive lists. However, no list can provide all possible half-reactions, and it is not practical to carry lists in your pocket for instant reference. The practical alternative is to learn to make your own half-reaction equations. There is only one prerequisite for this approach: you must know the oxidation state of the oxidized and reduced forms of the substances involved in the electron-transfer reaction. In Chapter 8 you learned the charges on the ions of the most common elements; now we should review the method of determining the oxidation state of an element when it is combined in a radical.

PROBLEM:
What is the charge of Cr in the $Cr_2O_7^=$ ion?

SOLUTION:
It will be convenient for you to remember that, whenever oxygen is combined with other elements, it *always* has a charge of −2 unless it is in a peroxide (in which case it is −1). Similarly, it will be useful to know that, whenever hydrogen is combined with other elements, it *always* has a charge of +1 unless it is a hydride (in which case it is −1). Peroxides and hydrides are not common.

The total charge (C) on an ion is the sum of the charges of the atoms that compose it. If we let Z be the charge of a given element in the ion, and n be the number of atoms of that element in the ion, then

$$C = n_1Z_1 + n_2Z_2 + \ldots = \Sigma n_iZ_i$$

If we apply this to the $Cr_2O_7^=$ ion whose charge is −2, we have

$$-2 = n_{Cr}Z_{Cr} + n_OZ_O$$
$$-2 = (2)(Z_{Cr}) + (7)(-2)$$
$$(2)(Z_{Cr}) = 14 - 2 = 12$$
$$Z_{Cr} = +6 = \text{charge on Cr}$$

Often the oxidized form of an atom is combined with oxygen, whereas in the reduced state it is combined with less oxygen, or none. Chromium is a typical example; in the Cr^{+6} state it is combined as $Cr_2O_7^=$, but in the Cr^{+++} state it is uncombined (except for hydration). In working out a suitable half-reaction equation, the student must decide what to do with the oxygen atoms. The answer is simple: he may use H^+, OH^-, and H_2O on either side of his equations for balancing, so long as he complies with the actual state of acidity of his solutions. If his solution is acidic, he must not use OH^- on either side of his equation for balancing; he must use H^+ and/or H_2O.

PROBLEM:
Write a balanced half-reaction equation for the oxidation of metallic gold to its highest oxidation state.

SOLUTION:
Like all elements in the uncombined state, metallic gold (Au) has a charge of 0. There is no simple way in which you can reason out the fact that gold's highest oxidation state is +3; presumably you learned this in Chapter 8. Now, knowing the two oxidation states involved, you can write your half-reaction as

$$3\ e^- + Au^{+++} \rightleftarrows Au$$

As a check we note that the sum of the electrical charges on each side is zero.

PROBLEM:
Write a balanced half-reaction equation for the oxidation of Mo^{+++} to MoO_2^+ in acid solution.

SOLUTION:
You are not expected to know about the chemistry of Mo, but once given the reactant and product there is no problem. For the MoO_2^+ ion,

$$C = +1 = (1)(Z_{Mo}) + (2)(Z_o)$$
$$1 = Z_{Mo} + (2)(-2)$$
$$Z_{Mo} = +5$$

Since the oxidized form contains oxygen and the reduced form does not, and since the solution is acidic, the most direct approach here is to add sufficient H^+ to combine with the oxygen atoms to form water; in this case 4 H^+ would combine with $2O^=$ to make 2 H_2O. Note that the H^+ and $O^=$ do not involve any changes in charge. We write

$$2\ e^- + 4\ H^+ + MoO_2^+ \rightleftarrows 2\ H_2O + Mo^{+++}$$

As a check, we note that the sum of the electrical charges on each side is +3.

Using the principles we have just discussed, you should be able to write a balanced equation for any electron-transfer reaction. However, unless you knew the positions of your half-reactions in a table of standard electrode potentials (such as Table 15-1) you would not be able to predict whether the reaction would actually occur spontaneously. Realizing that it is not practical to have an extensive table of $E°$ values always at hand, we have organized Table 16-1, which associates natural groups of common chemicals with approximate values of $E°$.

If you study this table carefully, you will note that the first five groups correspond to reactivities and general information with which you are familiar; the association of approximate $E°$ values with each group is very helpful. Group 7 contains all the common metals that do not displace H^+ (as H_2 gas) from acids. Again, the range and degree of reactivity is related to common experience; from a chemist's standpoint it is very helpful to associate an approximate value of the electrode potential with each.

Likewise, group 8 contains elements often associated with each other because of position in the periodic table and similarity in chemical properties. The emphasis in this group is on the oxidized form of the element (the halogen X_2). The same reduced forms of the halogen occur again in group 9, but associated in a different way. There is no problem with F_2 because it is the strongest oxidizing agent of all chemicals—it lies at the bottom of every list. The student needs to concern himself only with the approximate $E°$ values of the first three; the order is the same as that in the periodic table.

TABLE 16-1.
Electron Transfer Table by Chemical Groups. In each group the elements are listed from left to right in decreasing strength as reducing agents. Half-cell potentials correspond to unit activity and 25°C. The number in parentheses after the symbol of some elements denotes the electrical charge of the oxidized form.

Chemical Groups	$E°$ in volts
1. Alkali metals ($e^- + M^+ \rightleftarrows M$) Li, Cs, Rb, K, Na	-3.04 to -2.71
2. Alkaline earth metals ($2\ e^- + M^{++} \rightleftarrows M$) Ba, Sr, Ca, Mg	-2.90 to -2.36
3. Active metals ($n\ e^- + M^{+n} \rightleftarrows M$) Al, Zn, Cr(III), Fe(II), Cd	-0.76 to -0.40 (Al is -1.66)
4. Medium active metals ($n\ e^- + M^{+n} \rightleftarrows M$) Co(II), Ni(II), Sn(II), Pb(II)	-0.40 to -0.13
5. Midrange: $2\ e^- + 2\ H^+ \rightleftarrows H_2$ $2\ e^- + 2\ H^+ + S \rightleftarrows H_2S$	0.00 0.14
6. Second stage oxidation ($n\ e^- + $ -ic $\rightleftarrows$ -ous) Sn^{++}, Fe^{++}, Hg_2^{++} H_2SO_3, H_3AsO_3, HNO_2	0.15, 0.77, 0.91 0.17, 0.56, 0.94
7. Jewelry metals ($n\ e^- + M^{+n} \rightleftarrows M$) Cu(II), Ag(I), Hg(II), Pt(II), Au(III)	0.34, 0.80, 0.85, 1.20, 1.50
8. Halogens ($2\ e^- + X_2 \rightleftarrows 2\ X^-$) I^-, Br^-, Cl^-, F^-	0.54, 1.07, 1.36, 2.87
9. Oxidizing negative ions and others	

$$
n\ e^- + H^+ + \begin{bmatrix} NO_3^- \rightleftarrows NO \\ IO_3^- \rightleftarrows I^- \\ O_2 \rightleftarrows H_2O \\ MnO_2 \rightleftarrows Mn^{++} \\ Cr_2O_7^= \rightleftarrows Cr^{+++} \\ BrO_3^- \rightleftarrows Br^- \\ ClO_3^- \rightleftarrows Cl^- \\ PbO_2 \rightleftarrows Pb^{++} \\ MnO_4^- \rightleftarrows Mn^{++} \\ H_2O_2 \rightleftarrows H_2O \end{bmatrix} + H_2O
$$

	$E°$ in volts
$NO_3^- \rightleftarrows NO$	0.96
$IO_3^- \rightleftarrows I^-$	1.20
$O_2 \rightleftarrows H_2O$	1.23
$MnO_2 \rightleftarrows Mn^{++}$	1.28
$Cr_2O_7^= \rightleftarrows Cr^{+++}$	1.33
$BrO_3^- \rightleftarrows Br^-$	1.44
$ClO_3^- \rightleftarrows Cl^-$	1.45
$PbO_2 \rightleftarrows Pb^{++}$	1.45
$MnO_4^- \rightleftarrows Mn^{++}$	1.51
$H_2O_2 \rightleftarrows H_2O$	1.78

Groups 6 and 9 lie farthest from the previous experience of students and require more study than the others. Most chemists think of the oxidizing agents of group 9 as being *very strong* oxidizing agents and would classify a large number of them as having $E°$ values of about 1.35 to 1.50. It pays to know that the $E°$ of HNO_3 (giving NO) is approximately 1.0, because it is so common. Note that almost none of the half-reactions in the first eight groups overlap with group 9. Another help with group 9 is to note that all the halogen-containing oxidizing agents are reduced to the halide forms. You should probably think of the first seven groups in terms of the relative strengths of the *reducing* agents (those chemicals on the righthand side of the half-reaction), and think of groups 8 and 9 in terms of the relative strengths of the *oxidizing* agents (those chemicals on the lefthand side of the half-reactions).

Two major oversimplifications are involved in Table 16-1: (a) only metals are shown lying above H_2, and (b) no reactions in basic solution are shown. You will probably spend far less time memorizing this table, or something similar to it, than you will spend trying to get comparable information in some other intuitive, haphazard manner, and the correctness of your prediction will be superior and rewarding. This table is not practical for making predictions about reactions composed of half-reactions whose $E°$ values lie very close together.

Balancing Equations Containing Organic Compounds

The oxidation and reduction of most organic compounds do not fit neatly into an "approximate electron-transfer table," yet the principles involved are much the same. The information most likely to be lacking is the nature of the products, but assuming these to be given, the balancing of the chemical equation is relatively simple. Let us look at a couple of examples.

PROBLEM:
Write a balanced ionic equation for the oxidation of $C_2O_4^=$ by MnO_4^- in acid solution, given that CO_2 is the oxidation product from $C_2O_4^=$.

SOLUTION:
The reducing half-reaction must be

$$2\ e^- + 2\ CO_2 \rightleftarrows C_2O_4^=$$

since we must oxidize 2 C atoms from +3 (in $C_2O_4^=$) to +4 (in CO_2). The oxidizing half-reaction is one with which we are familiar:

$$5\ e^- + 8\ H^+ + MnO_4^- \rightleftarrows 4\ H_2O + Mn^{++}$$

Multiply the first half-reaction by 5 and the second half-reaction by 2 to give an equal loss and gain of 10 electrons. Then subtract the first half-reaction from the second, to get

$$5\ C_2O_4^= + 2\ MnO_4^- + 16\ H^+ \rightleftarrows 10\ CO_2 + 2\ Mn^{++} + 8\ H_2O$$

PROBLEM:
Write a balanced ionic equation for the oxidation of C_2H_3OCl by $Cr_2O_7^=$ in acid solution, given that CO_2 and Cl_2 are the two main oxidation products of the organic compound.

SOLUTION:
We can arbitrarily assign charges of +1 and −2 to H and O in the organic compound, but we still are in a quandary about what charges should be assigned to C and Cl. *For purposes of balancing the equation* it makes no difference what charges are assigned, as long as the whole organic molecule retains a net charge of zero. As a matter of convenience, let us assume that Cl has the same charge as it has in the product (Cl_2); i.e., zero. With this assumption, the charge on 2 C atoms must be equal and opposite in sign to the net charge on the 3 H and 1 O, namely, −1, or an effective charge of $-\frac{1}{2}$ per C atom. This would give us a trial half-reaction of

$$9\ e^- + 9\ H^+ + \tfrac{1}{2}\ Cl_2 + 2\ CO_2 \rightleftarrows 3\ H_2O + C_2H_3OCl$$

since each C atom is being oxidized from $-\frac{1}{2}$ to +4. It is undesirable to use a coefficient of $\frac{1}{2}$, so we multiply through by 2 to get our final half-reaction,

$$18\ e^- + 18\ H^+ + Cl_2 + 4\ CO_2 \rightleftarrows 6\ H_2O + 2\ C_2H_3OCl$$

We subtract this half-reaction from our familiar $Cr_2O_7^=$ half-reaction (taken three times in order to accept 18 e^-),

$$6\ e^- + 14\ H^+ + Cr_2O_7^= \rightleftarrows 7\ H_2O + 2\ Cr^{+++}$$

to give our balanced electron-transfer equation,

$$2\ C_2H_3OCl + 3\ Cr_2O_7^= + 24\ H^+ \rightarrow 4\ CO_2\uparrow + Cl_2\uparrow + 6\ Cr^{+++} + 15\ H_2O$$

In the actual subtraction process, there were 42 H^+ and 6 H_2O on the left side and 18 H^+ and 21 H_2O on the right side; this was simplified by subtracting 6 H_2O and 18 H^+ from both sides to give the equation shown.

EFFECT OF ACID CONCENTRATION

The $E°$ values of all the half-reactions involving H^+ depend on having H^+ concentration of unit activity; for all other half-reactions, the $E°$ values are independent of the H^+ concentration. For those that do involve H^+, the actual half-reaction potentials, and therefore the strengths as oxidizing or reducing agents, can be greatly affected by the H^+ concentration. On p. 195, we showed that in pure water, where $[H^+] = 10^{-7}\ M$,

$$E_{H_2-H^+} = -0.41\ \text{volts}$$

Immediately we see that none of the medium active metals (group 4) are able to react with water to displace H_2, and most of those in the active metal group (group 3) would react with little energy change.

PROBLEM:
Calculate the electrode potential for the Cr^{+++}-$Cr_2O_7^=$ half-reaction if the $Cr_2O_7^=$ and Cr^{+++} ions are kept at unit activity and the H^+ concentration is (a) raised to an activity of 5 M and (b) lowered to 10^{-7} M as it is in distilled water.

SOLUTION:
Our fundamental equation for the electrode potential (p. 187) gives

$$E_{Cr^{+++}-Cr_2O_7^=} = E^\circ_{Cr^{+++}-Cr_2O_7^=} - \frac{0.0591}{6} \log \frac{[Cr^{+=}]^2}{[Cr_2O_7^=][H^+]^{14}}$$

With unit activity for Cr^{+++} and $Cr_2O_7^=$, and an E° value of 1.36 volts from Table 15-1, we get

$$E_{Cr^{+++}-Cr_2O_7^=} = 1.36 + \frac{(0.0591)(14)}{6} \log [H^+]$$

 (a) If $[H^+] = 5$ M, then $E_{Cr^{+++}-Cr_2O_7^=} = +1.46$ volts. There is only a modest increase in oxidizing power by making a five-fold increase in the activity of the $[H^+]$.
 (b) If $[H^+] = 10^{-7}$ M, $E_{Cr^{+++}-Cr_2O_7^=} = +0.39$ volts. This result shows that one of the potentially strongest oxidizing agents, $Cr_2O_7^=$, is *greatly reduced* in potency when it is merely dissolved in water (with the H^+ activity decreased 10 million-fold); it is about as strong an oxidizing agent as Cu^{++} at unit activity.

PROBLEM:
Calculate the electrode potential for the NO-NO_3^- half-reaction for an aqueous solution of KNO_3 in which the NO_3^- is at unit activity, the NO pressure is 1 atm, and the $[H^+] = 10^{-7}$ M (distilled water).

SOLUTION:

$$E_{NO-NO_3^-} = E^\circ_{NO-NO_3^-} - \frac{0.0591}{3} \log \frac{[NO]}{[H^+]^4[NO_3^-]}$$

$$= +0.96 + \frac{(0.0591)(4)}{3} \log (10^{-7})$$

$$= +0.41 \text{ volts.}$$

This result shows that *in a neutral aqueous solution* the NO_3^- ion is a relatively weak oxidizing agent. It is a common error to attempt to make an electron-transfer reaction using NO_3^- ion in a neutral aqueous solution; you should not confuse the weak oxidizing power of the NO_3^- ion under these conditions with its strong oxidizing power in strong acid solution. Nitrate salts are commonly used in the laboratory because they are so soluble and readily available, and because they are not appreciably oxidizing in neutral aqueous solution.

We have already seen that strongly acid solutions will make the oxygen-containing negative ions even stronger oxidizing agents; the nitrate ion in an acid solution is no exception. When concentrated HNO_3 is used, the half-cell potential is comparable to that of MnO_4^-, but NO is no longer the gaseous product; it is NO_2. For concentrated HNO_3, the half-reaction that must be used is

$$e^- + 2\ H^+ + NO_3^- \rightleftarrows H_2O + NO_2$$

It is a common error to think that a metal, if it reacts with an acid, will produce H_2 gas. This is not true if the acid is HNO_3; the gaseous products will be NO or NO_2, depending on the concentration. Under certain circumstances, usually involving dilute solution and a strong reducing agent such as Zn, the reduction products of HNO_3 may actually be N_2 or NH_3.

Likewise, it is a common error to think that a sulfide, if it reacts with an acid, will always produce H_2S. This is not true if the acid is HNO_3; the gaseous products will be NO or NO_2 (depending on the concentration), and sulfur will be formed. The NO_3^- half-reaction is below the H_2S (or $S^=$) half-reaction. Concentrated HNO_3 is so powerful an oxidizing agent that almost all the really difficultly soluble sulfides can be dissolved through oxidation, even though the same sulfides remain untouched by those strong acids which would lead to the formation of H_2S.

It is worth noting that a large number of reducing agents (most of those in the first eight groups of the abbreviated electron-transfer table!) are unstable in solution if not protected from the air. Even in neutral water solution the effect of air oxidation may be very marked, because O_2 is a strong oxidizing agent. In some cases the effects may be complicated because of low O_2 pressure or unusual hydration effects.

By now it may have become a matter of some concern to you that aqueous solutions of such strong oxidizing agents as $KMnO_4$ and $Na_2Cr_2O_7$ are stable for indefinite periods of time. In fact, it would appear that no oxidizing agent which lies below the half-reaction

$$4\ e^- + 4\ H^+ + O_2 \rightleftarrows 2\ H_2O \qquad E^\circ = 1.23\ \text{volts}$$

could exist in water without decomposing the water to liberate O_2 gas. Since water is not decomposed by these substances it is evident that, although the equilibrium position of such reactions does lie far in the direction favoring the evolution of O_2, the activation energy must be extremely high and the *rate* be vanishingly small. The detailed reasons why this should be so are still unknown, in spite of enormous research effort.

Molecular Equations

After we have developed an ionic equation for an electron-transfer reaction, we frequently need to show the molecules involved in the solutions; that is, the substances that are initially put into the solution and those that are obtained from it after the reaction has occurred. We must have such molecular equations if stoichiometric calculations are to be made.

We can use the ionic equation to write the molecular equation of the same reaction, keeping in mind that every ion of the original substances was obtained from some acid, base, or salt, and that every ion in the products must be shown as the salt, base, or acid that would be obtained if the solution were evaporated to dryness.

PROBLEM:
Metallic copper is oxidized by dilute nitric acid. Write ionic and molecular equations for the reaction.

SOLUTION:
The two half-reactions are (Table 15-1)

$$3\ e^- + 4\ H^+ + NO_3^- \rightleftarrows NO + 2\ H_2O$$

$$2\ e^- + Cu^{++} \rightleftarrows Cu$$

Balancing electrons and subtracting the second half-reaction from the first gives:

$$6\ e^- + 8\ H^+ + 2\ NO_3^- \rightleftarrows 2\ NO + 4\ H_2O$$

$$6\ e^- + 3\ Cu^{++} \rightleftarrows 3\ Cu$$

$$3\ Cu + 2\ NO_3^- + 8\ H^+ \rightleftarrows 3\ Cu^{++} + 2\ NO + 4\ H_2O$$

To write the molecular equation, we use 8 HNO_3 to furnish the required 8 H^+. In the products Cu^{++} appears as $Cu(NO_3)_2$. We have

$$3\ Cu + 8\ HNO_3 \rightleftarrows 3\ Cu(NO_3)_2 + 2\ NO + 4\ H_2O$$

PROBLEMS A   Answers on page 346

1. Find the electrical charge on each atom in the following molecules and ions:
   (a) CO_2 (d) $Fe_2(SO_4)_3$ (g) $C_2O_4^=$ (j) UO_2^{++}
   (b) $AgNO_3$ (e) $Ca_2P_2O_7$ (h) $PtCl_6^=$
   (c) BaO_2 (f) LiH (i) $B_4O_7^=$

2. Write balanced ionic half-reaction equations for the oxidation of each of the following reducing agents in acid solution:
   (a) HNO_2 (c) Al (e) Hg_2^{++} (g) I^-
   (b) H_3AsO_3 (d) Ni (f) H_2O_2

3. Write balanced ionic half-reaction equations for the reduction of each of the following oxidizing agents in acid solution.
   (a) PbO_2 (c) Co^{+++} (e) BrO^- (g) F_2
   (b) NO_3^- (d) ClO_4^- (f) Ag^+ (h) Sn^{++}

4. Write balanced ionic equations for the following reactions.
   (a) $Zn + Cu^{++} \rightarrow$
   (b) $Zn + H^+ + SO_4^= \rightarrow$
   (c) $Cr_2O_7^= + I^- + H^+ \rightarrow$
   (d) $MnO_4^- + Cl^+ + H^+ \rightarrow$
   (e) $ClO_3^- + Br^- + H^+ \rightarrow$
   (f) $MnO_4^- + H_2O_2 + H^+ \rightarrow$
   (g) $MnO_2 + H^+ + Cl^- \rightarrow$
   (h) $Ag + H^+ + NO_3^-$ (dilute) $\rightarrow$
   (i) $PbO_2 + Sn^{++} + H^+ \rightarrow$
   (j) $MnO_4^- + H_2C_2O_4 + H^+ \rightarrow$
   (k) $H_2O_2 + HNO_2 \rightarrow$
   (l) $Fe^{++} + Zn \rightarrow$
   (m) $Fe^{+++} + I^- \rightarrow$
   (n) $ClO_4^- + H_3AsO_3 \rightarrow$
   (o) $H_2S + ClO^- \rightarrow$

5. Write balanced molecular equations for the reactions in Problem 4. Use potassium salts of the negative ions and sulfate salts of the positive ions for the reactants given.

6. Complete and balance the following ionic equations for reactions that occur in aqueous solution. The nature of the solution, acidic or basic, is indicated for each reaction.
   (a) $CH_2O + Ag_2O \rightarrow Ag + HCO_2^-$ (basic)
   (b) $C_2H_2 + MnO_4^- \rightarrow CO_2$ (acidic)
   (c) $C_2H_3OCl + Cr_2O_7^= \rightarrow CO_2 + Cl_2$ (acidic)
   (d) $Ag^+ + AsH_3 \rightarrow Ag + H_3AsO_3$ (acidic)
   (e) $CN^- + Fe(CN)_6^{\equiv} \rightarrow CNO^- + Fe(CN)_6^{\equiv}$ (basic)
   (f) $C_2H_4O + NO_3^- \rightarrow NO + C_2H_4O_2$ (acidic)

7. Write balanced equations for the reactions that occur in each of the following situations.
   (a) Metallic zinc and dilute nitric acid produce ammonium nitrate as one of the products.
   (b) Sodium thiosulfate, $Na_2S_2O_3$, is used to titrate iodine in quantitative analysis; one of the products is sodium tetrathionate, $Na_2S_4O_6$.
   (c) A sample of potassium iodide contains some potassium iodate as impurity. When sulfuric acid is added to a solution of this sample, iodine is produced, as shown by a blue color that appears when a little starch solution is added. Give the equation for the formation of the iodine.
   (d) When copper is heated in concentrated sulfuric acid, an odor of sulfur dioxide is noted.
   (e) A classical operation in quantitative analysis is the use of a Jones Reductor—a column of granulated zinc. A solution of ferric salts is passed through this column prior to titration with potassium permanganate. Give the equation for the reaction in the column.
   (f) Pure hydriodic acid cannot be prepared by adding concentrated sulfuric acid to sodium iodide and distilling off the hydriodic acid because of side reactions. One side reaction yields hydrogen sulfide as noted by the odor. Give the equation for this side reaction.
   (g) A solution of sodium hypochlorite is heated. One of the products is sodium chlorate.

PROBLEMS B   No answers given

8. Find the oxidation number of each atom in the following molecules and ions.
   (a) SO_3 (d) Na_2O_2 (g) $V_2O_7^{\equiv}$ (j) BiO^+
   (b) H_2SO_3 (e) $(NH_4)_3PO_4$ (h) $SiO_3^=$
   (c) $Zn(IO_3)_2$ (f) NaH (i) $S_2O_3^=$

9. Write balanced ionic half-reaction equations for the oxidation of each of the following reducing agents in acid solution:
   (a) Sn (c) $H_2C_2O_4$ (e) Br^-
   (b) Sn^{++} (d) H_2S (f) Ba

10. Write balanced ionic half-reaction equations for the reduction of each of the following oxidizing agents in aqueous acid solution:
    (a) MnO_2 solid   (c) ClO_2^- (e) H_2O_2 (g) Fe^{++}
    (b) $Cr_2O_7^=$ (d) IO_3^- (f) Br_2 (h) Cd^{++}

11. Write balanced ionic equations for the following reactions.
    (a) $Fe + Ag^+ \rightarrow$
    (b) $Cd + H^+ \rightarrow$
    (c) $Cd + H^+ + NO_3^- \rightarrow$
    (d) $Cl_2 + HNO_2 \rightarrow$
    (e) $MnO_4^- + H_2S + H^+ \rightarrow$
    (f) $ClO_3^- + Sn^{++} + H^+ \rightarrow$
    (g) $Br_2 + Fe^{++} \rightarrow$
    (h) $H_2O_2 + H_2SO_3 \rightarrow$
    (i) $H_2O_2 + ClO_3^- \rightarrow$
    (j) $H_3AsO_3 + Ce^{++++} \rightarrow$
    (k) $Sn^{++} + ClO_2^- + H^+ \rightarrow$
    (l) $Sn^{++} + Mg \rightarrow$
    (m) $PbO_2 + HNO_2 + H^+ \rightarrow$
    (n) $Cr_2O_7^= + H_2C_2O_4 + H^+ \rightarrow$
    (o) $Cl_2 + I^- \rightarrow$

12. Write balanced molecular equations for the reactions in Problem 11. Use potassium salts of the negative ions and sulfate salts of the positive ions for the reactants given.

13. Complete and balance the following ionic equations for reactions that occur in aqueous solution. The nature of the solution, acidic or basic, is indicated for each reaction.
    (a) $C_3H_8O + MnO_4^- \rightarrow CO_2$ (acidic)
    (b) $CN^- + MnO_4^- \rightarrow CNO^- + MnO_2$ (basic)
    (c) $CH_2O + Ag(NH_3)_2^+ \rightarrow Ag + HCO_2^-$ (basic)
    (d) $CHCl_3 + MnO_4^- \rightarrow Cl_2 + CO_2$ (acidic)
    (e) $MnO_4^- + C_2H_3OCl \rightarrow CO_2 + Cl_2$ (acidic)
    (f) $C_2H_6O + Cr_2O_7^= \rightarrow HC_2H_3O_2$ (acidic)

14. Write balanced equations for the reactions that occur in each of the following situations.
    (a) Chlorine gas is bubbled through a solution of ferrous bromide.
    (b) In the final step of producing bromine from sea water, a mixture of sodium bromide and sodium bromate is treated with sulfuric acid.

(c) A microchemical procedure uses a cadmium amalgam (cadmium dissolved in metallic mercury) to reduce iron salts to their lowest valence state prior to titration with standard ceric sulfate. Give equation for the reaction involving the cadmium.

(d) When zinc is heated with concentrated sulfuric acid, hydrogen sulfide is evolved.

(e) Pure hydrobromic acid cannot be prepared by treating sodium bromide with concentrated sulfuric acid and distilling off the hydrobromic acid because some sulfur dioxide is produced at the same time, as noted by the odor. Give the equation for the production of sulfur dioxide.

(f) Sodium perchlorate is prepared by carefully heating solid sodium chlorate.

(g) The Marsh Test for the detection of arsenic depends on the reaction of an arsenic compound, such as H_3AsO_4, with metallic zinc in acid solution to give arsine, AsH_3.

Stoichiometry IV.
Equivalent Weight and Normality

The stoichiometric calculations of Chapter 12 are based on the mole as the fundamental chemical unit in reactions. An alternative method of calculation utilizes the *equivalent* as a fundamental chemical unit. Unlike the mole, there are two kinds of equivalents, the type depending on the reaction in question; we shall refer to them as acid-base equivalents (or simply as equivalents) and electron-transfer equivalents (or E-T equivalents).

Acid-Base Equivalents

The acid-base equivalent is defined as the weight, in grams, of a substance that will provide, react with, or be equivalent to 1 mole of H^+. Whenever another weight unit is used, the equivalent may be referred to as a ton-equivalent, a pound-equivalent, a milliequivalent, and so on, just as is done with the mole. Thus, 1 mole or 98.1 g of H_2SO_4 contains $2H^+$, and its acid-base equivalent weight is 49.05 g per equivalent. One mole (171.4 g/mole) of $Ba(OH)_2$ contains $2OH^-$ and will react with $2H^+$; its acid-base equivalent weight is therefore 85.70 g/equiv. For $La_2(SO_4)_3$, there are 6 equivalents per mole, because the $2La^{+++}$ ions provide the same total positive charge as $6H^+$; the equivalent weight is $\frac{1}{6}$ the mole weight.

Care must be taken to avoid the ambiguity* that may arise when a substance may have more than one equivalent weight. For example, Na_2CO_3 might react with HCl in either of two ways:

$$2HCl + Na_2CO_3 \rightarrow 2NaCl + CO_2 + H_2O$$

$$HCl + Na_2CO_3 \rightarrow NaHCO_3 + NaCl$$

In the first reaction, Na_2CO_3 acts as though it has two equivalents per mole; in the second it acts as though it had one. The ambiguity is avoided by stating which reaction is involved.

When we prepare a solution that contains 1 equivalent weight per liter of solution, its concentration is said to be 1 normal, designated as $1N$. A $0.20N$ $Ba(OH)_2$ solution contains 0.20 equivalents per liter, or $0.20 \dfrac{\text{equiv}}{\text{liter}} \times 85.7$ $\dfrac{\text{g}}{\text{equiv}} = 17.14$ grams per liter. The *normality* of the solution is said to be 0.20 equivalents per liter.

Calculations involving equivalents, milliequivalents, normalities, and volumes of solutions are made in just the same way as those involving molarities of solutions. The unique and useful feature about the use of equivalents is that for *any* chemical reaction, when reactant A has just exactly consumed reactant B, we can say

$$\text{equivalents of A} = \text{equivalents of B}$$

regardless of the number of moles of A required to react with a mole of B. At the endpoint in a titration, for example, we can say that

$$\text{endpoint equivalents} = (N_{acid})(V_{acid}) = (N_{base})(V_{base})$$

regardless of the acid or base that is used. Or, similarly, if a weighed sample of acid were titrated with a standard base solution,

$$\text{endpoint equivalents} = \frac{\text{grams of acid}}{\dfrac{\text{grams}}{\text{equiv}}} = \frac{W}{E} = (N_{base})(V_{base})$$

PROBLEM:
A solution of HCl is standardized by titration of a known weight of pure Na_2CO_3. Find N of HCl, given that

$$\text{vol of HCl} = 4.13 \text{ ml}$$

$$\text{wt of } Na_2CO_3 = 0.215 \text{ g}$$

---

* Because of possible ambiguities about equivalent weights, many chemists prefer to base all stoichiometric calculations on the mole. The concept of normality is, however, widely used, and every student needs to understand it.

SOLUTION:
A mole of Na_2CO_3 is 106.0 g. Therefore, a gram equivalent is $106.0/2 = 53.0$ g:

$$\text{g equiv of } Na_2CO_3 = \frac{0.215 \text{ g}}{53.0 \frac{\text{g}}{\text{equiv}}} = 0.00406$$

Since equivalents of A are equal to equivalents of B, we have in 41.3 ml of HCl the same number of equivalents, or 0.00406. Normality = equiv per liter or

$$N = \frac{0.00406 \text{ equiv}}{0.0413 \text{ liter}} = 0.0985 \frac{\text{equiv}}{\text{liter}}$$

If we prefer to use mg, meq, and ml, the solution is:

$$\text{meq of } Na_2CO_3 = \frac{215 \text{ mg}}{53 \frac{\text{mg}}{\text{meq}}} = 4.06$$

$$\text{meq of HCl} = 4.06$$

$$N \text{ of HCl} = \frac{4.06 \text{ meq}}{41.3 \text{ ml}} = 0.0985 \frac{\text{meq}}{\text{ml}} = 0.0985N$$

PROBLEM:
A 0.500-g sample of impure $CaCO_3$ is dissolved in 50.0 ml of 0.0985N HCl. After the reaction is complete, the excess HCl is titrated by 6.0 ml of 0.105N NaOH. Find the percentage purity of the $CaCO_3$ (assume no other constituent of the sample reacts with HCl).

SOLUTION:
In this problem part of the HCl reacts with $CaCO_3$, the remainder with NaOH.

$$\text{equiv of HCl} = \text{equiv of } CaCO_3 + \text{equiv of NaOH}$$

We are given

$$\text{equiv of HCl} = NV = 0.050 \text{ liter} \times 0.0985 \frac{\text{equiv}}{\text{liter}} = 0.004925 \text{ equiv}$$

$$\text{equiv of NaOH} = 0.0060 \text{ liter} \times 0.105 \frac{\text{equiv}}{\text{liter}} = 0.00063 \text{ equiv}$$

$$\text{equiv of } CaCO_3 = \text{equiv of HCl} - \text{equiv of NaOH}$$

$$= 0.004925 - 0.00063 = 0.004295$$

$$\text{A g-equiv of } CaCO_3 \text{ is } \frac{100.1 \frac{\text{g}}{\text{mole}}}{2 \frac{\text{equiv}}{\text{mole}}} = 50.0 \text{ g/equiv}$$

$$\text{wt of } CaCO_3 = 50.0 \frac{\text{g}}{\text{equiv}} \times 0.004295 \text{ equiv} = 0.215 \text{ g}$$

$$\text{per cent of } CaCO_3 = \frac{0.215 \text{ g}}{0.500 \text{ g}} \times 100 = 43.0\%$$

PROBLEM:

An organic chemist synthesized a new compound, X, with acidic properties. A 0.7200-gram sample required 30.00 ml of 0.2000M Ba(OH)$_2$ for titration. What is the equivalent weight of X?

SOLUTION:

$$\text{equivalents of Ba(OH)}_2 = \left(0.2000 \ \frac{\text{mole}}{\text{liter}}\right)\left(2 \ \frac{\text{equiv}}{\text{mole}}\right)(0.03000 \text{ liter})$$

$$= 0.0120 \text{ equivalents of } X$$

$$\text{equivalent weight of } X = \frac{0.7200 \text{ g of } X}{0.0120 \text{ equiv of } X} = 60 \text{ g/equiv}$$

This knowledge, together with the empirical formula and the molecular weight of X, will help the chemist elucidate the structure of the compound.

Electron-Transfer Equivalents

The electron-transfer equivalent weight was defined in Chapter 15 as the weight of material oxidized or reduced by one mole of electrons. This quantity is easily calculated by dividing the coefficient of each component of the half-reaction that is involved in the reaction in question by the number of moles of electrons (n) in the half-reaction. For example,

$$e^- + \tfrac{1}{2}\,Zn^{++} \rightleftarrows \tfrac{1}{2}\,Zn$$

$$e^- + Fe^{+++} \rightleftarrows Fe^{++}$$

For the first half-reaction, 0.5 mole of Zn (or Zn^{++}) is oxidized (or reduced) per mole of electrons, and in the second one mole of Fe^{++} (or Fe^{+++}) is oxidized (or reduced) per mole of electrons. With our definition for E-T equivalent weight as

$$\text{E-T equiv wt} = \frac{\text{grams}}{\text{mole of electrons}}$$

the E-T equivalent weight for zinc can be found from its molecular weight as follows:

$$\text{E-T equiv wt} = \left(65.4 \ \frac{\text{g Zn}}{\text{mole Zn}}\right)\left(0.5 \ \frac{\text{mole Zn}}{\text{mole of electrons}}\right)$$

$$= 32.7 \ \frac{\text{g Zn}}{\text{mole of electrons}} = 32.7 \ \frac{\text{g Zn}}{\text{equiv}}$$

As is pointed out on p. 197, many substances have more than one E-T equivalent weight; so one must state the reaction involved when specifying the E-T equivalent weight for a given substance.

The terms normal and normality are defined and applied as they were for acid-base equivalents. Again, the unique property of normality is that for any electron-transfer reaction, when the reducing agent has just exactly consumed the oxidizing agent,

$$\text{reducing equivalents} = \text{oxidizing equivalents}$$

regardless of the number of moles of each involved. Moreover, all the statements made about the endpoint in acid-base titrations also apply to the endpoint in electron-transfer titrations.

PROBLEM:
A solution of $KMnO_4$ is standardized by titrating a known weight of $Na_2C_2O_4$. From the data compute the normality of the $KMnO_4$. Find the molarity of the $KMnO_4$. The data are:

$$\text{wt of } Na_2C_2O_4 = 0.280 \text{ g}$$

$$\text{purity of } Na_2C_2O_4 = 99.8\%$$

$$\text{vol of } KMnO_4 = 40.0 \text{ ml}$$

SOLUTION:
The reaction is

$$2MnO_4^- + 16H^+ + 5C_2O_4^- = 2Mn^{++} + 8H_2O + 10CO_2$$

Each molecule of $Na_2C_2O_4$ loses $2e^-$. Therefore, using equiv to mean E-T equiv,

$$\text{equiv wt of } Na_2C_2O_4 = \frac{134.0 \frac{g}{\text{mole}}}{2 \frac{\text{equiv}}{\text{mole}}} = 67.0 \text{ g/equiv}$$

$$\text{wt of pure } Na_2C_2O_4 = 0.280 \text{ g} \times 0.998 = 0.2794 \text{ g}$$

$$\text{no. of equiv of } Na_2C_2O_4 = \frac{0.2794 \text{ g}}{67.0 \frac{g}{\text{equiv}}} = 0.00417 \text{ equiv}$$

$$\text{no. of equiv of } KMnO_4 = 0.00417 \text{ equiv}$$

$$N \text{ of } KMnO_4 = \frac{\text{equiv}}{\text{liter}} = \frac{0.00417 \text{ equiv}}{0.0400 \text{ liter}} = 0.104 \frac{\text{equiv}}{\text{liter}}$$

A mole of $KMnO_4$ contains 5 equivalents. Therefore the molarity is

$$M \text{ of } KMnO_4 = \frac{0.104 \frac{\text{equiv}}{\text{liter}}}{5 \frac{\text{equiv}}{\text{mole}}} = 0.0208 \text{ mole/liter}$$

PROBLEM:

A solution of $FeSO_4$ is titrated by a standard solution of $KMnO_4$ in acid solution. From the data, find the normality of the $FeSO_4$:

$$\text{vol of } FeSO_4 = 45.3 \text{ ml}$$

$$\text{vol of } KMnO_4 = 42.6 \text{ ml}$$

$$N \text{ of } KMnO_4 = 0.105$$

SOLUTION:

Using equiv to mean E-T equiv, we compute equivalents of $KMnO_4$ as follows:

$$\text{equiv of } KMnO_4 = \text{normality} \times \text{volume}$$

$$= 0.105 \text{ equiv per liter} \times 0.0426 \text{ liter}$$

$$= 0.00448 \text{ equiv}$$

$$\text{equiv of } FeSO_4 = 0.00448$$

$$N \text{ of } FeSO_4 = \frac{\text{equiv}}{\text{liter}} = \frac{0.00448 \text{ equiv}}{0.0453 \text{ liter}} = 0.0988 \frac{\text{equiv}}{\text{liter}}$$

PROBLEM:

A sample of iron ore is dissolved in HCl and the iron reduced to Fe^{++}. The solution is then titrated by $KMnO_4$. From the data compute the percentage of iron as Fe_2O_3 in the sample:

$$\text{wt of sample} = 0.446 \text{ g}$$

$$\text{vol of } KMnO_4 = 38.6 \text{ ml}$$

$$N \text{ of } KMnO_4 = 0.105N$$

SOLUTION:

Here we are given normality and volume, from which we can compute the equivalents of $KMnO_4$. Again, using equiv to mean E-T equiv,

$$\text{equiv of } KMnO_4 = 0.105 \frac{\text{equiv}}{\text{liter}} \times 0.0386 \text{ liter} = 0.00405 \text{ equiv}$$

$$\text{equiv. of } Fe_2O_3 = 0.00405$$

$$\text{equiv wt of } Fe_2O_3 = \frac{\text{mole wt } Fe_2O_3}{2 \frac{\text{equiv}}{\text{mole}}} = \frac{159.6 \frac{\text{g}}{\text{mole}}}{2 \frac{\text{equiv}}{\text{mole}}} = 79.8 \text{ g/equiv}$$

(Each Fe atom loses $1e^-$ when oxidized. Since the Fe_2O_3 molecule contains 2 Fe atoms, the equivalent weight is half the molecular weight.)

$$\text{wt of } Fe_2O_3 = 79.8 \frac{\text{g}}{\text{equiv}} \times 0.00405 \text{ equiv} = 0.324 \text{ g}$$

$$\text{percentage of } Fe_2O_3 \text{ in sample} = \frac{0.324 \text{ g}}{0.446 \text{ g}} \times 100 = 72.5\%$$

PROBLEM:

A radiochemist isolated 8.6 micrograms (γ) of a chloride of neptunium (atomic weight = 237), which he proved had the formula $NpCl_3$. In trying to find the possible valence states of Np he found that titration of his 8.6 γ sample required 37.5 λ (microliters) of $0.002N$ $KMnO_4$ solution. To what electrical charge must the Np have been oxidized?

SOLUTION:

Using equiv to mean E-T equiv, we have

$$\frac{8.6 \ \gamma \ NpCl_3}{343.5 \ \dfrac{\gamma \ NpCl_3}{\mu\text{-mole } NpCl_3}} = 0.025 \ \mu\text{-mole of } NpCl_3$$

require

$$\left(0.002 \ \frac{\mu\text{-equiv}}{\lambda}\right)(37.5 \ \lambda) = 0.075 \ \mu\text{-equiv of } MnO_4^-$$

Therefore,

$$\frac{0.075 \ \mu\text{-equiv}}{0.025 \ \mu\text{-mole}} = 3 \ \frac{\text{equiv}}{\text{mole}}$$

With Np^{+++} showing a change in electrical charge of 3 on oxidation by MnO_4^-, its new electrical charge must be +6.

PROBLEMS A   Answers on page 348

1. Tell how you would prepare each of the following solutions:
   (a) 250 ml of $0.1N$ sulfamic acid from solid $H(NH_2)SO_3$
   (b) 450 ml of $0.2N$ $Ba(OH)_2$ from solid $Ba(OH)_2 \cdot 8H_2O$
   (c) 750 ml of $2.50N$ $Al(NO_3)_3$ from solid $Al(NO_3)_3 \cdot 9H_2O$

2. Tell how you would prepare each of the following solutions:
   (a) 2 liters of $0.060N$ $Ba(OH)_2$ from $1.86M$ $Ba(OH)_2$
   (b) 75 ml of $0.30N$ $AlCl_3$ from $4.3M$ $AlCl_3$
   (c) 150 ml of $0.032M$ $CdSO_4$ from $0.100N$ $CdSO_4$

3. What volume of $6M$ H_2SO_4 should be added to 10 liters of $2N$ H_2SO_4 in order to get 20 liters of $3M$ H_2SO_4 on dilution with water?

4. How much of each of the following substances is needed to react with 40 ml of $0.1N$ H_2SO_4: (a) milliliters of $0.15N$ KOH; (b) milligrams of KOH; (c) milligrams of Na_2CO_3; (d) milligrams of CaO; (e) milligrams of Zn?

5. How many milliliters of $0.0150M$ $Ba(OH)_2$ will react with 40.00 ml of $0.1000M$ H_2SO_4?

6. A 31.00-ml sample of 0.5000N H_2SO_4 reacts with 48.00 ml of NH_3 solution. What is the normality of the NH_3?

7. What is the normality of a $KMnO_4$ solution if 30.00 ml of it is needed for the titration of 45.00 ml of 0.3100N $Na_2C_2O_4$ solution? The reaction is

$$2MnO_4^- + 5C_2O_4^= + 16H^+ \rightarrow 2Mn^{++} + 10CO_2 + 8H_2O$$

8. What is the normality of a ceric sulfate solution if 46.35 ml is required for the titration of a 0.2351-g sample of $Na_2C_2O_4$ that is 99.60% pure? The reaction is

$$2Ce^{++++} + C_2O_4^= \rightarrow 2Ce^{+++} + 2CO_2$$

9. How many milligrams of $H_2C_2O_4 \cdot 2H_2O$ are needed to decolorize 35.00 ml of 0.2605N $KMnO_4$? (See Problem 7 for the equation.)

10. What is the normality of a $K_2Cr_2O_7$ solution if 35.00 ml of it is equivalent to 0.3216 g of 99.7% pure Fe? (The iron is first reduced to Fe^{++}.) The reaction is

$$Cr_2O_7^= + 6Fe^{++} + 14H^+ \rightarrow 2Cr^{+++} + 6Fe^{+++} + 7H_2O$$

11. (a) A 0.500-g sample of pure Na_2CO_3 is titrated by 46.7 ml of HCl. What is the normality of the HCl? (b) If 45.0 ml of this titrates 43.2 ml of NaOH, what is the normality of the NaOH?

12. What volume of 10N HCl is needed to prepare 12.7 liters of CO_2 at 735 torr and 35°C? The reaction is

$$CaCO_3 + 2HCl \rightarrow CaCl_2 + CO_2 + H_2O$$

13. (a) What weight of MnO_2 and (b) what volume of 12N HCl are needed for the preparation of 750 ml of 2M $MnCl_2$? (c) What volume of Cl_2 at 745 torr and 23°C will be formed? The reaction is

$$MnO_2 + 4HCl \rightarrow MnCl_2 + Cl_2 + 2H_2O$$

14. How many milliliters of 0.12N KOH are needed to neutralize 40.00 ml of KHC_2O_4 solution if 32.56 ml of the oxalate solution is equivalent to 39.27 ml of 0.1000N $KMnO_4$? (See Problem 7 for the equation.)

15. A soda-lime sample is 85% NaOH and 15% CaO. (a) If 3.5 g is dissolved in a volume of 300 ml., what is the total normality of the solution? (b) How many milliliters of 0.5000N H_2SO_4 would be required to titrate 1 ml of the solution?

16. The amount of limestone in a sample may be determined by dissolving the sample and precipitating the Ca as CaC_2O_4. This precipitate may be washed and dissolved in H_2SO_4, and the $C_2O_4^=$ titrated with standard $KMnO_4$ solution. Following this procedure, the CaC_2O_4 from a 0.4526-g sample of limestone was dissolved in H_2SO_4 The resulting solution required 41.60 ml of 0.1000N $KMnO_4$ for titration. What is the percentage of $CaCO_3$ in the original sample?

PROBLEMS B   No answers given

17. Tell how you would prepare each of the following solutions:
    (a) 5.5 liters of $0.6N$ LiOH from solid LiOH
    (b) 2.8 liters of $0.01N$ $Sr(OH)_2$ from solid $Sr(OH)_2$
    (c) 350 ml of $3.5N$ $FeCl_3$ from solid $FeCl_3 \cdot 6H_2O$

18. Tell how you would prepare each of the following solutions:
    (a) 5.32 liters of $1.20M$ $CdCl_2$ from $6.85N$ $CdCl_2$
    (b) 600 ml of $2.5N$ H_3PO_4 from $12M$ H_3PO_4
    (c) 6.3 liters of $0.003M$ $Ba(OH)_2$ from $0.1N$ $Ba(OH)_2$
    (d) 15.8 liters of $0.32N$ NH_3 from $8M$ NH_3

19. Tell how you would prepare each of the following solutions (weight percentage given):
    (a) 500 ml of $0.2N$ $HClO_4$ from a 50% solution whose density is 1.410 g/ml
    (b) 2.5 liters of $1.5N$ H_3PO_4 from an 85% solution whose density is 1.689 g/ml
    (c) 750 ml of $0.25N$ $Cr_2(SO_4)_3$ from a 35% solution whose density is 1.412 g/ml

20. How many milliliters of $0.0800M$ NaOH solution will react with 30.00 ml of $0.1100M$ H_2SO_4?

21. If 20.0 ml of H_2SO_4 is titrated by 40.0 ml of $0.1N$ NaOH, what is the normality of the H_2SO_4?

22. A 35.45-ml sample of an NaOH solution is needed to titrate a 2.0813-g sample of benzoic acid, $HC_7H_5O_2$. What is the normality of the NaOH solution?

23. How much of each of the following is needed to react with 45 ml of $0.1N$ H_2SO_4: (a) milliliters of $0.25N$ NH_3; (b) milligrams of NaOH; (c) milligrams of $CaCO_3$; (d) milligrams of Mg; (e) milligrams of BaO?

24. What is the normality of a $K_2C_2O_4$ solution if 35.00 ml of it is needed for the titration of 47.65 ml of $0.0632M$ $KMnO_4$ solution? The reaction is

$$2MnO_4^- + 5C_2O_4^= + 16H^+ \rightarrow 2Mn^{++} + 10CO_2 + 8H_2O$$

25. What is the normality of a ceric sulfate solution if 39.65 ml is required for the titration of a 0.3215-g sample of As_2O_3 that is 99.7% pure? The reaction is

$$4Ce^{++++} + As_2O_3 + 5H_2O \rightarrow 4Ce^{+++} + 2H_3AsO_4 + 4H^+$$

26. A 35.00-ml sample of HCl solution is required to react with 0.1967 g of pure Na_2CO_3. (a) What is the normality of the acid? (b) If 33.46 ml of the acid reacts with 39.65 ml of NaOH, what is the normality of the NaOH?

27. What volume of $12N$ HCl is needed to prepare 3 liters Cl_2 at 730 torr and 25°C by the reaction

$$2KMnO_4 + 16HCl \rightarrow 2MnCl_2 + 5Cl_2 + 8H_2O + 2KCl$$

28. (a) What volume of $6M$ HNO_3 and what weight of copper are needed for the production of 1.5 liters of a $0.50M$ $Cu(NO_3)_2$ solution? (b) What volume of NO, collected over water at 745 torr and 18°C, will be produced at the same time? The reaction is

$$3Cu + 8HNO_3 \rightarrow 3Cu(NO_3)_2 + 2NO + 4H_2O$$

29. The amount of calcium in blood samples may be determined by precipitating it as CaC_2O_4. This precipitate may then be dissolved in H_2SO_4 and the $C_2O_4^=$ titrated with standard $KMnO_4$ solution. Following this procedure, 10.00 ml of blood were diluted to 100.00 ml, and then from a 10.00-ml portion of this diluted sample the Ca^{++} was precipitated as CaC_2O_4. This precipitate, when washed and dissolved in H_2SO_4, required 1.53 ml of $0.0075N$ $KMnO_4$. How many milligrams of Ca^{++} were in the 10-ml sample of blood?

Colligative Properties of Solutions

The colligative properties of solutions are those properties that depend upon the number of dissolved molecules or ions, irrespective of their kind. They are: the lowering of the vapor pressure, the depression of the freezing point, the elevation of the boiling point, and osmotic pressure. These properties may be used in determining molecular weights of dissolved substances.

Vapor Pressure

When a nonvolatile substance is dissolved in a liquid, the vapor pressure is lowered. Raoult's law gives the mathematical relation

$$P = P_0 X$$

where P is the vapor pressure of the solution, P_0 is the vapor pressure of the pure solvent, and X is the mol fraction of solvent in the solution. The lowering of vapor pressure is not as widely used as some other colligative properties for experimental determinations of molecular weights, because it is difficult to make the measurements precisely.

PROBLEM:

The vapor pressure of liquid A (mol wt 120) is 70 torr at 25°C. What is the vapor pressure of a solution containing 10 g $C_6H_4Cl_2$ in 30 g of A?

SOLUTION:

To use Raoult's law, we must compute the mol fraction of A in the solution (see page 157):

$$\text{moles of A} = \frac{30 \text{ g}}{120 \frac{g}{mole}} = 0.25 \text{ mole}$$

$$\text{moles of } C_6H_4Cl_2 = \frac{10 \text{ g}}{147 \frac{g}{mole}} = 0.068 \text{ mole}$$

$$\text{mol fraction of A} = \frac{0.25 \text{ mole}}{(0.25 + 0.068) \text{ mole}} = 0.785$$

$$P = 70 \text{ torr} \times 0.785 = 55 \text{ torr}$$

PROBLEM:

How many grams of $C_6H_4Br_2$ must be added to 30 g of liquid A to give a vapor pressure of 55 torr at 25°C?

SOLUTION:

By application of Raoult's law, we have

$$55 \text{ torr} = 70 \text{ torr} \times \text{mol fraction A}$$

$$\text{mol fraction A} = \frac{55 \text{ torr}}{70 \text{ torr}} = 0.785$$

Since

$$\text{mol fraction A} = \frac{\text{moles A}}{\text{total moles}}$$

we have

$$0.785 = \frac{0.25}{\text{total moles}}$$

Then

$$\text{total moles} = \frac{0.25}{0.785} = 0.318 \text{ mole}$$

$$\text{moles of } C_6H_4Br_2 = 0.318 \text{ mole} - 0.25 \text{ mole A} = 0.068 \text{ mole}$$

$$\text{mol wt of } C_6H_4Br_2 = 236 \text{ g/mole}$$

$$\text{wt of } C_6H_4Br_2 = 236 \frac{g}{mole} \times 0.068 \text{ mole} = 16 \text{ g}$$

Freezing and Boiling Points of Solutions

The lowering of the vapor pressure by a solute also brings about other changes; the freezing point is lowered and the boiling point is raised.* The amount of the change, ΔT_F or ΔT_B, is determined by the molality (m) of the solution. The relationship is

$$\Delta T_F = K_F m$$

$$\Delta T_B = K_B m$$

Each solvent has its own characteristic freezing-point constant, K_F, and boiling-point constant, K_B, the changes caused by 1 mole of solute in 1 kilogram of solvent. Selected constants are given in Table 18-1.

TABLE 18-1.
Boiling- and Freezing-point Constants for Selected Compounds

Liquid	K_B	Normal BP,°C	Liquid	K_F	FP, °C
Benzene	2.53	80.15	Benzene	5.12	5.48
Carbon disulfide	2.34	46.13	Camphor	37.7	178.4
Water	0.51	100.00	Urethane	5.14	49.7
Camphor	5.95	208.25	Diphenyl	8.0	70.0
Chloroform	3.63	60.19	Water	1.86	0
Carbon tetrachloride	5.03	76.50	Barium chloride	108	962
Iodobenzene	8.53	188.47	Stannic bromide	28	31

PROBLEM:
Calculate the freezing point of a solution that contains 2.0 g $C_6H_4Br_2$ in 25 g of benzene (mol wt of $C_6H_4Br_2 = 236$).

SOLUTION:
First we find the weight of solute in 1 kg of benzene. Since there are 2.0 g in 25 g of benzene, the weight in 1 kg is

$$\text{wt of solute in } 1{,}000 \text{ g} = \frac{2.0 \text{ g of solute}}{25 \text{ g of benzene}} \times \frac{1{,}000 \text{ g of benzene}}{\text{kg benzene}}$$

$$= 80 \text{ g/kg benzene}$$

$$\text{moles of solute in 1 kg of benzene} = \frac{80 \dfrac{\text{g}}{\text{kg benzene}}}{236 \dfrac{\text{g}}{\text{mole}}} = 0.34 \text{ mole/kg}$$

We find in Table 18-1 that K_F for benzene, the freezing-point depression caused by 1 mole of solute per kg, is 5.12°C:

---

* See your textbook for a more detailed discussion of these effects.

$$\text{freezing-point lowering} = 5.12 \, \frac{\text{deg}}{\text{mole/kg}} \times 0.34 \, \text{mole/kg} = 1.74°$$

Since the freezing point of pure benzene is 5.48°C (Table 18-1), the freezing point of the solution is $5.48° - 1.74° = 3.74°C$.

The FP or BP constants can be used in making experimental measurements of molecular weights. The principle is the same for both, but the FP depression is used more frequently than the other because the FP constants are larger than the BP constants and because it is easier to make accurate measurements of freezing points than of boiling points.

The experimental determination of a molecular weight is illustrated in the following problem.

PROBLEM:

The freezing point of a solution that contains 1 g of compound B in 10 g of benzene is found to be 2.07°C. Calculate the molecular weight of B.

SOLUTION:

We find in Table 18-1 that benzene freezes at 5.48°C and that the molal constant is 5.12°C. The FP depression for our solution is $5.48° - 2.07° = 3.41°$. The molality of the solution (moles per kg of benzene) is

$$\text{molality} = \frac{3.41 \, \text{deg}}{5.12 \, \dfrac{\text{deg}}{\text{mole/kg}}} = 0.665$$

Since molality is the number of moles of solute in 1 kg of solvent, we compute the weight of B that would be present in 1 kg of benzene. This is 1,000 times the weight of B per gram of benzene:

$$\frac{\text{wt of B}}{\text{kg of benzene}} = 1,000 \times \frac{1 \, \text{g of B}}{10 \, \text{g of benzene}} = \frac{100 \, \text{g of B}}{\text{kg of benzene}}$$

The freezing point lowering shows that this amount is 0.665 moles. Therefore,

$$\text{wt of 1 mole B} = \frac{100 \, \dfrac{\text{g B}}{\text{kg benzene}}}{0.665 \, \dfrac{\text{moles B}}{\text{kg benzene}}} = \frac{150 \, \text{g B}}{\text{mole B}}$$

The use of boiling-point elevation to determine molecular weights is based upon the same type of calculation, using K_B instead of K_F.

We recall from Chapter 10 that the percentages of the elements in a compound can be used to compute the simplest formula for the compound. When the substance is soluble in some suitable liquid, we can combine the empirical formula with a molecular-weight determination by FP depression to get the true formula.

PROBLEM:
Compound B of the preceding problem is found by analysis to have the composition

$$C = 49.0\%, \; H = 2.7\%, \; Cl = 48.3\%$$

Find the formula and the exact molecular weight.

SOLUTION:
Using the methods of Chapter 10, we find the empirical formula to be C_3H_2Cl. If this were the true formula, the molecular weight would be 73.5. The FP depression gives a molecular weight of approximately 150. This is not an accurate value, for the experimental measurement is subject to some error, but it indicates that the true molecular weight is near 150.

Knowing the empirical formula to be C_3H_2Cl, we know that the real formula is $(C_3H_2Cl)_n$, where n is some small integer. If n is 1, a mole is 73.5 g; if n is 2, a mole is 147 g; if n is 3, a mole is 221 g. Our experimental molecular weight of 150 enables us to decide that the true value of n is 2, since 147 is the molecular weight nearest our experimental value. Thus the formula is $(C_3H_2Cl)_2$ or $C_6H_4Cl_2$.

Osmotic Pressure

When solvent and solution are separated by a semipermeable membrane that permits solvent molecules to pass, an osmotic pressure is developed in the solution. This pressure, π, is defined as the pressure that must be applied to the solution to prevent solvent molecules from diffusing into it.

For water solutions the relation between π and the molal concentration, m, is given by the equation

$$\pi = 0.082mT$$

π is in atmospheres and T is the absolute temperature. A 1-molal solution at $0°C$ ($273°K$) would have an osmotic pressure of

$$\pi = (0.082)(1)(273) = 22.4 \text{ atm}$$

A 0.0001-molal solution at 25°C would have an osmotic pressure of

$$\pi = (0.082)(1 \times 10^{-4})(298) = 2.44 \times 10^{-3} \text{ atm}$$

$$\pi = 2.44 \times 10^{-3} \text{ atm} \times 760 \text{ torr per atm} = 1.85 \text{ torr}$$

PROBLEM:
The osmotic pressure at 25°C of a solution containing 1.35 g of a protein per 100 g of water was found to be 9.12 torr. Estimate the mole weight of the protein.

SOLUTION:

The osmotic pressure is

$$\pi = 9.12 \text{ torr} \times \frac{1}{760 \text{ torr per atm}} = 0.012 \text{ atm}$$

The molality of the solution is

$$m = \frac{\pi}{(0.082)(T)} = \frac{0.012}{(0.082)(298)} = 4.91 \times 10^{-4} \text{ mole/kg } H_2O$$

The 1.35 g of protein per 100 g H_2O is equivalent to 13.5 g of protein per kg of water. This 13.5 g of protein must contain 4.91×10^{-4} mole; thus the mole weight is given by the equation

$$\text{mol wt} = \frac{13.5 \text{ g of protein}}{4.91 \times 10^{-4} \text{ mole}} = 27,500 \frac{\text{g protein}}{\text{mole}}$$

Abnormal Colligative Properties*

If a mole of NaCl is dissolved in 1 kg of water, the freezing point is not $-1.86°C$, as it would be for a mole of sugar or other nonelectrolyte. Rather, the freezing point is about $-3.5°C$, a depression almost twice as great as we should expect. The theory of ionization provides an explanation for this. When NaCl is dissolved, it breaks up into Na^+ and Cl^- ions, so that there are twice as many particles in solution as there would be if the dissociation did not occur. The water does not know whether the particles are molecules or ions insofar as the colligative properties are concerned. For every mole of NaCl dissolved we have 2 moles in solution—a mole of Na^+ ions and a mole of Cl^- ions. Thus we get an abnormal freezing-point depression.

According to modern theory, many strong electrolytes are completely dissociated in dilute solutions. The freezing-point lowering, however, does not indicate complete dissociation; for NaCl it is not twice the amount calculated on the basis of the number of moles added. In the solution the ions attract one another to some extent, and therefore they do not behave as completely independent particles, as they would if they were nonelectrolytes. From the colligative properties, therefore, we can compute only the "apparent degree of dissociation" of a strong electrolyte in solution.

To illustrate, let us consider the FP depression that occurs when we put 1 mole of NaCl into 1 kg of water. NaCl dissociates according to the equation

$$NaCl \rightarrow Na^+ + Cl^-$$

Let us assume that α is the fraction of NaCl molecules that dissociate and that $1 - \alpha$ is the fraction that act as if they were still combined as NaCl

---

*This section should not be used until the theory of ionization and the properties of electrolytes have been studied.

molecules. Remember that we are talking about our apparent degree of dissociation, as measured by the colligative properties. Then we have, if we start with n moles of NaCl,

$$n(1 - \alpha) \text{ mole of undissociated molecules,}$$
$$n\alpha \text{ mole of Na}^+ \text{ ions, and}$$
$$n\alpha \text{ mole of Cl}^- \text{ ions.}$$

Adding, we get

$$\text{total moles in solution} = n(1 - \alpha + 2\alpha) = n(1 + \alpha)$$

This we can use to compute the value of α from the FP lowering. As mentioned previously, we find that a solution of 1 mole of NaCl in 1 kg of H_2O freezes at $-3.5°C$. Here n is 1; so we have in solution $1 + \alpha$ moles. The freezing-point lowering tells us that we have 3.5/1.86 or 1.88 moles. Equating these two gives

$$1 + \alpha = 1.88$$
$$\alpha = 0.88$$

Thus, according to these measurements, the apparent degree of dissociation for 1-molal NaCl is 0.88, or 88%.

If we have an electrolyte such as $CaCl_2$, we find a different expression for α. Suppose we put into solution 1 mole, and it behaves as if α moles had dissociated. Then

$$\begin{array}{ccc} 1 - \alpha & \alpha & 2\alpha \\ CaCl_2 \rightarrow & Ca^{++} + & 2Cl^- \end{array}$$

We have in solution

$$1 - \alpha \text{ mole of CaCl}_2$$
$$\alpha \text{ mole of Ca}^{++}$$
$$2\alpha \text{ moles of Cl}^-$$

The total number in solution is $1 - \alpha + \alpha + 2\alpha = 1 + 2\alpha$. From this, and the measured FP depression, we can evaluate α as we did for NaCl.

PROBLEMS A   Answers on page 348

1. A solution is prepared by dissolving 1.28 g of naphthalene, $C_{10}H_8$, in 10.0 g of benzene.
   (a) What is the lowering of the freezing point of benzene?
   (b) What is the freezing point of the solution?
   (c) What is the mole fraction of benzene in the solution?

(d) The vapor pressure of pure benzene is 100 torr at room temperature. What is the vapor pressure of this solution at the same temperature?

(e) What is the boiling point of the solution?

2. A student uses a thermometer on which he can read temperatures to the nearest 0.1 degree. In a laboratory experiment he observes a freezing point of 5.4°C for pure benzene. He then dissolves 0.75 g of an unknown in 15.0 g of benzene and finds the freezing point of the solution to be 2.8°C. What is the molecular weight of the unknown?

3. What will be the freezing point of a 10% solution of sucrose, $C_{12}H_{22}O_{11}$, in water?

4. A 10-g sample of naphthalene mothballs, $C_{10}H_8$, is added to 50 ml of benzene (density $= 0.879$ g/ml). What will be the boiling point of this solution?

5. Expensive special thermometers are usually needed for determinations of freezing-point depression and boiling-point elevation, but when camphor is used a common lab thermometer may be employed. A student mixes 0.1032 g of camphor and 7.32 mg of an unknown compound, and finds the melting point to be 159.3°C. The melting point of the camphor before mixing was 175.1°C. What is the mole weight of the compound?

6. When 2.832 g of sulfur is dissolved in 50.0 ml of CS_2 (density $=1.263$ g/ml), the solution boils 0.411°C higher than the pure CS_2. What is the molecular formula of sulfur?

7. When 0.532 g of a certain solid organic compound was dissolved in 16.8 g of urethane, whose freezing point was 49.50°C, the freezing point of the solution was lowered to 48.32°C. Chemical analysis showed this compound to be 69.5% C, 7.25% H, and 23.25% O. Determine the true formula of this compound.

8. A 0.356-g sample of a solid organic compound, when dissolved in 9.15 ml of carbon tetrachloride (density $= 1.595$ g/ml), raised the boiling point of the carbon tetrachloride by 0.56°C. When analyzed, this compound was found to be 55.0% C, 2.75% H, 12.8% N, and 29.4% O. What is the true formula of this compound?

9. United States silver coins, which are 10% Cu in Ag, melt completely at 875°C, pure silver melts at 960°C. What is the molal freezing-point constant for Ag?

10. Calculate the mole fraction and the molality of the first-named component in each of the following solutions:

(a) 10% solution of glycerine, $C_3H_8O_3$, in water

(b) 5% solution of water in acetic acid

(c) 5.0 g CCl_4 and 25 g of C_2H_5OH

(d) 0.1 g CN_2H_4O in 5 g of water

(e) 8.0 g CH_3OH in 57.0 g C_6H_6

11. How many quarts of ethylene glycol, $C_2H_6O_2$ (density = 1.116 g/ml), will have to be added to 5 gal of water (density = 1 g/ml) to protect an automobile radiator down to a temperature of $-10°F$? (Ethylene glycol is a "permanent" antifreeze agent.)

12. A beaker containing 100 g of 10% sucrose ($C_{12}H_{22}O_{11}$) solution and another containing 150 g of a 30% sucrose solution are put under a bell jar at 25°C and allowed to stand until equilibrium is attained. What weight of solution will each beaker contain when equilibrium has been reached?

13. By how many torr will the vapor pressure of water at 27°C be lowered if 50 g of urea, CH_4N_2O, is dissolved in 50 g of water?

14. A carbon-hydrogen-oxygen analysis of benzoic acid gave 68.8% C, 5.0% H, 26.2% O. A 0.1506-g sample of benzoic acid dissolved in 100 g of water gave a solution whose freezing point was $-0.023°C$. A 2.153-g sample of benzoic acid in 50 g of benzene, C_6H_6, gave a solution whose freezing point was 4.58°C. The freezing point of the pure benzene was 5.48°C. Determine the molecular formula for benzoic acid in these two solutions, and explain any difference observed.

15. A 75-g sample of glucose, $C_6H_{12}O_6$, is dissolved in 250 g of water at 27°C. What will be the osmotic pressure of the solution?

16. A solution containing 1.346 g of a certain protein per 100 g of water was found to have an osmotic pressure of 9.69 cm *of water* at 25°C. Calculate the mole weight of the protein.

17. A 0.5% solution of a certain plant polysaccharide (complex plant sugar) has an osmotic pressure of 5.40 torr at 27°C. Of how many simple sucrose sugar units, $C_{12}H_{22}O_{11}$, must this polysaccharide be composed?

18. If an aqueous sucrose ($C_{12}H_{22}O_{11}$) solution has an osmotic pressure of 12.5 atm at 23°C, what will be the vapor pressure of this solution at 23°C?

19. A 0.1-molal solution of acetic acid in water freezes at $-0.190°C$. Calculate the percentage of ionization of acetic acid at this temperature.

20. A 0.5-molal KBr solution freezes at $-1.665°C$. What is its apparent percentage of ionization at this temperature?

21. A 0.01-molal solution of NH_3 is 4.15% ionized. What will be the freezing point of this solution?

22. A 0.2-molal solution of $Mg(NO_3)_2$ in water freezes at $-0.956°C$. Calculate the apparent percentage of ionization of $Mg(NO_3)_2$ at this temperature.

23. A solution prepared by dissolving 1 g of $Ba(OH)_2 \cdot 8H_2O$ in 200 g of water has a freezing point of $-0.0833°C$. Calculate the apparent percentage of ionization of $Ba(OH)_2$ at this temperature.

PROBLEMS B   No answers given

24. Caclulate the mole fraction and the molality of the first-named component in each of the following solutions:
    (a) 15% solution of ethylene glycol, $C_2H_6O_2$, in water
    (b) 7% solution of water in acetic acid
    (c) 7.0 g of chloroform, $CHCl_3$, and 30.0 g of methyl aclohol, CH_3OH
    (d) 40.0 g of ethyl alcohol, C_2H_5OH, and 60.0 g of acetone, C_3H_6O
    (e) 50 g of formamide, $HCONH_2$, and 50 g of water

25. What will be the freezing point of a 20% solution of glucose, $C_6H_{12}O_6$, in water?

26. A 20-g sample of p-dichlorobenzene mothballs, $C_6H_4Cl_2$, is added to 65 ml of benzene (density $= 0.879$ g/ml). What will be the boiling point of this solution?

27. A brass sample composed of 20% Zn and 80% Cu melts completely at 995°C; pure Cu melts at 1,084°C. What is the molal freezing-point constant for copper?

28. What is the boiling point of a solution of 10 g of diphenyl, $C_{12}H_{10}$, and 30 g of naphthalene, $C_{10}H_8$, in 60 g of benzene?

29. A common way to check the purity of a compound is to take its melting point, since any soluble impurity always lowers the melting point. What is the molal concentration of impurity in urethane if its melting point is 47.7°C?

30. When 1.645 g of phosphorus is dissolved in 60 ml of CS_2 (density $= 1.263$ g/ml), the solution boils at 46.709°C; CS_2 alone boils at 46.300°C. What is the molecular formula of phosphorus?

31. By analysis, a certain solid organic compound was found to be 40.0% C, 6.7% H, and 53.3% O. When a 0.650-g sample of this compound was dissolved in 27.80 g of diphenyl, the freezing point was lowered by 1.56°C. Determine the true formula of this compound.

32. A certain solid organic compound is analyzed as 18.3% C, 0.51% H, and 81.2% Br. When 0.793 g of this compound is dissolved in 14.80 ml of chloroform (density $= 1.485$ g/ml), whose boiling point is 60.3°C, the solution is found to boil at 60.63°C. What is the true formula of this compound?

33. How many quarts of ethylene glycol, $C_2H_6O_2$ (density $= 1.115$ g/ml), will have to be added to 5 gal of water (density $= 1$ g/ml) to protect an automobile radiator down to a temperature of $-15°F$? (Ethylene glycol is a "permanent" antifreeze agent.)

34. What is the vapor pressure at 25°C of an aqueous solution containing 100 g glycerine, $C_3H_8O_3$, in 150 g of water?

35. What is the osmotic pressure of a solution in which 50 g of sucrose, $C_{12}H_{22}O_{11}$, is dissolved in 150 g of water at 21°C?

36. If an aqueous glucose ($C_6H_{12}O_6$) solution has an osmotic pressure of 50 atm at 35°C, what will be its vapor pressure at the same temperature?

37. A solution containing 1.259 g of a certain protein fraction per 100 g of water was found to have an osmotic pressure of 8.32 cm of *water* at 28°C. Calculate the mole weight of the protein fraction.

38. A 0.7% solution of a certain plant polysaccharide (complex plant sugar) has an osmotic pressure of 6.48 torr at 21°C. Of how many simple sucrose sugar units, $C_{12}H_{22}O_{11}$, must this polysaccharide be composed?

39. A 0.02-molal aqueous solution of picric acid, $HC_6H_2O_7N_3$, freezes at −0.0656°C. Picric acid ionizes to a certain extent in water, as

$$HC_6H_2O_7N_3 \rightarrow H^+ + C_6H_2O_7N_3^-$$

Calculate the percentage of ionization of picric acid at this temperature.

40. A 0.2-molal $NaNO_3$ solution freezes at −0.665°C. What is its apparent percentage of ionization?

41. A 0.1-molal solution of HNO_2 in water is 6.5% ionized. What will be the freezing point of this solution?

42. A 0.2-molal solution of $Ni(NO_3)_2$ freezes at −0.982°C. Calculate the apparent percentage of ionization of $Ni(NO_3)_2$ at this temperature.

43. If a 0.5-molal Na_2SO_4 solution has an apparent ionization of 72%, at what temperature will the solution freeze?

44. *o*-Chlorobenzoic acid has a composition of 53.8% C, 3.2% H, 20.4% O, and 22.6% Cl. A 0.1236-g sample of this acid dissolved in 100 g of water gave a solution whose freezing point was −0.0147°C. A 3.265-g sample of *o*-chlorobenzoic acid dissolved in 60 g of benzene gave a solution whose freezing point was 4.59°C. Determine the molecular formula of *o*-chloro-benzoic acid in these two solutions, and explain any difference in results.

Hydrogen-Ion Concentration and pH

Chemical reactions in aqueous solutions, including the chemistry of life processes, very often depend on the concentration of hydrogen ion in the solution. As we shall later see, we may deal with hydrogen-ion concentrations varying from greater than $1M$ to less than $10^{-14}M$. Consequently it is convenient to express these concentrations on a logarithmic basis; for this purpose the terms "pH" and "pOH," which we shall discuss in this chapter, have been introduced.

Hydrogen and Hydroxide Concentrations in Water

Water is a *weak* electrolyte, ionizing slightly as *

$$H_2O \rightarrow H^+ + OH^-$$

In pure water at room temperature the concentration of H^+ is 10^{-7} mole per liter. Rather than write this out we normally use the notation

$$[H^+] = 10^{-7}M$$

---

*Actually the hydrogen ion is solvated, forming chiefly the H_3O^+ ion. Since other ions may also be hydrated, we shall omit the bound H_2O in the formulas for all ions.

where the square brackets indicate that we mean the moles per liter of the ion or molecule inside the brackets.

Since the dissociation of a water molecule gives both a H^+ ion and an OH^- ion, it follows that in pure water

$$[OH^-] = 10^{-7}M$$

As we shall see in the following chapters, it can be shown that there is a fixed relation between $[H^+]$ and $[OH^-]$ in water. This very important relation is given by the equation

$$[H^+][OH^-] = 10^{-14} = K_w$$

This tells us that in any water solution at room temperature the product of the moles per liter of H^+ and the moles per liter of OH^- is constant, having the value 10^{-14}. This constant is conventionally represented by the symbol K_w. It follows that, if we add an acid to water, thereby increasing the H^+ concentration, there must be a corresponding decrease in the OH^- concentration; or, conversely, if we increase the $[OH^-]$, the $[H^+]$ is decreased.

To illustrate, consider a $0.1M$ HCl solution. As we know, HCl is a strong acid and almost completely dissociated. Nearly all the HCl molecules in the solution have broken up into H^+ and Cl^- ions. This means that in a $0.1M$ HCl solution

$$[H^+] = 10^{-1}M$$

Now, since $[H^+][OH^-] = 10^{-14}$, it follows that $[OH^-] = 10^{-13}M$ or is one millionth of the concentration in pure water. Likewise, in $0.1M$ NaOH solution, the OH^- concentration is $0.1M$ or $10^{-1}M$, since NaOH also is a strong electrolyte. Therefore $[H^+]$ in $0.1M$ NaOH solution is $10^{-13}M$.

Definition of pH and pOH

The wide range in the hydrogen-ion concentrations of aqueous solutions makes it difficult to plot these values on a linear scale. As a convenience, we use a logarithmic scale introduced by Sorensen many years ago. Hydrogen-ion concentrations are represented by "pH" and hydroxide ions by "pOH," defined by the relations

$$pH = -\log [H^+]$$
$$pOH = -\log [OH^-]$$

In keeping with this usage, we also use

$$pK_w = -\log K_w$$

You recall that $\log AB = \log A + \log B$. Therefore, since $[H^+][OH^-] = K_w$,

$$pH + pOH = pK_w = 14$$

Relations between pH and pOH for solutions of HCl and NaOH are shown in Table 19-1. Note that in each solution the sum of pH and pOH values is 14. In the neutral solution, containing neither acid nor base, the pH is 7.

TABLE 19-1.
Relations Between Concentrations of Strong Acid and Strong Base to the pH of the Solution

Concentration of HCl	10^{-1}	10^{-3}	10^{-5}	0			
pH	1	3	5	7	9	11	13
pOH	13	11	9	7	5	3	1
Concentration of NaOH				0	10^{-5}	10^{-3}	10^{-1}

Calculation of pH from $[H^+]$

We need to know how to calculate the pH of any hydrogen-ion or hydroxide-ion concentration. This is done by means of the relation

$$pH = -\log [H^+]$$

A few examples will show the procedure.

PROBLEM:
What is the pH corresponding to a hydrogen-ion concentration of 5×10^{-4}?

SOLUTION:

$$\log (5 \times 10^{-4}) = \log 5 + \log 10^{-4} = 0.70 - 4 = -3.30$$
$$pH = -\log [H^+] = -(-3.30) = 3.30$$

If you are asked to compute the pH of a strong acid solution, remember that the acid is almost completely ionized and that $[H^+]$ is the same as the concentration of acid in the solution.

PROBLEM:
Find the pH of a $0.02M$ HCl solution.

SOLUTION:
$[H^+] = 0.02M$. Write $[H^+]$ in standard exponential form (review Chapter 2, if necessary):

$$[H^+] = 2 \times 10^{-2}M$$
$$\log [H^+] = \log 2 + \log 10^{-2} = 0.30 - 2 = -1.70$$
$$pH = 1.70$$

PROBLEM:
What is the pH of a $0.02M$ NaOH solution?

SOLUTION:
First find the pOH. Since NaOH is completely dissociated,

$$[OH^-] = 0.02 = 2 \times 10^{-2}M$$
$$pOH = -\log (2 \times 10^{-2}) = -(0.30 - 2) = -(-1.70) = 1.70$$
$$pH = 14 - pOH = 14 - 1.70 = 12.30$$

PROBLEM:
If 25 ml of $0.16M$ NaOH are added to 50 ml of $0.1M$ HCl, what is the pH?

SOLUTION:
When acid and base are mixed, they react to form H_2O. To find the amount of acid or base remaining in the solution after reaction occurs, we first compute the moles of acid and of base put into the solution:

$$\text{moles of HCl} = V \times M = 0.050 \text{ liter} \times 0.1 \text{ mole/liter} = 0.005 \text{ mole}$$
$$\text{moles of NaOH} = V \times M = 0.025 \text{ liter} \times 0.16 \text{ mole/liter}$$
$$= 0.004 \text{ mole}$$

Since acid is in excess, all the base is used up. (The salt formed—NaCl—has no effect on the pH of the solution.) So we subtract moles of base from moles of acid to find the amount of unneutralized acid:

$$0.005 \text{ mole HCl} - 0.004 \text{ mole NaOH} = 0.001 \text{ mole HCl not neutralized}$$

The concentration of the HCl remaining after NaOH is added is

$$\text{concentration} = \frac{\text{no. of moles}}{\text{vol}} = \frac{0.001}{0.050 + 0.025} = \frac{0.001}{0.075} = 0.0133M$$
$$\text{concentration of } H^+ = 0.0133 = 1.33 \times 10^{-2} \text{ mole/liter}$$
$$pH = -\log (1.33 \times 10^{-2}) = -(0.12 - 2) = -(-1.88) = 1.88$$

PROBLEM:
What is the pH after 25 ml of $0.2M$ NaOH are added to 50 ml of $0.1M$ HCl?

SOLUTION:
Solve as in the preceding problem:

$$\text{moles of HCl} = 0.050 \times 0.1 = 0.005$$
$$\text{moles of NaOH} = 0.025 \times 0.2 = 0.005$$

Since moles of acid = moles of base, the solution is neutral. The pH is that of water, or 7.0. (This is true only for a salt of a strong acid and a strong base.)

PROBLEM:

Find the pH after 26 ml of 0.2M NaOH are added to 50 ml of 0.1M HCl.

SOLUTION:

As before,

$$\text{moles of HCl} = 0.050 \times 0.1 = 0.0050$$

$$\text{moles of NaOH} = 0.026 \times 0.2 = 0.0052$$

Here NaOH is in excess, so

$$\text{moles of NaOH} - \text{moles of HCl} = 0.0052 - 0.0050 = 0.0002$$

$$\text{concentration of NaOH} = \frac{0.0002}{0.050 + 0.026} = \frac{0.0002}{0.076}$$

$$= \frac{2 \times 10^{-4}}{0.76 \times 10^{-1}} = 2.63 \times 10^{-3}M$$

$$pOH = -\log (2.63 \times 10^{-3}) = -(.42 - 3) = 2.58$$

$$pH = 14 - 2.58 = 11.42$$

The preceding problems have illustrated the computation of the pH for acidic, basic, and neutral solutions. We also need to understand the reverse calculation, or how to go from the pH or pOH to the actual concentration of acid or base in a solution.

PROBLEM:

What is the hydrogen-ion concentration in a solution of pH 4.3?

SOLUTION:

Since $pH = -\log [H^+]$,

$$[H^+] = \text{antilog} (-pH) = 10^{-4.3}M$$

The decimal part of this exponent must be changed to a positive number to allow us to look up the antilog. Therefore we rewrite as $10^{0.7} \times 10^{-5}$. The antilog of 0.7 is 5. This gives $[H^+] = 5 \times 10^{-5}M$.

PROBLEM:

What is the hydroxide-ion concentration in a solution of pH 8.40?

SOLUTION:

Since we want $[OH^-]$, we convert first to pOH:

$$pOH = 14 - pH = 5.60$$

$$[OH^-] = 10^{-5.60} = 10^{0.40} \times 10^{-6} = 2.5 \times 10^{-6}M$$

PROBLEM:

How many grams of HCl must be added to 200 ml of water to give a solution
of pH 2.70?

SOLUTION:

$$[H^+] = 10^{-2.70} = 10^{0.30} \times 10^{-3} = 2 \times 10^{-3} M$$

$$[HCl] = 2 \times 10^{-3} M$$

$$\text{number of moles} = V \times M = 0.2 \text{ liter} \times 2 \times 10^{-3} \text{ mole/liter}$$
$$= 0.4 \times 10^{-3} \text{ mole}$$

$$\text{wt of HCl} = 0.4 \times 10^{-3} \text{ mole} \times 36.5 \text{ g/mole} = 14.6 \times 10^{-3} \text{ g}$$
$$= 1.46 \times 10^{-2} \text{ g}$$

PROBLEMS A   Answers on page 349

1. Calculate the pH of solutions with the following H^+ concentrations in
   moles per liter: (a) 10^{-4}; (b) 10^{-6}; (c) 10^{-8}; (d) 10; (e) 0.012; (f) 8.9×10^{-2};
   (g) 3.7×10^{-5}; (h) 6.5×10^{-8}; (i) 3.5; (j) 0.5.

2. Calculate the pH of solutions with the following OH^- concentrations in
   moles per liter: (a) 10^{-4}; (b) 10^{-6}; (c) 10^{-8}; (d) 10; (e) 0.025; (f) 7.91×10^{-2};
   (g) 4.65×10^{-5}; (h) 2.56×10^{-8}; (i) 6.5; (j) 0.72.

3. Calculate the H^+ concentration for each of the solutions with the following values for pH: (a) 3.61; (b) 7.52; (c) 13.43; (d) 0.77; (e) 6.45; (f) 8.96;
   (g) 0; (h) 2.80; (i) −0.6; (j) 14.8.

4. Calculate the pOH for each of the solutions in Problem 1.

5. A student puts 50 ml of 0.1M HCl in a beaker and then adds increments of
   0.1M NaOH solution. Compute the pH after the addition of each of the
   following volumes of the NaOH solution:

   (a) No NaOH added   (e) 50 ml NaOH added
   (b) 10 ml NaOH added   (f) 51 ml NaOH added
   (c) 25 ml NaOH added   (g) 55 ml NaOH added
   (d) 49 ml NaOH added

   Plot the results, showing pH on the y-axis and volume of NaOH added on
   the x-axis.

6. The following volumes of 0.1M HCl are added to a beaker that initially
   contains 50 ml of 0.1M NaOH solution: 0 ml, 10 ml, 25 ml, 49 ml, 50 ml,
   51 ml, 55 ml. Compute the pH after each addition of HCl, and plot pH on
   the y-axis versus volume of HCl on the x-axis; compare the pH curve of
   Problem 5 with that of Problem 6.

PROBLEMS B   No answers given

7. Calculate the pH of solutions with the following H^+ concentrations in moles per liter: (a) 10^{-3}; (b) 10^{-5}; (c) 10^{-9}; (d) 0.0056; (e) 10; (f) 7.6×10^{-4}; (g) 4.3×10^{-6}; (h) 8.3×10^{-9}; (i) 5.6; (j) 0.35.

8. Calculate the pH of solutions with the following OH^- concentrations in moles per liter: (a) 10^{-2}; (b) 10^{-4}; (c) 10^{-10}; (d) 10; (e) 0.077; (f) 8.5×10^{-3}; (g) 1.67×10^{-6}; (h) 4.73×10^{-16}; (i) 7.65; (j) 0.22.

9. Calculate the H^+ concentration for each of the solutions with the following values for pH: (a) 6.35; (b) 2.78; (c) 12.91; (d) 0.55; (e) 10.47; (f) 7.32; (g) 15.21; (h) 0.76; (i) 0; (j) −0.36.

10. Calculate the pOH for each of the solutions in Problem 7.

11. Calculate the pH of a solution made by dissolving 20.0 g of sulfamic acid, $H(NH_2)SO_3$, and diluting to 200 ml with water. (Sulfamic acid is a strong acid like HCl.)

12. Calculate the pH of a solution made by diluting 50 ml of a HNO_3 solution (56% by weight, and density $= 1.350$ g/ml) to 2.75 liters.

Acid-Base Equilibria in Solution

In Chapter 14 we applied the fundamental general equilibrium expression to gaseous equilibrium reactions. In this chapter we apply the same expression to the equilibria that involve weak acids and bases in aqueous solution, the principal difference being that all concentrations will be expressed in moles per liter instead of in atmospheres as was done for gases. All the general conclusions that we reached in Chapter 14, and that are summarized in the Principle of Le Chatelier, apply to equilibria in solutions as well as in gases.

Ionization Constant

When we put a strong electrolyte, such as HCl, into solution, almost all the molecules dissociate to ions, in this case, H^+ and Cl^-. But when we put into solution a weak electrolyte, such as acetic acid $(HC_2H_3O_2)$, only a small fraction of the molecules dissociate. The equation is

$$HC_2H_3O_2 \rightleftarrows H^+ + C_2H_3O_2^-$$

Since this reaction is at equilibrium, we can apply the mathematical expression

$$\frac{[H^+][C_2H_3O_2^-]}{[HC_2H_3O_2]} = K_e$$

The equilibrium constant for the ionization of a weak electrolyte is usually designated as K_i, which we call the ionization constant.

Ionization constants are determined by experimental measurements of equilibrium concentrations. For example, to determine K_i for acetic acid, we prepare a solution of known concentration and by any of several methods measure the H^+ concentration or the pH. The method most widely used today is measuring with a pH meter, which gives a direct dial reading for the pH. We find experimentally that in a 0.10M solution of acetic acid the pH is 2.88. From this we calculate the concentrations in the solution and use these to evaluate K_i. Starting with $[H^+]$, we have

$$[H^+] = 10^{-2.88} = 10^{0.12} \times 10^{-3} = 1.32 \times 10^{-3} \text{ mole/liter}$$

Since each molecule that ionizes yields a H^+ ion and a $C_2H_3O_2^-$ ion, the concentration of $C_2H_3O_2^-$ is also 1.32×10^{-3}. We have put 0.10 mole/liter of $HC_2H_3O_2$ in solution. Since 1.32×10^{-3} mole/liter has dissociated, there remains $0.10000 - 0.00132 = 0.09868$ mole/liter of undissociated molecules. Substituting these molar concentrations into the mathematical equation for equilibrium gives

$$\frac{(1.32 \times 10^{-3})(1.32 \times 10^{-3})}{(0.09868)} = 1.8 \times 10^{-5} = K_i$$

Experimental values for selected ionization constants are given in Table 20-1. In using such values it should be kept in mind that although some of them, such as the constant for acetic acid, are reliable to at least two significant figures, others may be in error, some as much as tenfold. For example, the dissociation constant for the HS^- ion is listed in different tables with values ranging from 10^{-13} to 10^{-15}. This uncertainty is due to the difficulty of determining the concentrations of ions, such as that of $S^=$, that are present in very low concentration. In general the second dissociation constant for a diprotic acid is less accurately known than the one for the first stage. In using tables of dissociation constants one must, therefore, keep in mind the limitations of many of the computations based on them. It must also be remembered that, even if a constant is accurately known, computations based on it are really accurate only when used in conjunction with activities, rather than concentrations, of the various ions present.

TABLE 20-1.
Equilibrium Constants at 25°C

Name	Reaction	Constant
Weak acids		
Acetic	$HC_2H_3O_2 \rightleftarrows H^+ + C_2H_3O_2^-$	1.8×10^{-5}
Boric	$H_3BO_3 \rightleftarrows H^+ + H_2BO_3^-$	5.8×10^{-10}
Carbonic	$H_2CO_3 \rightleftarrows H^+ + HCO_3^-$	$K_1 = 4.5 \times 10^{-7}$
	$HCO_3^- \rightleftarrows H^+ + CO_3^=$	$K_2 = 6 \times 10^{-11}$
Cyanic	$HCNO \rightleftarrows H^+ + CNO^-$	2×10^{-4}
Formic	$HCHO_2 \rightleftarrows H^+ + CHO_2^-$	2×10^{-4}
Hydrazoic	$HN_3 \rightleftarrows H^+ + N_3^-$	1.9×10^{-5}
Hydrocyanic	$HCN \rightleftarrows H^+ + CN^-$	4×10^{-10}
Hydrofluoric	$HF \rightleftarrows H^+ + F^-$	7.2×10^{-4}
Hydrogen sulfide	$H_2S \rightleftarrows H^+ + HS^-$	$K_1 = 1.0 \times 10^{-7}$
	$HS^- \rightleftarrows H^+ + S^=$	$K_2 = 1.3 \times 10^{-13}$
Nitrous	$HNO_2 \rightleftarrows H^+ + NO_2^-$	4.5×10^{-4}
Oxalic	$H_2C_2O_4 \rightleftarrows H^+ + HC_2O_4^-$	$K_1 = 5.9 \times 10^{-2}$
	$HC_2O_4^- \rightleftarrows H^+ + C_2O_4^=$	$K_2 = 6.4 \times 10^{-5}$
Phosphoric	$H_3PO_4 \rightleftarrows H^+ + H_2PO_4^-$	$K_1 = 7.5 \times 10^{-3}$
	$H_2PO_4^- \rightleftarrows H^+ + HPO_4^=$	$K_2 = 2 \times 10^{-7}$
	$HPO_4^= \rightleftarrows H^+ + PO_4^\equiv$	$K_3 = 1 \times 10^{-12}$
Phosphorous	$H_3PO_3 \rightleftarrows H^+ + H_2PO_3^-$	$K_1 = 1.7 \times 10^{-2}$
Bisulfate ion	$HSO_4^- \rightleftarrows H^+ + SO_4^=$	$K_2 = 1.2 \times 10^{-2}$
Sulfurous	$H_2SO_3 \rightleftarrows H^+ + HSO_3^-$	$K_1 = 1.2 \times 10^{-2}$
	$HSO_3^- \rightleftarrows H^+ + SO_3^=$	$K_2 = 1 \times 10^{-7}$
Weak bases		
Ammonia	$NH_3 + H_2O \rightleftarrows NH_4^+ + OH^-$	1.8×10^{-5}
Methylamine	$CH_3NH_2 + H_2O \rightleftarrows CH_3NH_3^+ + OH^-$	4.4×10^{-4}
Ethylamine	$C_2H_5NH_2 + H_2O \rightleftarrows C_2H_5NH_3^+ + OH^-$	5.6×10^{-4}
Dimethylamine	$(CH_3)_2NH + H_2O \rightleftarrows (CH_3)_2NH_2^+ + OH^-$	7.5×10^{-4}
Trimethylamine	$(CH_3)_3N + H_2O \rightleftarrows (CH_3)_3NH^+ + OH^-$	7.4×10^{-5}
Aniline	$C_6H_5NH_2 + H_2O \rightleftarrows C_6H_5NH_3^+ + OH^-$	3.8×10^{-10}
Pyridine	$C_5H_5N + H_2O \rightleftarrows C_5H_5NH^+ + OH^-$	1.4×10^{-9}
Water	$H_2O \rightleftarrows H^+ + OH^-$	1.0×10^{-14}

Uses of ionization constants to compute concentrations of the ions present in solution and the pH of the solution are illustrated in the following problems. First we will consider the dissociation of a monoprotic acid, using acetic acid as an example; later, the dissociation of a diprotic acid, H_2S, in connection with precipitation of metal sulfides.

PROBLEM:

What is the pH of a $0.05M$ acetic acid solution?

SOLUTION:

Write the chemical equation, and above each term of the equation write the molar concentration. Since we are not given the H^+ and $C_2H_3O_2^-$ concentrations, and since the two are the same, we represent this value by x. This gives a concentration of $(0.05 - x)$ mole/liter of undissociated $HC_2H_3O_2$ molecules, since 0.05 mole of acid is put into solution and x moles dissociate:

$$\overset{0.05-x}{HC_2H_3O_2} \rightleftarrows \overset{x}{H^+} + \overset{x}{C_2H_3O_2^-}$$

Substitute the molar concentrations into the K_i equation:

$$\frac{x^2}{0.05 - x} = 1.8 \times 10^{-5}$$

To solve an equation of this type, we usually first assume x to be so small that $0.05 - x$ may be considered as 0.05 (in other words, subtraction of x from 0.05 does not appreciably change the value). This gives, as the simplified equation,

$$\frac{x^2}{0.05} = 1.8 \times 10^{-5}$$

$$x^2 = 9.0 \times 10^{-7}$$

Since we want the pH, it is convenient to take logs at this stage, rather than extract the square root and then take the log:

$$\log x^2 = 0.95 - 7 = -6.05$$

$$-\log x^2 = 6.05$$

$$-\log x = \frac{6.05}{2} = 3.02 = pH$$

This pH corresponds to a hydrogen-ion concentration of $10^{-3.02}M$. Since this figure also represents the number of moles of $HC_2H_3O_2$ that have dissociated, we see that x is much smaller than the original concentration, 0.05 M, and that we were justified in making the assumption that $0.05 - x = 0.05$ (neglecting the value of x subtracted from the larger number). In general when the computation gives a value of x that does not exceed 10 per cent of the number from which it is subtracted or to which it is added, we can consider it permissible to make the approximation.

Weak Bases

When weak bases are dissolved in water, a few of the molecules accept protons from water, leaving OH^- ions in the solution to make it slightly basic. For many years it was said that such solutions contained the *hydrated* form of the base (instead of the base itself) and that the hydrated base then subsequently dissociated to a slight degree. For ammonia it was said that NH_3 first reacts with water to form NH_4OH, which then dissociated slightly as a weak base:

$$NH_3 + H_2O \rightleftharpoons NH_4OH$$

$$NH_4OH \rightleftharpoons NH_4^+ + OH^-$$

Since most of the dissolved base probably does not exist in the hydrated form in solution, it is now more acceptable to write the chemical equilibrium equation as the sum of the two equations above:

$$NH_3 + H_2O \rightleftharpoons NH_4^+ + OH^-$$

Just as it is customary to consider the concentration of water to be constant (or at unit activity) in dilute solutions of weak acids, so we shall consider that the water concentration remains constant in dilute solutions of weak bases.

PROBLEM:
What is the pH of a $0.1M$ NH_3 solution?

SOLUTION:
First, write the chemical equation,

$$NH_3 + H_2O \rightleftharpoons NH_4^+ + OH^-$$

Second, write the K_i expression, assuming H_2O to be at unit activity and obtaining the K_i value from Table 20-1;

$$K_i = \frac{[NH_4^+][OH^-]}{[NH_3]} = 1.8 \times 10^{-5}$$

Third, write what you know and do not know. Since you are asked for the pH of a *basic* solution, you will first have to find the $[OH^-]$; let $[OH^-] = x$. As before, NH_4^+ and OH^- ions are formed in equal amounts, so $[NH_4^+] = x$; x moles of original 0.1 moles of NH_3 will be used up, leaving $[NH_3] = 0.1 - x$. Substituting these values in the K_i expression gives

$$\frac{[x][x]}{[0.1 - x]} = 1.8 \times 10^{-5}$$

and, for simplicity, we try neglecting x compared to 0.1, giving:

$$x^2 = (0.1)(1.8 \times 10^{-5}) = 1.8 \times 10^{-6}$$

$$x = [OH^-] = 1.34 \times 10^{-3} = 10^{1.13} \times 10^{-3} = 10^{-2.87}$$

$$pOH = 2.87$$

$$pH = 14 - pOH = 14 - 2.87$$

$$= 11.13$$

Common-ion Effect

If we add some sodium acetate or other source of $C_2H_3O_2^-$ ion to a solution of $HC_2H_3O_2$, we shift the equilibrium

$$HC_2H_3O_2 \rightleftarrows H^+ + C_2H_3O_2^-$$

to the left, or use up some of the H^+ ions to form more undissociated $HC_2H_3O_2$. We may use the regular K_i expression for acetic acid to compute the pH of such a mixture of acetic acid and sodium acetate.

PROBLEM:
What is the pH after 1.0 g of $NaC_2H_3O_2$ is added to 150 ml of 0.05M $HC_2H_3O_2$?

SOLUTION:
First determine the concentrations of the substances put into the solution. Let $x = [H^+]$, for which we are to solve; $[HC_2H_3O_2] = 0.05 - x$, since we start with 0.05M acetic acid solution. The concentration of acetate ion is not given, but we can compute it from the amount of sodium acetate added to the solution. Remember that $NaC_2H_3O_2$ is a salt and is therefore essentially completely dissociated.

$$\text{moles of } NaC_2H_3O_2 = \frac{1 \text{ g}}{82 \frac{\text{g}}{\text{mole}}} = 0.0122 \text{ moles}$$

$$\text{molarity of } NaC_2H_3O_2 = \frac{0.0122 \text{ mole}}{0.150 \text{ liter}} = 0.0815M$$

Since x moles per liter of $HC_2H_3O_2$ dissociate,

$$[C_2H_3O_2^-] = 0.0815 + x$$

Writing the chemical equation and placing above each substance its concentration in the solution, we have

$$\begin{array}{ccc} 0.05 - x & x & 0.0815 + x \\ HC_2H_3O_2 & \rightleftarrows H^+ & + C_2H_3O_2^- \end{array}$$

The mathematical equation is

$$\frac{(x)(0.0815 + x)}{(0.05 - x)} = 1.8 \times 10^{-5}$$

Assuming that x is negligible in comparison with 0.05 and 0.0815, we have

$$\frac{x(0.0815)}{0.05} = 1.8 \times 10^{-5}$$

$$x = 1.1 \times 10^{-5}$$

(We see that our assumption that x could be neglected was sound.) Therefore

$$p\text{H} = -\log(1.1 \times 10^{-5}) = -(0.04 - 5) = 4.96$$

When we compare this pH with the pH of 0.05M $HC_2H_3O_2$ in the preceding problem, we see that the addition of the salt $NaC_2H_3O_2$ has greatly repressed the ionization of the acid, changing the $[H^+]$ about 100-fold, from $10^{-3.02}$ to $10^{-4.96}$. The shift in equilibrium caused by adding a substance with a common ion is known as the "common-ion effect."

Buffers

A solution that contains a weak acid or base plus a salt of the acid or base is known as a buffer. This name is given because of the regulating action a buffer mixture has on the pH of the solution. If we add an acid or base to a buffer, there is only a small change in pH.

To illustrate, consider the equilibrium in a buffer containing sodium acetate and acetic acid:

$$\frac{[H^+][C_2H_3O_2^-]}{[HC_2H_3O_2]} = K_i$$

Since most of the $C_2H_3O_2^-$ ion is provided by the $NaC_2H_3O_2$ (completely dissociated), we can write

$$\frac{[H^+][salt]}{[acid]} = K_i$$

or

$$[H^+] = K_i \times \frac{[acid]}{[salt]}$$

If we add a small amount of base to this buffer, it reacts with acid to form more salt, but the ratio of acid to salt is not greatly changed. Likewise, if we add acid, it reacts with salt to give more acid. In neither case does the pH change appreciably.

PROBLEM:
A buffer solution contains 1 mole per liter each of acetic acid and sodium acetate. Find the pH.

SOLUTION:

$$[H^+] = K_i \times \frac{[acid]}{[salt]} = 1.8 \times 10^{-5} \times \frac{1}{1} = 1.8 \times 10^{-5}M$$

$$pH = -\log(1.8 \times 10^{-5}) = -(0.26 - 5) = 4.74$$

PROBLEM:
What is the pH if 0.2 mole of HCl is added to 1 liter of the buffer in the preceding problem?

SOLUTION:

The 0.2 mole of HCl added reacts with the salt,

$$H^+ + C_2H_3O_2^- \rightleftharpoons HC_2H_3O_2$$

This gives 1.2 moles of $HC_2H_3O_2$ in the solution and $1 - 0.2 = 0.8$ mole of $NaC_2H_3O_2$:

$$[H^+] = 1.8 \times 10^{-5} \times \frac{1.2}{0.8} = 2.7 \times 10^{-5} M$$

$$pH = -\log(2.7 \times 10^{-5}) = -(0.43 - 5) = 4.57$$

PROBLEM:

How many moles of sodium acetate must be added to 1 liter of $0.2M$ $HC_2H_3O_2$ solution to make a buffer of pH 5?

SOLUTION:

We know that

$$[H^+] = 1.8 \times 10^{-5} \times \frac{[acid]}{[salt]}$$

Since we specify a pH of 5, $[H^+] = 10^{-5}M$. Substituting this gives

$$10^{-5} = 1.8 \times 10^{-5} \times \frac{[acid]}{[salt]}$$

Solving, we get

$$\frac{[salt]}{[acid]} = \frac{1.8 \times 10^{-5}}{1 \times 10^{-5}} = 1.8$$

The problem gives the concentration of acid as $0.2M$; therefore

$$[salt] = 1.8 \times 0.2M = 0.36M$$

Since we have a liter of solution, we add 0.36 mole of $NaC_2H_3O_2$ to obtain a salt concentration of $0.36M$.

PROBLEM:

If 20 ml of $0.2M$ NaOH is added to 50 ml of $0.1M$ $HC_2H_3O_2$, what is the pH?

SOLUTION:

We have 0.050 liter $\times$ 0.1 mole/liter $= 0.005$ mole of $HC_2H_3O_2$, and 0.020 liter $\times$ 0.2 mole/liter $= 0.004$ mole of NaOH. Since the acid is in excess, all the NaOH is neutralized to form 0.004 mole of $NaC_2H_3O_2$, leaving 0.001 mole of unneutralized $HC_2H_3O_2$.

$$[HC_2H_3O_2] = \frac{0.001}{0.070} - x$$

$$[C_2H_3O_2^-] = \frac{0.004}{0.070} + x$$

$$[H^+] = x$$

It is convenient to leave the solution volume, 0.070 liters, in the expressions for the concentrations of the acetic acid and acetate ions rather than to solve, since we will be able to cancel the volume later.

Substituting these concentrations into the equilibrium-constant expression for acetic acid gives

$$\frac{(x)\left(\dfrac{0.004}{0.070} + x\right)}{\left(\dfrac{0.001}{0.070} - x\right)} = 1.8 \times 10^{-5}$$

Neglecting x added to or subtracted from a larger number, and canceling the volume, 0.070, which appears in both numerator and denominator, we have:

$$\frac{(x)(0.004)}{(0.001)} = 1.8 \times 10^{-5}$$

$$x = 0.45 \times 10^{-5} = 4.5 \times 10^{-6}$$

$$pH = -\log (4.5 \times 10^{-6}) = -(0.65 - 6) = 5.35$$

Hydrolysis*

Pure water has a pH of 7. If we add a salt of a strong base and a strong acid, such as NaCl, it does not affect the pH, since neither the Na^+ ion nor the Cl^- ion can react with the H^+ ion or the OH^- ion of water.

If we add to water a salt whose ions come from a weak acid or base, some of the salt reacts with water, or hydrolyzes. An ion of the salt ties up some of the H^+ or OH^- ions of the water, leaving the other ion in excess. The reaction of water with sodium acetate is:

$$HOH \rightleftarrows H^+ + OH^-$$
$$+ \qquad +$$
$$NaC_2H_3O_2 \rightleftarrows C_2H_3O_2^- + Na^+$$
$$\updownarrow \qquad \updownarrow$$
$$HC_2H_3O_2 \qquad NR \text{ (no reaction)}$$

By tying up some of the H^+ ions of water, the solution is left with an excess of OH^- ions and is basic.

Similarly, NH_4Cl gives by hydrolysis an acid solution, since it removes OH^- ion to form the weak base NH_3, thereby leaving an excess of H^+ ions:

---

* Hydrolysis effects due to dissociation of hydrated metal ions [for example, $Al(H_2O)_6^{+++}$] are not included in the simplified discussion of this chapter. See Chapter 22 for this material.

$$\text{HOH} \rightleftharpoons \text{H}^+ + \text{OH}^-$$

$$+ \qquad +$$

$$\text{NH}_4\text{Cl} \rightleftharpoons \text{Cl}^- + \text{NH}_4^+$$

$$\Updownarrow \qquad \Updownarrow$$

$$\text{NR} \quad \text{NH}_3 + \text{H}_2\text{O}$$

We have used here a form of equation that emphasizes the ions of the salt that tie up ions of water. Ordinarily, we do not write hydrolysis equations in this way. The common forms of the equations above are:

$$\text{C}_2\text{H}_3\text{O}_2^- + \text{H}_2\text{O} \rightleftharpoons \text{HC}_2\text{H}_3\text{O}_2 + \text{OH}^-$$

$$\text{NH}_4^+ + \text{H}_2\text{O} \rightleftharpoons \text{NH}_3 + \text{H}_3\text{O}^+$$

As one ion is removed from water, the increase in concentration of the other ion lowers the concentration of the reacting ion and brings about a state of equilibrium. Consequently we can apply the mathematical relation (taking the hydrolysis of acetate ion as an example):

$$\frac{[\text{HC}_2\text{H}_3\text{O}_2][\text{OH}^-]}{[\text{C}_2\text{H}_3\text{O}_2^-][\text{H}_2\text{O}]} = K_e$$

Since $[\text{H}_2\text{O}]$ remains constant at unit activity, we combine this with K_e to obtain

$$\frac{[\text{HC}_2\text{H}_3\text{O}_2][\text{OH}^-]}{[\text{C}_2\text{H}_3\text{O}_2^-]} = [\text{H}_2\text{O}] \times K_e = K_h$$

K_h is called the hydrolysis constant. Unlike ionization constants, which are tabulated in extensive tables, values of K_h are never tabulated because they are easily calculated from values of K_i. For example, if we take the product of the expressions for K_i and K_h for acetic acid and the acetate ion, we get

$$K_i \times K_h = \frac{[\text{H}^+][\text{C}_2\text{H}_3\text{O}_2^-]}{[\text{HC}_2\text{H}_3\text{O}_2]} \times \frac{[\text{HC}_2\text{H}_3\text{O}_2][\text{OH}^-]}{[\text{C}_2\text{H}_3\text{O}_2^-]} = [\text{H}^+][\text{OH}^-] = K_w = 10^{-14}$$

from which we may solve for K_h:

$$K_h = \frac{K_w}{K_i}$$

This equation is a general expression, in which K_i is the ionization constant for the weak acid or weak base that is *formed* on hydrolysis. For the hydrolysis of the $\text{C}_2\text{H}_3\text{O}_2^-$ ion, we get

$$K_h = \frac{1.0 \times 10^{-14}}{1.8 \times 10^{-5}} = 5.5 \times 10^{-10} = \frac{[HC_2H_3O_2][OH^-]}{[C_2H_3O_2^-]}$$

For the hydrolysis of the NH_4^+ ion, we get

$$K_h = \frac{1.0 \times 10^{-14}}{1.8 \times 10^{-5}} = 5.5 \times 10^{-10} = \frac{[NH_3][H^+]}{[NH_4^+]}$$

If the salt has ions of both a weak acid and a weak base,

$$K_h = \frac{K_w}{K_a K_b}$$

We may use the hydrolysis constant to compute the pH of a salt solution, as illustrated in the following problems.

PROBLEM:
Compute the pH of $0.05M$ $NaC_2H_3O_2$.

SOLUTION:
Write the chemical equation, showing the concentrations above the symbols:

$$\overset{0.05-x}{C_2H_3O_2^-} + H_2O \rightleftharpoons \overset{x}{HC_2H_3O_2} + \overset{x}{OH^-}$$

As in previous problems, we represent the unknown concentration of OH^- ion by x. Since $HC_2H_3O_2$ is formed simultaneously with OH^- and in equimolar amount, its concentration is also x. The $C_2H_3O_2^-$ concentration is $0.05 - x$. This gives

$$\frac{x^2}{0.05-x} = K_h = \frac{K_w}{K_a} = \frac{1 \times 10^{-14}}{1.8 \times 10^{-5}} = 5.5 \times 10^{-10}$$

Neglecting x when subtracted from 0.05, we have:

$$\frac{x^2}{0.05} = 5.5 \times 10^{-10}$$

$$x^2 = 2.75 \times 10^{-11}$$

$$-\log x^2 = -(0.44 - 11) = -(-10.56) = 10.56$$

$$-\log x = -\log [OH^-] = pOH = 5.28$$

$$pH = 14 - 5.28 = 8.72$$

PROBLEM:
If 25 ml of $0.2M$ NaOH is added to 50 ml of $0.1M$ $HC_2H_3O_2$, what is the pH of the resulting solution?

SOLUTION:
We have 0.025 liter $\times$ 0.2 mole/liter $= 0.005$ mole of NaOH, and 0.050 liter

$\times$ 0.1 mole/liter = 0.005 mole of $HC_2H_3O_2$. Since the moles of acid are just equal to the moles of base, neutralization is essentially complete, and the solution contains 0.005 moles of $NaC_2H_3O_2$. The concentration is

$$\frac{0.005}{V} = \frac{0.005}{0.075} = 0.067M$$

The hydrolysis equilibrium is

$$\overset{0.067-x}{C_2H_3O_2^-} + H_2O \rightleftarrows \overset{x}{HC_2H_3O_2} + \overset{x}{OH^-}$$

Thus

$$\frac{x^2}{0.067-x} = K_h = \frac{10^{-14}}{1.8 \times 10^{-5}} = 5.5 \times 10^{-10}$$

$$x^2 = 0.367 \times 10^{-10} = 3.67 \times 10^{-11}$$

$$-\log x^2 = -(0.56 - 11) = 10.44$$

$$-\log x = pOH = 5.22$$

$$pH = 14 - 5.22 = 8.78$$

The pH of Aqueous Solutions (Summary)

We are now in a position to consider the various types of aqueous solutions we may deal with in chemistry and the computation of the pH of each. In these calculations we use the ionization constants of Table 20-1.

1. Pure water (see p. 237). The equilibrium reaction is:

$$H_2O \rightleftarrows H^+ + OH^-$$

$$[H^+] = [OH^-] = 1.0 \times 10^{-7}M$$

$$pH = 7$$

2. Strong acid (see p. 238). Because of complete dissociation of acid in dilute solution,

$$[H^+] = \text{molarity of acid.}$$

Thus, for $10^{-3}M$ HCl,

$$[H^+] = 10^{-3}M$$

$$pH = 3$$

3. Strong base (see p. 239). Because of complete dissociation of base in dilute solution,

$$[OH^-] = \text{molarity of base.}$$

Thus, for $10^{-3}M$ NaOH,

$$[OH^-] = 10^{-3}M$$

$$pOH = 3$$

$$pH = 14 - 3 = 11$$

4. Weak acid (see p. 246). The equilibrium reaction is

$$HA \rightleftarrows H^+ + A^-$$

$$[H^+] = [A^-]$$

$$[H^+][A^-] = [H^+]^2 = [HA]K_i$$

$$[H^+] = \sqrt{[HA]K_i}$$

5. Weak base (see p. 247). The equilibrium reaction is

$$B + H_2O \rightleftarrows BH^+ + OH^-$$

$$[BH^+] = [OH^-]$$

$$[OH^-][BH^+] = [OH^-]^2 = [B]K_i$$

$$[OH^-] = \sqrt{[B]K_i}$$

6. Weak acid and its salt, or weak base and its salt (see p. 248). This involves the common-ion effect; the solutions are buffers. The equilibria involved are the same as in (4) or (5) above, *except* that for the acid solution $[H^+] \neq [A^-]$, and for the basic solution $[OH^-] \neq [BH^+]$. After first calculating the salt concentration and the concentration of the corresponding acid or base,

$$[H^+] = \frac{[HA]K_i}{[A^-]} = \frac{[acid]K_i}{[salt]} \quad \text{for the acid solution}$$

$$[OH^-] = \frac{[B]K_i}{[BH^+]} = \frac{[base]K_i}{[salt]} \quad \text{for the base solution}$$

7. Salt of a strong acid and a strong base (see p. 251). Salts of this type do not hydrolyze, consequently the pH of such solutions is 7.0, the same as pure water.

8. Salt of a weak acid and a strong base (see p. 252). This is a hydrolysis equilibrium reaction:

$$A^- + H_2O \rightleftarrows HA + OH^-$$

$$[OH^-] = [HA]$$

$$[OH^-]^2 = [A^-]K_h = \frac{[A^-]10^{-14}}{K_i} = \frac{[salt]10^{-14}}{K_i}$$

$$[OH^-] = \sqrt{\frac{[A^-]10^{-14}}{K_i}} \quad \text{where } K_i \text{ is for the acid, HA}$$

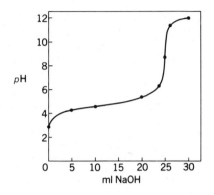

FIGURE 20-1.
Titration of acetic acid with sodium hydroxide: pH against ml of NaOH.

9. Salt of a weak base and a strong acid (see p. 253). This is a hydrolysis equilibrium reaction:

$$BH^+ + H_2O \rightleftharpoons B + H_3O^+$$

$$[H^+] = [B]$$

$$[H^+]^2 = [BH^+]K_h = \frac{[BH^+]10^{-14}}{K_i} = \frac{[salt]10^{-14}}{K_i}$$

$$[H^+] = \sqrt{\frac{[BH^+]10^{-14}}{K_i}} \quad \text{where } K_i \text{ is for the base, B}$$

Titration Curves

When an acid or base solution is titrated, each addition of reagent causes a change in pH. A plot of pH versus volume of reagent added is known as a titration curve. An example is shown in Figure 20-1, for titration of 25 ml of $0.1M$ $HC_2H_3O_2$ by $0.1M$ NaOH. The points used to plot this curve are computed by the methods of the preceding section. Four different types of calculations are involved:

1. Initial point, before any base is added. The solution contains only $HC_2H_3O_2$ (calculate as in No. 4).
2. Intermediate region. The solution contains excess acid plus its salt (calculate as in No. 6).
3. Equivalence point. The solution contains the salt of a weak acid and a strong base (calculate as in No. 8).
4. Beyond equivalence point. The solution contains an excess of strong base (calculate as in No. 3).

When an indicator is used to locate the endpoint of an acid-base titration, we select one whose color change occurs at or near the pH of the equivalence point. If the titration is done by use of a pH meter, the midpoint of the nearly vertical region of the curve is taken as the endpoint.

PROBLEMS A   Answers on page 350

1. Calculate the ionization constants for each of the weak electrolytes in the following table, using the experimental data given.

Electrolyte	Concentration	$[H^+]$	$[OH^-]$	K_i
(a) HCN	$1M$	2.0×10^{-5}		
(b) $HC_2H_3O_2$	$0.001M$	1.26×10^{-4}		
(c) NH_3	$0.0069M$		3.45×10^{-4}	
(d) $HC_2H_3O_2$	$0.2M$	1.9×10^{-3}		
(e) $HCHO_2$	$0.2M$	6.4×10^{-3}		

2. By pH meter or by color comparison with indicators the solutions listed below were found to have the pH values given. Calculate the ionization constant for each.
   (a) $0.01M$ $HC_2H_3O_2$ has a pH of 3.37
   (b) $0.05M$ HCN has a pH of 5.35
   (c) $0.04M$ NH_3 has a pH of 10.93
   (d) $0.02M$ HCNO has a pH of 2.70
   (e) $0.01M$ CH_3NH_2 has a pH of 11.34

3. Calculate the pH of each of the following solutions: (a) $0.5M$ HCNO; (b) $0.1M$ HN_3; (c) $0.01M$ H_3BO_3; (d) $0.25M$ $HC_2H_3O_2$; (e) $0.75M$ HNO_2; (f) $1.5M$ NH_3; (g) $0.5M$ $C_2H_5NH_2$; (h) $0.2M$ C_5H_5N; (i) $0.005M$ NH_3; (j) $1.25M$ $(CH_3)_3N$.

4. Calculate the percentage of ionization of each of the following in water solution: (a) $0.003M$ HCN; (b) $0.6M$ $HCHO_2$; (c) $1.25M$ $(CH_3)_2NH$; (d) $0.05M$ H_2S; (e) $0.001M$ C_5H_5N.

5. A 5-g sample of $HC_2H_3O_2$ is added to 500 ml of water. (a) What is the pH of the solution? (b) Now 5 g of $NaC_2H_3O_2$ are added to the solution. What is the pH of the solution now?

6. How many grams of $NaC_2H_3O_2$ must be added to 250 ml of a $0.1M$ $HC_2H_3O_2$ solution to give a pH of 6.5?

7. A 10-ml sample of $2M$ $HC_2H_3O_2$ is added to 30 ml of $1M$ $NaC_2H_3O_2$. What is the pH of the solution?

8. A 5-g sample of NH_4NO_3 is added to 100 ml of $0.1M$ NH_3. What is the pH of the solution?

9. How many grams of NH_4Cl must be added to 250 ml of $0.2M$ NH_3 in order that the solution shall have a pH of 7.2?

10. A 125-ml sample of $1M$ NH_4Cl is added to 50 ml of $2M$ NH_3. What is the pH of the resulting solution?

11. What is the pH of the solution that results from the addition of 50 ml of $0.1M$ $NaOH$ to 50 ml of $0.2M$ $HC_2H_3O_2$?

12. What is the pH of the solution that results from the addition of 50 ml of $0.1M$ NH_3 to 15 ml of $0.2M$ HCl?

13. The pH of each of the following solutions was determined experimentally by pH meter or by color comparison. Calculate the hydrolysis constant for each salt.
(a) $0.2M$ NH_4Cl has a pH of 4.94
(b) $0.01M$ KCN has a pH of 10.70
(c) $0.1M$ CH_3NH_3Cl has a pH of 5.86
(d) $0.1M$ $KCHO_2$ has a pH of 8.34
(e) $0.01M$ $LiC_2H_3O_2$ has a pH of 8.37

14. Calculate the pH of each of the following solutions: (a) $0.4M$ NaN_3; (b) $1.5M$ NH_4NO_3; (c) $0.1M$ KF; (d) $0.25M$ NaCl; (e) $0.25M$ $(CH_3)_2NH_2Cl$; (f) $0.1M$ KNO_3; (g) $0.5M$ $LiCHO_2$; (h) $10^{-9}M$ HCl.

15. Calculate the pH of the solution resulting from each of the following neutralization processes:
(a) 50 ml of $0.1M$ HCl + 50 ml of $0.1M$ NaOH
(b) 25 ml of $0.1M$ HCl + 50 ml of $0.05M$ NH_3
(c) 35 ml of $0.2M$ HCN + 35 ml of $0.2M$ LiOH
(d) 30 ml of $0.1M$ $HC_2H_3O_2$ + 10 ml of $0.3M$ NaOH

16. If 40 ml of $1M$ $HC_2H_3O_2$ is added to 50 ml of $2M$ $NaC_2H_3O_2$ solution, what will be the pH of the resulting solution?

17. How would you prepare a solution, using NH_4NO_3 and NH_3, so that its pH would be 8?

18. You have a buffer solution that contains 1 mole of NH_4Cl and 1 mole of NH_3 per liter. (a) Calculate the pH of this solution. (b) Calculate the pH of the solution after the addition of 0.1 mole of solid NaOH to a liter. (c) Calculate the pH of this solution after the addition of 0.1 mole of HCl gas to a separate 1-liter portion of the buffer. (d) Calculate the pH of a solution made by adding 0.1 mole of solid NaOH to 100 ml of water. (e) Calculate the pH of a solution made by adding 0.1 mole of HCl gas to 100 ml of H_2O. (f) Compare answers (a)-(e), and think about their significance.

19. Compute the pH for each of the points shown in Figure 20-1. The titration involves the successive addition of 0, 5, 10, 20, 24, 24.9, 25.0, 25.1, 26, and 30 ml of $0.10M$ NaOH to 25.0 ml of $0.10M$ $HC_2H_3O_2$.

20. Compute pH values and construct a titration curve for HCl versus NaOH. Assume that an initial volume of 25.0 ml of $0.10M$ HCl is used and titrated by $0.10M$ NaOH. Make calculations for additions of 0, 5, 10, 24, 24.9, 25.0, 25.1, 26, and 30 ml. Plot on same graph as the $HC_2H_3O_2$-NaOH curve. Note that the two plots coincide when NaOH is somewhat in excess.

PROBLEMS B   No answers given

21. Experimental measurement shows that each of the following is ionized in water as shown. Calculate the ionization constant for each.
    (a) $0.1M$ HF is 8.5% ionized   (d) $0.25M$ NH_3 is 0.84% ionized
    (b) $0.2M$ HNO_2 is 4.75% ionized   (e) $0.2M$ $HAsO_2$ is 0.0055% ionized
    (c) $0.1M$ $HC_2H_3O_2$ is 1.3% ionized

22. By pH meter or by color comparison with indicators, the solutions listed below were found to have the pH values given. Calculate the ionization constant for each.
    (a) $0.01M$ HNO_2 has a pH of 2.71
    (b) $0.001M$ HN_3 has a pH of 3.85
    (c) $0.005M$ HCN has a pH of 5.85
    (d) $0.2M$ $(CH_3)_2NH$ has a pH of 12.08
    (e) $0.002M$ $C_6H_5NH_2$ has a pH of 7.98

23. Calculate the pH of each of the following solutions: (a) $0.002M$ HCN; (b) $0.75M$ $HCHO_2$; (c) $1.20M$ HF; (d) $0.002M$ H_3BO_3; (e) $0.1M$ H_2S; (f) $2.5M$ CH_3NH_2; (g) $0.5M$ NH_3; (h) $1.75M$ $C_6H_5NH_2$; (i) $0.75M$ $(CH_3)_2NH$; (j) $0.02M$ NH_3.

24. Calculate the percentage of ionization of each of the following in water solution: (a) $0.005M$ C_5H_5N; (b) $0.80M$ HN_3; (c) $0.65M$ HCNO; (d) $1.25M$ NH_3; (e) $2.5M$ HF.

25. A 10-g sample of formic acid is added to 200 ml of water. (a) What is the pH of the solution? (b) Now 20 g of sodium formate is added to the solution. What is the pH of the solution now?

26. How many grams of $NaNO_2$ must be added to 300 ml of a $0.2M$ HNO_2 solution to give a pH of 4.7?

27. A 20-ml sample of $1M$ HNO_2 is added to 50 ml of $0.6M$ $NaNO_2$. What is the pH of the solution?

28. A 25-g sample of NH_4Cl is added to 200 ml of $0.5M$ NH_3. What is the pH of the solution?

29. How many grams of NH_4NO_3 must be added to 100 ml of $3M$ NH_3 in order that the solution shall have a pH of 8.5?

30. A 75-ml sample of $2M$ NH_4NO_3 is added to 100 ml of $1M$ NH_3. What is the pH of the resulting solution?

31. What is the pH of the solution that results from the addition of 25 ml of 0.2M KOH to 50 ml of 0.15M HNO_2?

32. What is the pH of the solution that results from the addition of 25 ml of 0.2M NH_3 to 30 ml of 0.15M HCl?

33. The pH of each of the following solutions was determined experimentally by pH meter or by color comparison. Calculate the hydrolysis constant for each salt.
    (a) 0.1M NH_4Cl has a pH of 5.13.
    (b) 0.01M $LiCHO_2$ has a pH of 7.86.
    (c) 0.1M $NaC_2H_3O_2$ has a pH of 8.86.
    (d) 0.01M $(CH_3)_2NH_2Cl$ has a pH of 6.43.
    (e) 0.1M NaCN has a pH of 11.2.

34. Calculate the pH of each of the following solutions: (a) 0.1M NaCNO; (b) 0.05M NH_4NO_3; (c) 1.50M KNO_2; (d) 0.35M $NaNO_3$; (e) 0.80M $C_2H_5NH_3Cl$; (f) 0.01M KCl; (g) 1.5M $KC_2H_3O_2$; (h) $10^{-9}M$NaOH.

35. Calculate the pH of the solution resulting from each of the following neutralization processes:
    (a) 25 ml of 0.2M HNO_3 + 25 ml of 0.2M KOH
    (b) 40 ml of 0.1M $HCHO_2$ + 20 ml of 0.2M NaOH
    (c) 35 ml of 0.1M HNO_3 + 35 ml of 0.1M NH_3
    (d) 20 ml of 0.25M $HC_2H_3O_2$ + 10 ml of 0.5M KOH

36. If 10 ml of 1M HNO_2 is added to 20 ml of 2M $NaNO_2$, what will be the pH of the resulting solution?

37. How would you prepare a solution, using $HC_2H_3O_2$ and $NaC_2H_3O_2$, so that the pH of the resulting solution would be 6?

38. You have a buffer solution that contains 1 mole of $NaC_2H_3O_2$ and 1 mole of $HC_2H_3O_2$ per liter. (a) Calculate the pH of this solution. (b) Calculate the pH of the solution after the addition of 0.1 mole of solid NaOH to a liter. (c) Calculate the pH of this solution after the addition of 0.1 mole of HCl gas to a separate 1-liter portion of the buffer. (d) Calculate the pH of a solution made by adding 0.1 mole of solid NaOH to 100 ml of water. (e) Calculate the pH of a solution made by adding 0.1 mole of HCl gas to 100 ml of H_2O. (f) Compare answers (a)-(e), and think about their significance.

39. Construct a titration curve for the titration of 40.0 ml of 0.20M NaOH with 0.20M HCl. Make pH calculations for the addition of 0, 5, 20, 30, 39, 39.9, 40.0, 40.1, 41, and 45 ml.

40. Calculate pH values and construct a titration curve for the titration of 25.0 ml of 0.10M NH_3 with 0.10M HCl. Make calculations for the addition of 0, 5, 10, 20, 24, 24.9, 25.0, 25.1, 26, and 30 ml.

Solubility Product and Precipitation

In the preceding chapter we discussed applications of equilibrium relations to the pH of aqueous solutions. Another important application of these principles involves precipitation reactions.

We tend to think of the precipitates we separate in chemical analysis as insoluble. Actually, all precipitates have some solubility. When we have a precipitate of AgCl, for example, in contact with supernatant liquid, the liquid is saturated with the ions of the precipitate, Ag^+ and Cl^-. The equation is

$$AgCl_{(s)} \rightleftharpoons Ag^+ + Cl^-$$

The symbol (s) designates a solid. At the surface of the precipitate, Ag^+ and Cl^- ions are constantly going into solution and redepositing from the solution. Since this is an equilibrium process, we can apply the mathematical relation

$$\frac{[Ag^+][Cl^-]}{[AgCl_{(s)}]} = K_e$$

Since the concentration of solid AgCl crystals is constant, we may simplify the equation by combining the two constants:

$$[Ag^+][Cl^-] = K_e \times [AgCl_{(s)}] = K_{sp}$$

When written in this form, to involve only the product of the concentrations of the ions, the constant K_{sp} is called the "solubility-product constant."

Evaluation of K_{sp}

It is found experimentally that in a saturated solution of AgCl at room temperature the concentration of dissolved salt is 0.00182 g/liter, or 1.27×10^{-5} mole per liter. When a molecule of AgCl dissolves, it dissociates to give an Ag^+ ion and a Cl^- ion. Therefore the concentrations of Ag^+ and Cl^- in the saturated solution are each $1.27 \times 10^{-5}M$. Substituting these concentrations gives

$$[Ag^+][Cl^-] = (1.27 \times 10^{-5})^2 = 1.6 \times 10^{-10} = K_{sp}$$

Computation of Solubility from K_{sp}

Table 21-1 gives experimentally measured K_{sp} values for some common precipitates. These can be used to compute the solubilities in grams.

PROBLEM:
The K_{sp} value for $AgC_2H_3O_2$ (mol wt = 167) is 2×10^{-3}. Find the approximate solubility in grams per 100 ml.

SOLUTION:
Let S = moles per liter of $AgC_2H_3O_2$ in a saturated solution. Since the salt is completely dissociated,

$$[Ag^+] = S$$
$$[C_2H_3O_2^-] = S$$

This gives

$$[Ag^+][C_2H_3O_2^-] = 2 \times 10^{-3}$$
$$[S]^2 = 2 \times 10^{-3}$$
$$S = \sqrt{2 \times 10^{-3}} = \sqrt{20 \times 10^{-4}} = 4.4 \times 10^{-2}$$

Therefore 4.4×10^{-2} mole $AgC_2H_3O_2$ dissolves per liter or $0.1 \times 4.4 \times 10^{-2}$ $= 4.4 \times 10^{-3}$ mole/100 ml. The weight in grams is $4.4 \times 10^{-3} \times 167 = 730 \times 10^{-3}$ $= 0.73$ g.

Precipitation

We can use the K_{sp} value to compute whether a precipitate will form when given quantities of Ag^+ and Cl^- are put together in a solution. We compute the molar concentrations of the ions, multiply these to obtain the product, and compare this value with the known K_{sp} value for a saturated solution.

TABLE 21-1.
Approximate Solubility Products (18-25°C)

Acetates		Hydroxides	
$AgC_2H_3O_2$	2×10^{-3}	$Al(OH)_3$	1×10^{-33}
		$Ca(OH)_2$	8×10^{-6}
Carbonates		$Cd(OH)_2$	1.2×10^{-14}
Ag_2CO_3	8×10^{-12}	$Cr(OH)_3$	1×10^{-30}
$BaCO_3$	5×10^{-9}	$Cu(OH)_2$	6×10^{-20}
$CaCO_3$	4.8×10^{-9}	$Fe(OH)_2$	1×10^{-15}
$CuCO_3$	1×10^{-10}	$Fe(OH)_3$	1×10^{-38}
$FeCO_3$	2×10^{-11}	$Mg(OH)_2$	1×10^{-11}
$MgCO_3$	1×10^{-5}	$Mn(OH)_2$	4×10^{-14}
$MnCO_3$	9×10^{-11}	$Pb(OH)_2$	1×10^{-16}
$PbCO_3$	1×10^{-13}	$Sn(OH)_2$	1×10^{-26}
$SrCO_3$	1×10^{-9}	$Zn(OH)_2$	1×10^{-17}

Chromates		Sulfates	
Ag_2CrO_4	1×10^{-12}	Ag_2SO_4	1.2×10^{-5}
$BaCrO_4$	2×10^{-10}	$BaSO_4$	1×10^{-10}
$PbCrO_4$	2×10^{-14}	$CaSO_4 \cdot 2H_2O$	2.4×10^{-5}
$SrCrO_4$	3.6×10^{-5}	Hg_2SO_4	6×10^{-7}
		$PbSO_4$	2×10^{-8}
Halides		$SrSO_4$	2.8×10^{-7}
$AgCl$	1.6×10^{-10}		
$AgBr$	4×10^{-13}	**Sulfides**	
AgI	1×10^{-16}	Ag_2S	10^{-51}
CaF_2	4×10^{-11}	Bi_2S_3	10^{-72}
Hg_2Cl_2	1×10^{-18}	CdS	10^{-28}
$PbCl_2$	1.7×10^{-5}	CoS	10^{-21}
PbI_2	9×10^{-9}	CuS	10^{-40}
SrF_2	4×10^{-9}	FeS	10^{-22}
		HgS	10^{-54}
Oxalates		MnS	10^{-16}
		NiS	10^{-21}
BaC_2O_4	1×10^{-7}	PbS	10^{-28}
CaC_2O_4	2×10^{-9}	SnS	10^{-28}
MgC_2O_4	9×10^{-5}	Tl_2S	10^{-22}
		ZnS	10^{-23}

PROBLEM:

If 0.01 mg of NaCl is added to 200 ml of $0.00002M$ $AgNO_3$, will we get a precipitate of AgCl?

SOLUTION:

First compute the concentrations of the Ag^+ and Cl^- ions. Since these come from salts that are almost completely dissociated, the ions have the same concentrations as the salts:

$$[Ag^+] = 0.00002M = 2 \times 10^{-5}M \text{ (the molarity of the}$$
$$\text{AgNO}_3 \text{ solution is given in the problem.)}$$

$$[Cl^-] = [NaCl] = 0.00001 \text{ g per 200 ml or } 0.00005 \text{ g/liter}$$

The molarity is $\dfrac{0.00005}{58.5} = \dfrac{50 \times 10^{-6}}{58.5} = 0.85 \times 10^{-6}M$.

The product $[Ag^+][Cl^-] = (2 \times 10^{-5})(0.85 \times 10^{-6}) = 1.7 \times 10^{-11}$. This is less than the K_{sp} value for a saturated solution (1.6×10^{-10}). Therefore no precipitation occurs.

PROBLEM:

If 1.0 g of $AgNO_3$ is added to 50 ml of $0.05M$ $HC_2H_3O_2$, will we get a precipitate of $AgC_2H_3O_2$?

SOLUTION:

The K_{sp} for $AgC_2H_3O_2$ is 2×10^{-3}. When we add 1.0 g of $AgNO_3$ to 50 ml, or $1.0 \times \dfrac{1,000}{50} = 20$ g/liter, the molarity of $AgNO_3 = \dfrac{20}{170} = 0.12M$. This is also the concentration of Ag^+ ion.

To compute the concentration of the $C_2H_3O_2^-$ ion, we must use the equation for the ionization of acetic acid:

$$\begin{matrix} 0.05 & & x & & x \\ HC_2H_3O_2 & \rightleftharpoons & H^+ & + & C_2H_3O_2^- \end{matrix}$$

$$\frac{x^2}{0.05} = 1.8 \times 10^{-5}$$

$$x^2 = 9 \times 10^{-7} = 90 \times 10^{-8}$$

$$x = \sqrt{90 \times 10^{-8}} = 9.5 \times 10^{-4}$$

Now we determine the ion-product value:

$$[Ag^+][C_2H_3O_2^-] = (0.12)(9.5 \times 10^{-4}) = 1.1 \times 10^{-4}$$

This value is less than the K_{sp} value for a saturated solution (2×10^{-3}), and consequently precipitation does not occur.

Solubility Products for Mixed-Charge Precipitates

We have so far in this chapter discussed only simple cases, where both ions of the precipitate have the same charge. The same principles apply to substances such as $Mg(OH)_2$ and $Al(OH)_3$, but the calculations are a little more complicated. We will discuss a few typical examples.

PROBLEM:
The solubility of $Mg(OH)_2$ is 0.0009 g per 100 ml at 18°C. Calculate the K_{sp} value.

SOLUTION:
A solubility of 0.0009 g per 100 ml is 0.009 g/liter. The molar solubility is

$$\frac{0.009}{58} = \frac{90 \times 10^{-4}}{58} = 1.5 \times 10^{-4}$$

The equation is

$$Mg(OH)_{2_{(s)}} \rightleftarrows Mg^{++} + 2\ OH^-$$

Each mole of $Mg(OH)_2$ that dissolves produces 1 mole Mg^{++} and 2 moles OH^- ion:

$$[Mg^{++}] = 1.5 \times 10^{-4}$$
$$[OH^-] = 2 \times 1.5 \times 10^{-4} = 3.0 \times 10^{-4}$$

The K_{sp} equation is

$$[Mg^{++}][OH^-]^2 = K_{sp}$$

(In the mathematical relation for the equilibrium constant, the concentration of each ion is raised to the power indicated by its coefficient in the chemical equation.) This gives

$$(1.5 \times 10^{-4})(3.0 \times 10^{-4})^2 = (1.5 \times 10^{-4})(9 \times 10^{-8})$$
$$= 13.5 \times 10^{-12} = 1.35 \times 10^{-11} = K_{sp}$$

PROBLEM:
The K_{sp} value for Ag_2CrO_4 is 1×10^{-12}. Calculate the solubility of Ag_2CrO_4.

SOLUTION:
Let S be the number of moles of Ag_2CrO_4 in a saturated solution. Then

$$[Ag^+] = 2S \text{ (each molecule gives 2 } Ag^+ \text{ ions)}$$
$$[CrO_4^=] = S \text{ (each molecule gives 1 } CrO_4^= \text{ ion)}$$

Substituting into the K_{sp} equation:

$$[Ag^+]^2[CrO_4^=] = 1 \times 10^{-12}$$
$$(2S)^2(S) = 1 \times 10^{-12}$$
$$4S^3 = 1 \times 10^{-12}$$
$$S^3 = 0.25 \times 10^{-12}$$
$$S = \sqrt[3]{0.25 \times 10^{-12}} = \sqrt[3]{250 \times 10^{-15}} = 6.3 \times 10^{-5} \text{ mole/liter}$$

To get the solubility in grams per liter we multiply by the molecular weight:

$$\text{grams per liter} = 6.3 \times 10^{-5} \text{ mole/liter} \times 332 \text{ g/mole} = 0.021 \text{ g/liter}$$

PROBLEM:
If 0.5 g of $MgSO_4$ and 2 g of NH_4Cl are added to 250 ml of $0.1M$ NH_3, will we get a precipitate of $Mg(OH)_2$?

SOLUTION:
We find the concentrations of the Mg^{++} and OH^- ions, substitute into the K_{sp} equation, and compare the product with the K_{sp} value.

The OH^- ion comes from NH_3, according to the equation

$$\begin{array}{cccc} 0.1 - x & 0.15 + x & & x \\ H_2O + NH_3 & \rightleftharpoons & NH_4^+ & + & OH^- \end{array}$$

$[NH_3]$ is $0.1M$, and $[OH^-]$ is x. $[NH_4^+]$ is calculated from the amount of NH_4Cl, as follows:

$$2 \text{ g of } NH_4Cl = \frac{8}{53.5} = 0.15M = [NH_4^+]$$

The K_i for NH_3 is 1.8×10^{-5}. Therefore

$$\frac{(0.15 + x)(x)}{(0.1 - x)} = 1.8 \times 10^{-5}$$

Neglecting x, as compared with 0.1 and 0.15, we have

$$\frac{0.15x}{0.1} = 1.8 \times 10^{-5}$$

$$x = 1.2 \times 10^{-5} = [OH^-]$$

The concentration of Mg^{++} is the molarity of $MgSO_4$:

$$0.5 \text{ g per 250 ml} = 2.0 \text{ g/liter}$$

$$\text{molarity of } MgSO_4 = \frac{2.0}{120} = 0.017M = [Mg^{++}]$$

The K_{sp} equation is

$$[Mg^{++}][OH^-]^2 = K_{sp} = 1.35 \times 10^{-11}$$

Substituting the given concentrations, we have

$$(0.017)(1.2 \times 10^{-5})^2 = (1.7 \times 10^{-2})(1.4 \times 10^{-10}) = 2.4 \times 10^{-12}$$

Since this is less than the K_{sp} value, no precipitation occurs. If we had not repressed the OH^- ion concentration from NH_3 by adding the NH_4Cl, the concentration of OH^- would have been sufficiently large to give a precipitate of $Mg(OH)_2$.

Precipitation by Hydrogen Sulfide

In qualitative analysis many of the metal ions are precipitated as sulfides by addition of H_2S to their solutions. This system has long been used because separations may be controlled by adjustment of the hydrogen-ion concentration of the solution.

H_2S is a diprotic acid, ionizing in two stages:

$$H_2S \rightleftarrows H^+ + HS^-$$

$$HS^- \rightleftarrows H^+ + S^=$$

The equilibrium expressions are;

$$K_1 = \frac{[H^+][HS^-]}{[H_2S]} = 1 \times 10^{-7}$$

$$K_2 = \frac{[H^+][S^=]}{[HS^-]} = 1.3 \times 10^{-13}$$

If we are primarily interested in the *acidity* of an H_2S solution, we consider only K_1, because the number of H^+ ions produced in the second stage is negligible compared to the number produced in the first. However, if we want to precipitate metal sulfides we must consider the second stage, since this is the *only* source of $S^=$ ions. In a simple aqueous solution of H_2S with no other source of H^+ or $S^=$, $[H^+] = [HS^-]$ as a result of the first stage of ionization; $[S^=]$ is calculated from the expression for K_2:

$$[S^=] = \frac{[HS^-](1.3 \times 10^{-13})}{[H^+]} \cong 1.3 \times 10^{-13} M$$

a concentration that is negligible compared to the $[HS^-]$ from which it came.

The variation and control of the $[S^=]$ by control of pH is best calculated from the overall ionization

$$H_2S \rightleftarrows 2H^+ + S^=$$

whose equilibrium constant is the product of the constants for the two stages; that is,

$$\frac{[H^+][HS^-]}{[H_2S]} \times \frac{[H^+][S^=]}{[HS^-]} = \frac{[H^+]^2[S^=]}{[H_2S]} = K_1 \times K_2 = 1.3 \times 10^{-20}$$

There is an important special form of this equilibrium expression for solutions *saturated* with H_2S where the H_2S concentration is about $0.1M$ at 25°C. For this common situation we have

$$[H^+]^2[S^=] = (0.1)(1.3 \times 10^{-20}) = 1.3 \times 10^{-21}$$

The following problem illustrates how two ions may be separated by controlling the $[S^=]$ through adjustment of the pH.

PROBLEM:
A solution containing Mn^{++} and Cd^{++} ions in $0.2M$ HCl is saturated with H_2S. Calculate the concentration of the two ions at equilibrium.

SOLUTION:
Substitute $0.2M$ for $[H^+]$ in the equilibrium expression for a saturated H_2S solution and solve for $[S^=]$:

$$[S^=] = \frac{1.3 \times 10^{-21}}{(0.2)^2} = 3.2 \times 10^{-20}M$$

The metal-ion concentrations in equilibrium with this $[S^=]$ are calculated from the respective K_{sp} values for the metal sulfides:

$$[Cd^{++}] = \frac{10^{-28}}{3.2 \times 10^{-20}} = 3.1 \times 10^{-9}M$$

$$[Mn^{++}] = \frac{10^{-16}}{3.2 \times 10^{-20}} = 3.1 \times 10^3M$$

These values show that K_{sp} for MnS is never reached in the solution and consequently no MnS precipitates. Most of the Cd^{++} ion of the solution is, however, precipitated as CdS until the final concentration left in solution is $3.1 \times 10^{-9}M$. If we wish to precipitate the Mn^{++} after removal of the CdS we must increase the sulfide-ion concentration enormously by decreasing the hydrogen-ion concentration. This is done by making the solution basic.

When a metal ion is precipitated as the sulfide, hydrogen ions are formed in the solution. For example, in the reaction

$$Cu^{++} + H_2S \rightleftharpoons CuS + 2\ H^+$$

2 moles of hydrogen ion are formed for each mole of CuS that is precipitated. As precipitation proceeds, therefore, the solution becomes more acidic and

the concentration of sulfide ion becomes less. In some precipitations it may be necessary to hold the hydrogen-ion concentration constant by buffering the solution in order to effect complete separations of metal sulfides.

Dissolving a Precipitate

When a precipitate is treated with a reagent that can react with one of its ions, its solubility is increased, since reduction of the concentration of one ion causes an increase in the concentration of the other. Two different types of reaction may be used to dissolve precipitates.

1. The negative ion of the precipitate may react with an added ion to form a weak acid or base.

2. The positive ion of the precipitate may react with an added substance to form a complex ion (see Chapter 22).

The most common example of the first type is the solubility of salts of weak acids in strong acids. Many carbonates, sulfides, phosphates, borates, oxalates, and salts of other weak acids may be dissolved in strong acids, even though the solubility in water is quite low. If the ionization constant of the weak acid and the solubility product of the salt are known, we may compute the solubility by consideration of the two equilibria that must be satisfied in the solution.

PROBLEM:
The solubility product for the salt MA is 10^{-8}. The negative ion A^- is from a weak acid, HA, whose ionization constant is 10^{-6}. If solid MA is added to 1 liter of $0.1M$ HCl, how many moles are dissolved?

SOLUTION:
The reaction is

$$MA_{(s)} + H^+ \rightleftarrows M^+ + HA$$

and the equilibrium constant is given by the relation

$$\frac{[M^+][HA]}{[H^+]} = K_e$$

Note that $[MA_{(s)}]$, which is constant, is included in the value for K_e and not shown in the equation.

To compute the concentration of the M^+ ion in solution we must know the value of the constant. This can be calculated from the known K_i and K_{sp} values, since the ionic concentrations in the solution must be such as to satisfy both equilibria. We have

$$[M^+][A^-] = K_{sp} = 10^{-8}$$

$$\frac{[H^+][A^-]}{[HA]} = K_i = 10^{-6}$$

Dividing K_{sp} by K_i gives the desired relation:

$$\frac{[M^+][HA]}{[H^+]} = \frac{K_{sp}}{K_i} = 10^{-2}$$

Let x = moles MA that dissolve in a liter of solution. Then

$$[HA] = x$$
$$[M^+] = x$$
$$[H^+] = 0.1 - x$$

Substituting these values into the equilibrium expression gives

$$\frac{x^2}{0.1 - x} = 10^{-2}$$

The magnitude of this constant makes us suspect that we should not neglect x in comparison to 0.1, so we solve the quadratic equation. This gives

$$x = 0.027 \text{ moles/liter}$$

PROBLEM:
CuS is treated with a liter of $1M$ HCl solution. How many grams of the salt are dissolved?

SOLUTION:
The reaction is

$$\text{CuS} + 2\text{ H}^+ \rightleftarrows \text{Cu}^{++} + \text{H}_2\text{S}$$

and K_e is

$$\frac{[\text{Cu}^{++}][\text{H}_2\text{S}]}{[\text{H}^+]^2} = K_e$$

To evaluate K_e we consider the simultaneous equilibria

$$[\text{Cu}^{++}][\text{S}^=] = K_{sp} = 10^{-40}$$
$$\frac{[\text{H}^+]^2[\text{S}^=]}{[\text{H}_2\text{S}]} = K_i = 1.3 \times 10^{-20}$$

Dividing K_{sp} by K_i, we have

$$\frac{[\text{Cu}^{++}][\text{H}_2\text{S}]}{[\text{H}^+]^2} = 0.77 \times 10^{-20}$$

Let x = moles of CuS that dissolve. Then, $x = [\text{Cu}^{++}] = [\text{H}_2\text{S}]$ and $1 - 2x = [\text{H}^+]$. Substituting these values gives

$$\frac{x^2}{(1-2x)^2} = 0.77 \times 10^{-20} = 77 \times 10^{-22}$$

Here we can neglect $2x$ when subtracted from 0.1. Solving, we have

$$x = \sqrt{77 \times 10^{-22}} = 8.8 \times 10^{-11} \text{ moles/liter}$$

$$\text{grams of CuS} = 8.8 \times 10^{-11} \frac{\text{moles}}{\text{liter}} \times 95.6 \frac{\text{g}}{\text{mole}}$$

$$= 8.4 \times 10^{-9} \text{ g/liter}$$

This calculation shows that CuS is so insoluble that no appreciable amount will dissolve in a strong acid. To dissolve it we must use an oxidizing agent, HNO_3, for example, that removes the sulfide ion completely by oxidation to free sulfur.

Sulfides whose K_{sp} values are considerably larger than that for CuS may be dissolved in strong acids. For example, if the K_{sp} value is taken as 10^{-24} and the calculation of the preceding section is repeated, we have

$$\frac{x^2}{(1-2x)^2} = \frac{10^{-24}}{1.3 \times 10^{-20}} = 0.77 \times 10^{-4}$$

$$x = \sqrt{77 \times 10^{-6}} = 8.8 \times 10^{-3} \text{ moles/liter}$$

Dissolving of precipitates by formation of a weak electrolyte is not restricted to salts of weak acids. Most oxides are soluble in acids, with formation of undissociated H_2O. Hydroxides behave similarly. Some hydroxides may be dissolved in solutions of ammonium salts, for example $Mg(OH)_2$:

$$Mg(OH)_2 + 2\,NH_4^+ \rightleftarrows Mg^{++} + 2\,NH_3 + 2\,H_2O$$

PROBLEMS A   Answers on page 351

1. The following solubilities in water have been determined by experiment. From these experimental data calculate the solubility products for the solids involved. The solubilities are given in grams per 100 ml of solution. (a) AgI, 2.88×10^{-7}; (b) $BaSO_4$, 2.4×10^{-4}; (c) $Cd(OH)_2$, 2.06×10^{-4}; (d) SrF_2, 1.22×10^{-2}; (e) Ag_3PO_4, 8.5×10^{-4}; (f) PbS, 8.6×10^{-14}; (g) As_2S_3, 5.17×10^{-5}; (h) $BaCO_3$, 1.4×10^{-3}.

2. Which of the following solids will dissolve readily in an excess of $1M$ HCl? (a) AgI; (b) $MnCO_3$; (c) $SrCrO_4$; (d) $Zn(OH)_2$; (e) $SrSO_4$; (f) BaC_2O_4.

3. In each of the following cases show whether a precipitate will form under the given conditions:
   (a) 1 ml of $0.1M$ $AgNO_3$ is added to 1 liter of $0.01M$ Na_2SO_4
   (b) 1 g of $Pb(NO_3)_2$ is put into 100 ml of $0.01M$ HCl
   (c) 1 ml of $1M$ NaOH is added to 1 liter of $10^{-4}M$ $Mg(NO_3)_2$
   (d) 1 ml of $1M$ NH_3 is added to 1 liter of $10^{-4}M$ $Mg(NO_3)_2$
   (e) 1 ml of $0.1M$ $Sr(NO_3)_2$ is added to 1 liter of $0.01M$ HF
   (f) 1 ml of $0.01M$ $Ba(NO_3)_2$ is added to 1 liter of $0.1M$ $NaHC_2O_4$
   (g) 1 mg of $CaCl_2$ and 1 mg of $Na_2C_2O_4$ are put in a liter of water

4. Calculate the solubility, in moles per liter, of each of the following compounds in water (neglect hydrolysis effects): (a) HgS; (b) $SrCrO_4$; (c) CaC_2O_4; (d) Ag_2S; (e) PbI_2; (f) $Zn(OH)_2$.

5. Calculate how many grams of CaC_2O_4 will dissolve in 1 liter of (a) water, (b) $0.1M$ $Na_2C_2O_4$, (c) $0.01M$ $CaCl_2$, (d) $0.1M$ $NaNO_3$.

6. Calculate how many grams of PbI_2 will dissolve in 250 ml of (a) water, (b) $0.01M$ $Pb(NO_3)_2$, (c) $0.01M$ CaI_2.

7. What concentration of sulfide ion is needed to commence precipitation of each of the following metals from solution as the sulfide? (a) $0.1M$ $CuCl_2$; (b) $10^{-4}M$ $AgNO_3$; (c) $0.02M$ $TlNO_3$; (d) $10^{-3}M$ $Bi(NO_3)_3$; (e) $10^{-6}M$ $Hg(NO_3)_2$.

8. What OH^- concentration is needed to commence precipitation of each of the following metals as the hydroxide, if each is present to the extent of 1 mg of metal ion per milliliter? (a) Cu^{++}; (b) Cr^{+++}; (c) Zn^{++}; (d) Sn^{++}; (e) Al^{+++}.

9. K_2CrO_4 is slowly added to a solution that is $0.02M$ in $Pb(NO_3)_2$ and $0.02M$ in $Ba(NO_3)_2$. (a) Which ion precipitates first? (b) What will be its concentration when the second ion begins to precipitate?

10. A 1-ml sample of $0.1M$ $Pb(NO_3)_2$ is added to 100 ml of a solution saturated with $SrCrO_4$. How many milligrams of $PbCrO_4$ will precipitate?

11. What concentration of $C_2O_4^=$ is needed to start precipitation of calcium from a saturated solution of $CaSO_4$?

12. An analyst wishes to determine the amount of calcium by precipitation as the oxalate. If his sample contains 50 mg Ca^{++} in 250 ml of solution, and if he adds $(NH_4)_2C_2O_4$ to give an oxalate ion concentration of $0.5M$, what percentage of his Ca^{++} will remain unprecipitated?

13. What will be the concentration of Cu^{++} remaining in solution when Cd^{++} just begins to precipitate as CdS from a solution that was originally $0.02M$ in Cu^{++} and Cd^{++}?

14. A solution is $0.02M$ in Mg^{++} and $0.1M$ in NH_4NO_3. What concentration of NH_3 must be attained in order to begin precipitation of $Mg(OH)_2$?

15. In order to prevent precipitation of $Mg(OH)_2$, what is the minimum number of grams of NH_4Cl that must be added to 500 ml of a solution containing 3 g of $Mg(NO_3)_2$ and 5 g of NH_3?

16. How many moles of $AgC_2H_3O_2$ will dissolve in 250 ml of $0.1M$ HNO_3?

17. What is the pH of a water solution saturated with $Fe(OH)_3$?

18. (a) To what final pH must a solution be adjusted in order to precipitate as much CdS as possible without precipitating any ZnS? (The solution is originally $0.02M$ with respect to each metal ion.) (b) How much Cd^{++} will be left in solution when the Zn^{++} begins to precipitate?

19. What must be the final pH of an H_2SO_4 solution in order to just dissolve 0.05 mole of ZnS in 1 liter of the solution?

20. What is the lowest pH that will permit the sulfide of each of the following metals to precipitate from a $10^{-4}M$ solution saturated with H_2S? (a) Pb^{++}; (b) Bi^{+++}; (c) Co^{++}; (d) Mn^{++}.

21. Can a suspension of AgI that is $5M$ in HCl be converted to Ag_2S by saturation with H_2S? Explain.

22. How many liters of a solution saturated with HgS would one have to take to find statistically one Hg^{++} ion? (Neglect hydrolysis of the sulfide ion.)

23. When a metal ion precipitates from solution on saturation with H_2S, the acidity of the solution increases as the metal removes the $S^=$. This in turn increases the solubility of the metal sufide. What will be the concentration of each of the following metals remaining after saturation of $0.1M$ aqueous solutions with H_2S? (a) Cu^{++}; (b) Cd^{++}; (c) Zn^{++}.

24. A $0.05M$ $ZnCl_2$ solution is buffered with 2 moles of $NaC_2H_3O_2$ and 1 mole of $HC_2H_3O_2$ per liter. What will be the Zn^{++} concentration remaining after saturation with H_2S?

25. A solution saturated with CO_2 at 1 atm and 25°C has a concentration of about $0.034M$. (a) Write an equilibrium expression, in terms of H^+ and $CO_3^=$, that could be usefully applied to the separation of metal ions as carbonates by saturation with CO_2. (b) At what pH must a solution be adjusted so as not to precipitate $BaCO_3$ from a $0.02M$ $Ba(NO_3)_2$ solution on saturation with CO_2? (c) What would be the concentration of Pb^{++} remaining in a solution adjusted to the pH in (b)? (d) Why can't saturation with CO_2 be applied in a practical way to the separation of metal ions in the same way that saturation with H_2S is?

26. Which will precipitate first when solid Na_2S is slowly added to a $0.01M$ $MnSO_4$ solution, MnS or $Mn(OH)_2$?

27. Calculate the solubility of (a) PbS and (b) Tl_2S in water. (Do not neglect hydrolysis of the sulfide ion.)

PROBLEMS B   No answers given

28. The following solubilities in water have been determined by experiment. From these experimental data calculate the solubility products for the solids involved. The solubilities are given in grams per 100 ml of solution. (a) HgBr, 3.9×10^{-6}; (b) AgCN, 2.2×10^{-5}; (c) $CaCO_3$, 6.93×10^{-4}; (d) PbI_2, 6×10^{-2}; (e) Ag_2S, 1.6×10^{-16}; (f) Ag_3PO_4, 6.7×10^{-4}; (g) CaF_2, 1.7×10^{-3}; (h) $MgNH_4PO_4$, 8.65×10^{-4}.

29. Which of the following solids will dissolve readily in an excess of $1M$ HCl? (a) CaF_2; (b) AgBr; (c) $MgCO_3$; (d) $BaCrO_4$; (e) $Cd(OH)_2$; (f) $BaSO_4$; (g) MgC_2O_4.

30. In each of the following cases show whether a precipitate will form under the given conditions:
    (a) 1 g of $Sr(NO_3)_2$ is added to 1 liter of $0.001M$ K_2CrO_4
    (b) 1 ml of $0.01M$ $Pb(NO_3)_2$ is added to 1 liter of $0.01M$ H_2SO_4
    (c) 1 ml of $0.1M$ NaOH is added to 1 liter of $10^{-5}M$ $MnSO_4$
    (d) 1 ml of $0.1M$ NH_3 is added to 1 liter of $10^{-5}M$ $MnSO_4$
    (e) 1 ml of $0.01M$ $Ca(NO_3)_2$ is added to 1 liter of $0.01M$ HF
    (f) 1 ml of $0.01M$ $Ca(NO_3)_2$ is added to 1 liter of $0.1M$ $NaHC_2O_4$
    (g) 1 mg of $Ba(NO_3)_2$ is added to 1 liter of $0.01M$ K_2CrO_4

31. Calculate the solubility, in moles per liter, of each of the following compounds in water (neglect hydrolysis effects): (a) BaC_2O_4; (b) $PbSO_4$; (c) AgBr; (d) Ag_2SO_4; (e) $Ca(OH)_2$.

32. Calculate how many grams of $BaSO_4$ will dissolve in a liter of (a) water, (b) $0.1M$ Na_2SO_4, (c) $0.1M$ $Ba(NO_3)_2$, (d) $0.01M$ KCl.

33. Calculate how many grams of $Cd(OH)_2$ will dissolve in 250 ml of (a) water, (b) $0.01M$ KOH, (c) $0.01M$ $CdCl_2$.

34. What concentration of Ag^+ is needed to initiate precipitation of the anions from each of the following solutions? Assume in each case that the anion has not reacted with the water. (a) $0.05M$ NaBr; (b) $0.1M$ K_2CO_3; (c) $0.001M$ $(NH_4)_2CrO_4$; (d) $2.0M$ H_2SO_4; (e) $0.5M$ $HC_2H_3O_2$.

35. What OH^- concentration is needed to commence precipitation of each of the following metals as the hydroxide if each is present in the amount of 1 mg/ml? (a) Cd^{++}; (b) Fe^{++}; (c) Fe^{+++}; (d) Mn^{++}; (e) Ca^{++}.

36. KI is slowly added to a solution that is $0.02M$ in $Pb(NO_3)_2$ and $0.02M$ in $AgNO_3$. (a) Which ion precipitates first? (b) What will be its concentration when the second ion begins to precipitate?

37. What is the pH of a water solution saturated with $Sn(OH)_2$?

38. A 1-ml sample of $0.5M$ K_2CrO_4 is added to 100 ml of a solution saturated with $PbCl_2$. How many milligrams of $PbCrO_4$ will precipitate?

39. What concentration of I^- is needed to start precipitation of silver from a saturated solution of AgCl?

40. An analyst wishes to determine the amount of Pb by precipitation as the sulfate. If his sample contains 40 mg Pb^{++} in 300 ml of solution, and if he adds H_2SO_4 to give a sulfate concentration of $0.3M$, what percentage of his Pb^{++} will remain unprecipitated?

41. What will be the concentration of Hg^{++} remaining in solution when Tl^+ first begins to precipitate as Tl_2S from a solution that was originally $0.1M$ in Hg^{++} and Tl^+?

42. A 100-ml sample of $0.25M$ NH_3 is added to 100 ml of $0.02M$ $Mn(NO_3)_2$ solution. What is the minimum number of grams of NH_4Cl that must be added to prevent the precipitation of $Mn(OH)_2$?

43. What volume of $3M$ NH_3 must be added to 100 ml of a solution containing 1 g of $Mn(NO_3)_2$ and 75 g of NH_4Cl in order to commence the precipitation of $Mn(OH)_2$?

44. How many grams of NaCN must be added to 1 liter of $0.01M$ $Mg(NO_3)_2$ in order to commence precipitation of $Mg(OH)_2$?

45. What will be the molar concentration of the following ions remaining in solutions saturated with H_2S and adjusted to a final pH of 2? (a) Ag^+; (b) Bi^{+++}; (c) Ni^{++}; (d) Mn^{++}; (e) Tl^+.

46. Is it possible to precipitate PbS from a suspension of PbI_2 in $2M$ HCl by saturation with H_2S? Explain.

47. (a) To what final pH must a solution saturated with H_2S be adjusted in order to precipitate SnS without precipitating any FeS? (This solution is originally $0.1M$ with respect to each metal ion.) (b) How much Sn^{++} will be left in solution when the Fe^{++} begins to precipitate?

48. What must be the final pH of an HCl solution in order to just dissolve 0.2 mole FeS in 1 liter of solution?

49. What is the lowest pH that will permit the sulfide of each of the following metals to precipitate from a $10^{-4}M$ solution saturated with H_2S? (a) Tl^+; (b) Cd^{++}; (c) Ni^{++}; (d) Sn^{++}.

50. What volume of saturated Ag_2S solution would you have to take in order to obtain, statistically, one Ag^+ ion? (Neglect hydrolysis of the sulfide ion.)

51. A 100-ml sample of a solution contains 10 g of $NaC_2H_3O_2$, 5 ml of $6M$ $HC_2H_3O_2$, and 50 mg of Mn^{++}. When the solution is saturated with H_2S, how many milligrams of Mn^{++} remain in solution?

52. A solution that is $0.05M$ in Cd^{++}, $0.05M$ in Zn^{++}, and $0.1M$ in H^+ is saturated with H_2S. Calculate the concentration of Zn^{++} and Cd^{++} remaining in solution after saturation. (Do not neglect the H^+ formed on precipitation.)

53. Calculate the solubility of (a) MnS and (b) Ag_2S in water. (Do not neglect hydrolysis of the sulfide ion.)

54. Which will precipitate first when solid Na_2S is slowly added to a $0.02M$ $CdCl_2$ solution, CdS or $Cd(OH)_2$? Explain.

Complex Ions

The last chapter emphasized controlling precipitation by controlling the way in which H^+ ions react with the *negative* ion of the solid. In this chapter we shall turn our attention to control by using the way in which reagents react with the *positive* ion to form a complex ion. A familiar example is the reaction of NH_3 with the Ag^+ in AgCl to form the complex ion, $Ag(NH_3)_2^+$.

All complex ions dissociate to some extent. The diammine silver(I) ion, for example, is in equilibrium with Ag^+ and NH_3,

$$Ag(NH_3)_2^+ \rightleftharpoons Ag^+ + 2\,NH_3$$

and the equilibrium constant,

$$\frac{[Ag^+][NH_3]^2}{[Ag(NH_3)_2^+]} = K_e = K_{inst}$$

is designated as the instability constant (note the formal analogy to the ionization constant for a weak acid). At room temperature the dissociation constant for the diammine silver (I) ion is 6×10^{-8}.

The instability constant is used to compute concentrations of ions and molecules in solution, as shown in the following problem.

PROBLEM:

What is the concentration of silver ion in a solution containing 0.1 mole of $AgNO_3$ in a liter of $0.5M$ ammonia?

SOLUTION:

Since NH_3 is in excess and K_{inst} is small, we know that almost all the silver ion is tied up as the complex. Let $x=[Ag^+]$, and $0.1-x=[Ag(NH_3)_2^+]$. Since each mole of complex contains 2 moles NH_3,

$$[NH_3] = 0.5 - 2(0.1 - x)$$

Substituting these values into the equilibrium expression gives

$$\frac{x[0.5 - 2(0.1 - x)]^2}{0.1 - x} = 6 \times 10^{-8}$$

Neglecting x when subtracted from much larger numbers, this simplifies to

$$\frac{(x)(0.3)^2}{0.1} = 6 \times 10^{-8}$$

Solving,

$$x = 6.7 \times 10^{-8}$$

TABLE 22-1.
Instability Constants of Complex Ions at 25°C

Name*	Equilibrium	K_{inst}
Tetrammine cadmium(II)	$Cd(NH_3)_4^{++} \rightleftarrows Cd^{++} + 4NH_3$	2×10^{-7}
Tetrammine copper(II)	$Cu(NH_3)_4^{++} \rightleftarrows Cu^{++} + 4NH_3$	5×10^{-14}
Diammine silver(I)	$Ag(NH_3)_2^+ \rightleftarrows Ag^+ + 2NH_3$	6×10^{-8}
Tetrammine zinc(II)	$Zn(NH_3)_4^{++} \rightleftarrows Zn^{++} + 4NH_3$	1×10^{-9}
Tetrachloro-mercurate(II)	$HgCl_4^= \rightleftarrows Hg^{++} + 4Cl^-$	1×10^{-16}
Dichloro-argentate(I)	$AgCl_2^- \rightleftarrows Ag^+ + 2Cl^-$	1×10^{-5}
Tetracyano-cadmate(II)	$Cd(CN)_4^= \rightleftarrows Cd^{++} + 4CN^-$	1×10^{-17}
Tetracyano-cuprate(I)	$Cu(CN)_4^\equiv \rightleftarrows Cu^+ + 4CN^-$	2×10^{-27}
Tetracyano-mercurate(II)	$Hg(CN)_4^= \rightleftarrows Hg^{++} + 4CN^-$	4×10^{-42}
Dicyano-argentate(I)	$Ag(CN)_2^- \rightleftarrows Ag^+ + 2CN^-$	4×10^{-19}
Tetrahydroxo-aluminate(III)	$Al(OH)_4^- \rightleftarrows Al(OH)_{3(s)} + OH^-$	2.5×10^{-2}
Tetrahydroxo-chromate(III)	$Cr(OH)_4^- \rightleftarrows Cr(OH)_{3(s)} + OH^-$	1×10^2
Trihydroxo-plumbate(II)	$Pb(OH)_3^- \rightleftarrows Pb(OH)_{2(s)} + OH^-$	50
Trihydroxo-stannate(II)	$Sn(OH)_3^- \rightleftarrows Sn(OH)_{2(s)} + OH^-$	2×10^3
Tetrahydroxo-zincate(II)	$Zn(OH)_4^= \rightleftarrows Zn(OH)_{2(s)} + 2OH^-$	10

*In the names of complex ions, the Roman numeral in parentheses after the name indicates the valence of the metal ion involved. *Di, tri, tetra-,* and so on are used to tell how many coordination partners are attached to the central metal ion. If the complex ion has a positive valence, the central metal ion uses only the name of the element, with no suffix. If the complex ion has a negative valence, the name of the central metal ion is always a common root of the element with the suffix -*ate*. The names of coordination partners usually end in -*o*; the common exception is NH_3 whose name here is *ammine.*

Some of the more commonly used instability constants, all based on experimental measurements, are given in Table 22-1. These constants, together with corresponding K_{sp} values for precipitates, may be used to compute solubilities of precipitates in solutions of complexing agents.

PROBLEM:

When AgCl is treated with an ammonia solution, some of the precipitate dissolves with formation of a complex ion. What weight of AgCl will dissolve in a liter of $1M$ NH_3 solution?

SOLUTION:

The reaction is

$$AgCl_{(s)} + 2NH_3 \rightleftharpoons Ag(NH_3)_2^+ + Cl^-$$

The equilibrium constant for this reaction is

$$\frac{[Ag(NH_3)_2^+][Cl^-]}{[NH_3]^2} = K_e$$

Note that $[AgCl_{(s)}]$, which is constant, is included in the value for K_e and not given in the equation.

We can evaluate K_e by considering the simultaneous equilibria of the solution, just as we previously did for the solution of salts of weak acids by strong acids. We have two equilibrium constants to satisfy:

$$[Ag^+][Cl^-] = K_{sp} = 1.6 \times 10^{-10}$$

$$\frac{[Ag^+][NH_3]^2}{[Ag(NH_3)_2^+]} = K_{inst} = 6 \times 10^{-8}$$

Dividing K_{sp} by K_{inst} gives

$$\frac{[Ag(NH_3)_2^+][Cl^-]}{[NH_3]^2} = \frac{1.6 \times 10^{-10}}{6 \times 10^{-8}} = 2.67 \times 10^{-3}$$

Let $x = [Ag(NH_3)_2^+]$; x is also the concentration of Cl^- ion, since these two ions are formed in equimolar amounts in the reaction. Then, $[NH_3] = 1 - 2x$. Substituting these values, we have

$$\frac{x^2}{(1 - 2x)^2} = 2.67 \times 10^{-3}$$

Taking square roots of both sides of the equation:

$$\frac{x}{1 - 2x} = 0.0517$$

$$x = 0.047 \text{ moles of AgCl dissolving}$$

$$\text{weight of AgCl} = 0.047 \frac{\text{moles}}{\text{liter}} \times 143 \frac{\text{g}}{\text{mole}} = 6.7 \text{ g/liter}$$

We find experimentally that AgI, whose K_{sp} value is 1×10^{-16}, will not appreciably dissolve in a solution of NH_3. Application of equilibrium calculations shows that the $Ag(NH_3)_2^+$ complex does not remove Ag^+ completely enough to compete with AgI for much of the Ag^+ ion.

PROBLEM:

How many grams of AgI will dissolve in a liter of $1M$ NH_3?

SOLUTION:

$$AgI_{(s)} + 2NH_3 \rightleftarrows Ag(NH_3)_2^+ + I^-$$

$$\frac{[Ag(NH_3)_2^+][I^-]}{[NH_3]^2} = K_e = \frac{K_{sp}}{K_{inst}} = \frac{1 \times 10^{-16}}{6.0 \times 10^{-8}} = 1.67 \times 10^{-9}$$

Let x = moles of AgI that dissolve. Then

$$[Ag(NH_3)_2^+] = x$$

$$[I^-] = x$$

$$[NH_3] = 1 - 2x$$

$$\frac{x^2}{(1 - 2x)^2} = 16.7 \times 10^{-10}$$

$$\frac{x}{1 - 2x} = 4.07 \times 10^{-5}$$

$$x = 4.07 \times 10^{-5}$$

grams of AgI per liter $= 4.07 \times 10^{-5} \times 235 = 9.55 \times 10^{-3}$ g

---

Comparison of this value with the AgCl solubility in the preceding example shows that the AgI is so insoluble in NH_3 that we can separate AgI and AgCl almost completely by dissolving the AgCl with NH_3 solution.

Aqueous solutions of the $+2$ and $+3$ metal salts of strong acids are always slightly acid. This would probably seem reasonable to you if you reasoned from our general statement about the hydrolysis of salts on p. 251, assuming that the metal ions come from weak bases (though they are actually "insoluble"). A more satisfactory explanation can be offered in terms of the hydrated aquo metal ions acting as weak acids. For example, hexaaquo iron(III) could dissociate to give

$$Fe(H_2O)_6^{+++} \rightleftarrows Fe(H_2O)_5(OH)^{++} + H^+$$

That the dissociation goes even farther is evident from the small bit of colloidal suspension of ferric hydroxide that is obtained by boiling a solution of $FeCl_3$ (you can see the beam of scattered light even though the solution is perfectly clear). This results from

$$Fe(H_2O)_6^{+++} \rightleftarrows Fe(H_2O)_3(OH)_3 + 3 H^+$$

The acidity, though slight, is easily detected by litmus paper. A similar experiment with $FeCl_2$ would show an appreciably less acid solution. A little thought easily rationalizes this difference: the $Fe(H_2O)_6^{+++}$, which has a much higher positive charge density than $Fe(H_2O)_6^{++}$, will exert more repulsion toward the H^+ on the ligand water molecules, the net result being more dissociation (and greater acidity). If we compared several aquo ions with the same charge, we would usually find that the smallest ion had the most dissociation (because it had the highest charge density) and the largest ion would have the least dissociation. We also find that strong-acid salts of singly charged ions, such as $Na(H_2O)_2^+$ in NaCl solution, have negligible dissociation. Table 22-2 lists values of K_i for a few aquo metal ions. Note that the aquo Al^{+++} ion has about the same strength as acetic acid, and that the aquo Hg^{++} and Fe^{+++} ions are appreciably stronger.

TABLE 22-2.
Ionization Constants of Aquo Metal Ions at 25°C

Dissociation Reaction	K_i
$Al(H_2O)_6^{+++} \rightleftarrows Al(H_2O)_5(OH)^{++} + H^+$	1.26×10^{-5}
$Cu(H_2O)_6^{++} \rightleftarrows Cu(H_2O)_5(OH)^+ + H^+$	1.00×10^{-8}
$Fe(H_2O)_6^{++} \rightleftarrows Fe(H_2O)_5(OH)^+ + H^+$	1.26×10^{-6}
$Fe(H_2O)_6^{+++} \rightleftarrows Fe(H_2O)_5(OH)^{++} + H^+$	3.96×10^{-3}
$Mg(H_2O)_6^{++} \rightleftarrows Mg(H_2O)_5(OH)^+ + H^+$	2.00×10^{-12}
$Hg(H_2O)_6^{++} \rightleftarrows Hg(H_2O)_5(OH)^+ + H^+$	2.00×10^{-3}
$Zn(H_2O)_6^{++} \rightleftarrows Zn(H_2O)_5(OH)^+ + H^+$	2.50×10^{-10}

PROBLEM:
What is the pH of a $0.1M$ $Al(NO_3)_3$ solution?

SOLUTION:
$Al(NO_3)_3$ is the salt of a strong acid, so we do not have to consider the effect of the NO_3^- ion; the pH will be determined by the dissociation of the hexaaquoaluminum(III) ion:

$$Al(H_2O)_6^{+++} \rightleftarrows Al(H_2O)_5(OH)^{++} + H^+$$

Application of the usual equilibrium expression, using the needed value of K_i from Table 22-2, gives

$$K_i = 1.26 \times 10^{-5} = \frac{[Al(H_2O)_5(OH)^{++}][H^+]}{[Al(H_2O)_6^{+++}]}$$

As we did on p. 246, we know that

$$[Al(H_2O)_5(OH)^{++}] = [H^+] = x$$

and

$$[Al(H_2O)_6^{+++}] = 0.1 - x \cong 0.1$$

Substitution gives

$$\frac{x^2}{0.1} = 1.26 \times 10^{-5}$$

$$x^2 = 1.26 \times 10^{-6}$$

$$[H^+] = x = 1.12 \times 10^{-3} = 10^{.05} \times 10^{-3} = 10^{-2.95}$$

$$pH = 2.95$$

Some insoluble hydroxides dissolve not only in acid but also in an excess of strong base. Those that do so are said to be *amphoteric*. This is illustrated by $Al(OH)_3$.

$$Al(OH)_{3(s)} + 3\ H^+ \rightleftarrows Al^{+++} + 3\ H_2O \text{ (in acid)}$$

$$Al(OH)_{3(s)} + OH^- \rightleftarrows Al(OH)_4^- \text{ (in base)}$$

The hydroxyl complex ions formed in this way have instability constants, just as ammonia or other complexes do. These instability constants are somewhat special, in that one of the products of the equilibrium is the insoluble amphoteric hydroxide. Thus, for aluminum hydroxide,

$$Al(OH)_4^- \rightleftarrows Al(OH)_{3(s)} + OH^-$$

the instability constant is

$$K_{inst} = \frac{[OH^-]}{[Al(OH)_4^-]} = 2.5 \times 10^{-2}$$

The $[Al(OH)_{3(s)}]$ is omitted, as usual, because it is a solid. Because the solids are a part of the equilibrium, the instability constants of these hydroxyl complex ions can be applied only to solutions that are saturated with respect to the solid.

PROBLEM:
Solid $Pb(OH)_2$ is added to 1 liter of $1M$ NaOH solution until no more dissolves. What weight of the solid goes into solution?

SOLUTION:
The reaction is

$$Pb(OH)_{2(s)} + OH^- \rightleftarrows Pb(OH)_3^-$$

The equilibrium constant (Table 22-1) is

$$\frac{[OH^-]}{[Pb(OH)_3^-]} = 50$$

Let $x = [Pb(OH)_3^-]$ and $1 - x = [OH^-]$. Substituting these values gives:

$$\frac{1-x}{x} = 50$$

$$x = 0.02 \text{ molar}$$

$$\text{weight of } Pb(OH)_2 = 0.02 \frac{\text{moles}}{\text{liter}} \times 241 \frac{g}{\text{mole}} = 4.8 \text{ g/liter}$$

PROBLEMS A   Answers on page 352

1. What is the concentration of Cu^{++} in a solution made by diluting 0.1 mole of $CuSO_4$ and 2 moles of NH_3 to 500 ml?

2. K_{sp} for MX (mol wt = 80) is 10^{-10}; K_{inst} for $M(NH_3)_2^+$ is 10^{-8}. If 0.1 mole of MNO_3 is added to 1 liter of $2M$ NH_3, what are the concentrations of (a) $M(NH_3)_2^+$; (b) NH_3; (c) M^+; (d) NO_3^-; (e) H^+; (f) OH^-?

3. If solid MX (see Problem 2) is added to 1 liter of $0.5M$ NH_3 solution, how many grams dissolve?

4. How many grams of AgBr will dissolve in 500 ml of $2M$ NH_3?

5. What must be the final NH_3 concentration to dissolve 2 g of AgCl in 250 ml of solution?

6. What must be the concentration of an HCl solution to dissolve 1 mg of AgCl in 100 ml of solution?

7. How many grams of HgS will dissolve in 100 ml of $1M$ NaCN? (Neglect hydrolysis of $S^=$ and CN^-.)

8. What must be the NH_3 concentration in a $0.1M$ $AgNO_3$ solution to prevent the precipitation of AgCl when enough NaCl has been added to raise its concentration to $0.5M$?

9. A 100-ml sample of a solution containing 5 mg of Cd^{++} was treated with 10 ml of $15M$ NH_3. Calculate the Cd^{++} concentration in the resulting solution.

10. What is the minimum concentration of CN^- that will prevent the precipitation of AgI from a solution that is $0.1M$ in I^- and $0.01M$ in Ag^+?

11. What must be the concentration of an NH_3 solution if 100 ml of it must dissolve 3 g of AgCl together with 3 g of $AgNO_3$?

12. Calculate the pH of (a) a $0.1M$ $Cu(NO_3)_2$ solution, and (b) a $0.1M$ $Fe(NO_3)_3$ solution.

13. The following questions involve the typical amphoteric hydroxides, $Pb(OH)_2$ and $Sn(OH)_2$. (a) Calculate the concentrations of Pb^{++},

$Pb(OH)_3^-$, and OH^- in a water solution saturated with $Pb(OH)_2$. (b) Calculate the Pb^{++} and $Pb(OH)_3^-$ concentrations in a solution which is saturated with $Pb(OH)_2$ and whose equilibrium NaOH concentration is $1.0M$. (c) Calculate the Sn^{++} and $Sn(OH)_3^-$ concentrations that result from the addition of excess solid $Sn(OH)_2$ to 500 ml of $0.5M$ NaOH. (d) What is the minimum volume of $0.5M$ NaOH needed to dissolve 0.1 mole of $Pb(OH)_2$?

PROBLEMS B   No answers given

14. What is the concentration of Cd^{++} in a solution made by diluting 0.1 mole of $CdSO_4$ and 2 moles of NH_3 to 250 ml?

15. K_{sp} for MX (mol wt $= 125$) is 10^{-12}; K_{inst} for $M(NH_3)_4^+$ is 10^{-10}. If 0.1 mole of MNO_3 is added to 500 ml of $3M$ NH_3 solution, what are the concentrations of (a) $M(NH_3)_2^+$; (b) NH_3; (c) M^+; (d) NO_3^-; (e) H^+; (f) OH^-?

16. If solid MX (see Problem 15) is added to 500 ml of $2M$ NH_3 solution, how many grams dissolve?

17. How many grams of AgBr will dissolve in 250 ml of a $6M$ NH_3 solution?

18. What must be the final NH_3 concentration to dissolve 5 g of AgCl in 500 ml of solution?

19. What must be the concentration of $CaCl_2$ solution to dissolve 3 g of AgCl in 500 ml of solution?

20. How many grams of $Cu(OH)_2$ will dissolve in 100 ml of $6M$ NH_3?

21. What must be the NH_3 concentration in order to prevent the precipitation of AgCl if the solution is $0.25M$ in $AgNO_3$ and $2M$ in NaCl?

22. A 100-ml sample of a solution containing 10 mg Cu^{++} is treated with 50 ml of $3M$ NH_3. Calculate the Cu^{++} concentration in the resulting solution.

23. What is the minimum concentration of CN^- that will prevent the precipitation of $Hg(OH)_2$ from a solution that is $0.5M$ in OH^- and $0.1M$ in Hg^{++}? [K_{sp} for $Hg(OH)_2$ is 2×10^{-22}.]

24. What volume of $6M$ NH_3 must be added to 250 ml of $0.1M$ $AgNO_3$ to prevent the precipitation of AgCl when 4 g of NaCl are added?

25. Calculate the pH of (a) a $0.1M$ $Mg(NO_3)_2$ solution, and (b) a $0.1M$ $Hg(NO_3)_2$ solution.

26. Answer the following questions involving the typical amphoteric hydroxides, $Zn(OH)_2$ and $Cr(OH)_3$. (a) Calculate the Zn^{++}, $Zn(OH)_4^=$, and OH^- concentrations in a water solution saturated with $Zn(OH)_2$. (b) Calculate the Cr^{+++} and $Cr(OH)_4^-$ concentrations that result when an excess of solid $Cr(OH)_3$ is added to 500 ml of $1.5M$ NaOH. (c) A saturated

$Zn(OH)_2$ solution is shown to be $0.25M$ in NaOH. Calculate the concentration of Zn^{++} and $Zn(OH)_4^=$ ions. (d) An excess of solid $Zn(OH)_2$ is added to 100 ml of $1M$ NaOH. How many moles of $Zn(OH)_2$ are dissolved? (e) Compare the minimum volume of $1M$ NH_3 required to dissolve 0.1 mole of $Zn(OH)_2$ to give $Zn(NH_3)_4^{++}$ with the minimum volume of $1M$ NaOH required to dissolve 0.1 mole of $Zn(OH)_2$ to give $Zn(OH)_4^=$.

27. It is proposed, by treatment with the minimum amount of NH_3 necessary to dissolve the AgCl, to separate a solid mixture of 0.50 g of AgCl and 0.75 g of AgBr. (a) What must be the NH_3 concentration in the final reaction mixture if all the AgCl is dissolved in a total volume of 200 ml? (b) What volume of $1M$ NH_3 should be added to the solids to just dissolve the AgCl? (c) How many grams of AgBr would also dissolve if the separation were performed as in (b)?

28. A 2.0000-g mixture of solids containing $NaNO_3$, NaCl, and NaBr was dissolved in water, and the Cl^- and Br^- precipitated with an excess of $AgNO_3$. This silver halide precipitate, which weighed 2.5000 g, was treated with 200 ml of $0.9M$ NH_3 that dissolved the AgCl. The remaining residue of AgBr weighed 0.9000 g. (a) Calculate the percentage of NaCl and NaBr in the original mixture. (b) By what percentage are the analyses in (a) in error as a result of the solution of some AgBr by the NH_3?

Thermochemistry

Thermochemistry is the study of energy changes in matter. The changes usually discussed in general chemistry are (1) changes in temperature, (2) changes in physical state, and (3) chemical reactions.

Changes in Temperature

If heat is applied to a substance, the temperature is raised; if heat is withdrawn, the temperature is lowered. The unit of heat is the calorie (cal), which is defined as the quantity of heat required to raise the temperature of 1 g of water 1 degree Celsius. In precise work we use the 15° calorie, the heat needed to raise 1 g of H_2O from 14.5°C to 15.5°C, or the mean calorie, $\frac{1}{100}$ of the heat needed to raise 1 g of H_2O from 0°C to 100°C.

The number of calories required to raise the temperature of an object 1°C is called the *heat capacity* of the object. The *molar heat capacity* is the number of calories needed to raise the temperature of a mole 1°C. The *atomic heat capacity* is the heat required to raise 1 g-atom of an element 1°C. The *specific heat* of any substance is the number of calories required to raise one gram 1°C. From our definition of the calorie it follows that the specific heat of water is 1 cal per gram-degree.

PROBLEM:

The specific heat of Fe_2O_3 is 0.151 cal per gram degree. How much heat is needed to raise the temperature of 200 g of Fe_2O_3 from 20°C to 30°C? What is the molar heat capacity of Fe_2O_3?

SOLUTION:

Since 0.151 cal raises 1 g of Fe_2O_3 1°C,

$$\text{total calories} = 200 \text{ g} \times \frac{0.151 \text{ cal}}{\text{g-degree}} \times \text{(temp change)}$$

$$= 200 \text{ g} \times \frac{0.151 \text{ cal}}{\text{g-degree}} \times 10° = 302 \text{ cal}$$

The molar heat capacity is the product of specific heat × mole weight in grams:

$$\text{molar heat capacity} = \frac{159.6 \text{ g}}{\text{mole}} \times \frac{0.151 \text{ cal}}{\text{g-degree}} = \frac{24 \text{ cal}}{\text{mol degree}}$$

Calorimetry

The amount of heat that is absorbed or liberated in a physical or chemical change can be measured in a well-insulated vessel called a calorimeter, such as that shown in Figure 23-1. Calorimetry is based on the principle that the observed temperature change which results from a chemical reaction can be simulated with an electrical heater. The electrical measurements of current (I), heater resistance (R), and duration (t) of heating make it possible to calculate how much heat is equivalent to the amount produced by the chemical change, using the formula $I^2Rt/4.184$. By using weighed quantities of reactants, one can calculate the heat change per gram or per mole.

The formula is derived as follows. Electrical current, measured in amperes, is the *rate* of flow of electrical charge (coulombs); by definition it is

$$\text{amperes} = \frac{\text{coulombs}}{\text{seconds}}$$

The common relationship between volts (E), amperes (I), and resistance (R) is known as Ohm's Law:

$$E = IR$$

Electrical energy is given by

$$\text{energy} = \text{(volts)(amperes)(seconds)} = \text{(volts)}\left(\frac{\text{coulombs}}{\text{second}}\right)\text{(seconds)}$$

$$= EIt = I^2Rt \text{ volt coulombs}$$

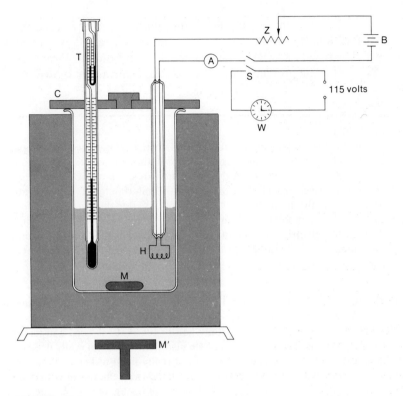

FIGURE 23-1
A calorimeter.

By definition,

1 joule = 1 volt coulomb

1 calorie = 4.184 joules (the so-called mechanical equivalent of heat)

By combining these laws and definitions, we get

$$\text{electrical energy} = \frac{I^2 R t}{4.184} \text{ calories}$$

The insulation surrounding the calorimeter minimizes heat loss or gain through the walls. The cover (C) supports the electrical heater (H); it has two holes, one to permit the insertion of a sensitive differential thermometer (T), and one through which reactants can be added. The solution is stirred by a Teflon-covered magnet (M) that is rotated by the motor-driven magnet (M'). The electrical heating is controlled and measured as follows. The double-pole switch (S) controls two things simultaneously: the timer (W), and the battery (B), which supplies a constant current to the immersion heater (H). The current (I) that flows through the heater is read from the ammeter (A), and the duration (t) of time that the current flows is read from the timer. The resistance (R) of the heater is known from a separate measurement. The following problems are based on the calorimeter shown in Figure 23-1.

PROBLEM:

A calorimeter contained 200 ml of 0.100M NaCl solution at room temperature (25.038°C according to the thermometer). With the magnetic stirrer going, exactly 10.00 ml of 1.000M AgNO$_3$ solution (also at room temperature) was added dropwise through the porthole of the calorimeter. The temperature, as noted by the thermometer, rose to 25.662°C. After the contents were cooled back to room temperature, a current of 0.700 ampere, passed for 3 min 26 sec through the electrical heater (whose resistance was 6.50 ohms), again raised the temperature to 25.662°C. Calculate the number of calories produced when one mole of AgCl precipitates from aqueous solution.

SOLUTION:

You will note first that in the calorimeter there is (0.200 liter) (0.100 mole/liter) =0.0200 mole of NaCl, and that we are adding only (0.0100 liter)(1.00 mole/liter) = 0.0100 mole of AgNO$_3$; i.e., only half of the NaCl is used and just 0.0100 mole of AgCl is formed. You will note also that we do not really need to know the intitial and final temperatures; all we really need to know is that the electrical heater was operated between the same two temperatures that were involved in the chemical change. The amount of heat generated by the electrical heater in 3 min 26 sec (i.e., 206 sec) is

$$\text{energy} = \frac{(0.700 \text{ amp})^2(6.50 \text{ ohms})(206 \text{ sec})}{4.184} = 156.8 \text{ cal}$$

$$\text{energy per mole} = \frac{156.8 \text{ cal}}{0.0100 \text{ mole}} = 15{,}680 \frac{\text{cal}}{\text{mole}} = 15.68 \frac{\text{kcal}}{\text{mole}}$$

If you want to measure the specific heat of a liquid, you need to know only the electrical energy needed to heat a known weight of the liquid and the measured temperature change, but you must use a *calibrated* calorimeter so that a correction can be made for the amount of electrical energy that was absorbed by the calorimeter walls rather than by the liquid.

PROBLEM:

A calorimeter requires a current of 0.800 amp for 4 min 15 sec to raise the temperature of 200.0 ml of H$_2$O by 1.100°C The same calorimeter requires 0.800 amp for 3 min 5 sec to raise the temperature of 200.0 ml of another liquid (whose density is 0.900 g/ml) by 0.950°C. The heater resistance is 6.50 ohms. Calculate the specific heat of the liquid.

SOLUTION:
The total heat energy produced by the electrical heater in water was used to raise the temperature of the water and the calorimeter walls by 1.100°C; it is

$$\text{total energy} = \frac{(0.800 \text{ amp})^2(6.50 \text{ ohms})(255 \text{ sec})}{4.184} = 254 \text{ cal}$$

The energy required to raise just the water by 1.100°C (assuming the density and specific heat of water are both 1.000) is

$$\text{energy for water} = (200.0 \text{ g})\left(1.000 \frac{\text{cal}}{\text{g deg}}\right)(1.100 \text{ deg}) = 220 \text{ cal}$$

The energy required to raise the temperature of the calorimeter walls is the difference between the total energy and that required for the water; i.e., it is 254 − 220 = 34 cal. The heat capacity of the calorimeter, i.e., calories required to raise that part of its walls in contact with the liquid by 1°C, is

$$\text{heat capacity of calorimeter} = \frac{34 \text{ cal}}{1.10 \text{ deg}} = 31 \frac{\text{cal}}{\text{deg}}$$

The total electrical energy produced when the heater was in the liquid is

$$\frac{(0.800 \text{ amp})^2(6.50 \text{ ohms})(185 \text{ sec})}{4.184} = 184 \text{ cal}$$

Of this total amount, the part required to raise the calorimeter walls by 0.950°C is

$$\text{energy for calorimeter} = (31 \text{ cal/deg})(0.950 \text{ deg}) = 29 \text{ cal}$$

The difference between 184 cal and 29 cal is 155 cal; this is the amount required to raise the temperature of the liquid by 0.95°C. The amount of heat required to raise 1 g of the liquid by 1°C, i.e., its *specific heat*, is

$$\text{specific heat of liquid} = \frac{155 \text{ cal}}{(200 \text{ ml})(0.900 \text{ g/ml})(0.950 \text{ deg})}$$

$$= 0.910 \frac{\text{cal}}{\text{g deg}}$$

Many years ago Dulong and Petit observed that the atomic heat capacity for most solid elements is approximately 6.2 cal/g-atom deg; i.e., the number of calories required to raise one mole of an element by 1°C is given by

$$\left(\frac{\text{g}}{\text{g-atom}}\right)\left(\frac{\text{cal}}{\text{g deg}}\right) = 6.2 \frac{\text{cal}}{\text{g-atom deg}}$$

This relation provides a simple way to find the approximate values of the atomic weights of solid elements. For example, if you had an unknown solid element in a finely divided state, you could put a weighed sample of it into your calibrated calorimeter and quickly find its specific heat; then, using the Dulong and Petit rule, you could find its approximate atomic weight.

PROBLEM:

A 50.0-g sample of a finely divided metal, insoluble and unreactive to water, was put into 200.0 ml of water in the calorimeter that was calibrated in the last problem. A current of 0.800 amp was passed for 15 min 50 sec in order to raise the temperature by 4.00°C. What is the specific heat of the metal?

SOLUTION:

The total electrical energy required to raise the water, the calorimeter, and the metal by 4.00°C is

$$\frac{(0.800 \text{ amp})^2(6.50 \text{ ohms})(950 \text{ sec})}{4.184} = 945 \text{ cal}$$

Together, the water and the calorimeter require 231 cal/deg, so the total energy required by them for 4.00°C is

$$\left(231 \frac{\text{cal}}{\text{deg}}\right)(4.00 \text{ deg}) = 924 \text{ cal}$$

The difference, $945 - 924 = 21$ cal, is required to raise the 50.0 g of metal by 4.00°C. Therefore, the specific heat of the metal is

$$\frac{21.0 \text{ cal}}{(50.0 \text{ g})(4.00 \text{ deg})} = 0.105 \frac{\text{cal}}{\text{g deg}}$$

$$\text{approximate atomic weight of the metal} = \frac{6.2}{0.105} = 59 \frac{\text{g}}{\text{g-atom}}$$

An alternate method of calorimetry that gives less accurate results but is simpler in concept uses only a single insulated container and a thermometer. Temperature changes in the calorimeter are brought about by adding hot (or cold) objects of known weight and temperature. Calculations are based on the principle that the heat lost by the added hot object is equal to that gained by the water in the calorimeter and the calorimeter walls. This simple approach is illustrated in the next two problems.

PROBLEM:

The temperature in a calorimeter containing 100 g of water is 22.7°C. Fifty grams of water are heated to boiling, 99.1°C at this location, and quickly poured into the calorimeter. The final temperature is 44.8°C. From these data calculate the heat capacity of the calorimeter.

SOLUTION:

The heat loss from the hot water is equal to the heat gain by the calorimeter and the water initially in it.

$$\text{heat lost by hot water} = (\text{wt of } H_2O)(\text{sp ht of } H_2O)(\text{temp change})$$

$$= (50 \text{ g})\left(1 \frac{\text{cal}}{\text{g deg}}\right)(99.1° - 44.8°)$$

$$= 2,715 \text{ cal}$$

heat gained by calorimeter water $=$ (wt of H_2O)(sp ht of H_2O)(temp change)

$$= (100 \text{ g})\left(1 \frac{\text{cal}}{\text{g deg}}\right)(44.8° - 22.7°)$$

$$= 2{,}210 \text{ cal}$$

heat gained by calorimeter $=$ (ht capacity of calorimeter)(temp change)

$$= (X \frac{\text{cal}}{\text{deg}})(44.8° - 22.7°)$$

$$= 22.1 \, X \text{ cal}$$

We equate the heat lost to the heat gained and solve for X, the heat capacity of the calorimeter:

$$2{,}715 \text{ cal} = 2{,}210 \text{ cal} + 22.1 \, X \text{ cal}$$

$$X = \frac{2{,}715 - 2{,}210}{22.1} = 22.8 \frac{\text{cal}}{\text{deg}}$$

PROBLEM:
The calorimeter of the preceding problem is used to measure the specific heat of a metal sample. A 100-g sample of water is put into the calorimeter at a temperature of 24.1°C. A 45.32-g sample of metal filings is put into a dry test tube that is immersed in a bath of boiling water until the metal is at the temperature of the latter, 99.1°C. The hot metal is then quickly poured into the calorimeter and the water stirred by a thermometer that is read at frequent intervals until the temperature reaches a maximum of 27.6°C. Compute the specific heat of the metal.

SOLUTION:
The heat lost by the metal sample is equal to the heat gained by the calorimeter.

heat gained by water $=$ (wt of H_2O)(sp ht of H_2O)(temp change)

$$= (100 \text{ g})(1 \frac{\text{cal}}{\text{g deg}})(27.6° - 24.1°)$$

$$= 350 \text{ cal}$$

heat gained by calorimeter $=$ (ht capacity of calorimeter)(temp change)

$$= \left(22.8 \frac{\text{cal}}{\text{deg}}\right)(27.6° - 24.1°)$$

$$= 80 \text{ cal}$$

total heat gained $= 350 \text{ cal} + 80 \text{ cal}$

$$= 430 \text{ cal}$$

heat lost by metal $=$ (wt of metal)(sp ht of metal)(temp change)

$$430 \text{ cal} = (45.32 \text{ g})\left(X \frac{\text{cal}}{\text{g deg}}\right)(99.1° - 27.6°)$$

$$X = \frac{430 \text{ cal}}{(45.32 \text{ g})(71.5 \text{ deg})}$$

$$= 0.133 \, \frac{\text{cal}}{\text{g deg}}$$

Changes in State

When substances melt or vaporize, they absorb energy but do not change temperature. Instead, this energy is used to overcome the mutual attraction of the molecules or ions and permit them to move more independently than they could in their former states; the new state with its added energy always has less molecular order. For example, liquid water at 0° is a less-ordered state than crystalline water at 0°, and water vapor at 100° is chaotic in its molecular organization compared to liquid water at 100°.

We can make these statements quantitative by saying that at a given temperature and pressure the molecules in "state 2" (say, vapor) have a heat content of H_2, in "state 1" (say, liquid) they have a heat content of H_1, and that the "heat of transition" (in this case, vaporization) is simply the "change in heat content" (ΔH_T):

$$\Delta H_T = H_2 - H_1 = \text{heat of transition}$$

Two heats of transition frequently used in chemistry are

$$\Delta H_f = \text{heat of fusion (melting) of ice at } 0°C = 1,435 \, \frac{\text{cal}}{\text{mole}}$$

$$\Delta H_v = \text{heat of vaporization of water at } 100°C = 9,713 \, \frac{\text{cal}}{\text{mole}}$$

To avoid the use of the rather ambiguous term "heat" in connection with "heat content," it is customary to use the term *enthalpy,* and correspondingly, the heat changes associated with chemical and physical changes at constant pressure are called changes in enthalpy; ΔH_T=enthalpy of transition, and 1,435 cal/mole and 9,713 cal/mole are the respective enthalpies of fusion and vaporization for water.

Energy is also involved in transitions from one allotropic form to another, or from one crystal form to another. To change a mole of red phosphorus to yellow phosphorus, we must supply 4.22 kilocalories ($\Delta H_T = +4.22$ kcal/mole), and when 1 mole of yellow silicon disulfide changes to white silicon disulfide, 3.11 kcal is liberated ($\Delta H_T = -3.11$ kcal/mole).

In the following problem we apply the principles involved in both specific heat and heats of transition.

PROBLEM:
What is the resulting temperature if 36 grams of ice at 0°C are put into 200 g of H_2O at 25°C contained in the calibrated calorimeter used in the last problem?

SOLUTION:
The heat required to melt the ice is supplied by the water and calorimeter walls, which are, of course, cooled. Let T be the final temperature; then

$$\text{heat required to melt} \atop \text{the ice at } 0° = \frac{(36 \text{ g})\left(1435 \dfrac{\text{cal}}{\text{mole}}\right)}{\left(18 \dfrac{\text{g}}{\text{mole}}\right)} = 2{,}870 \text{ cal}$$

$$\text{heat required to raise} \atop 36 \text{ g of } H_2O \text{ from } 0° \text{ to } T° = (36 \text{ g})\left(1 \dfrac{\text{cal}}{\text{g deg}}\right)(T \text{ deg}) = 36T \text{ cal}$$

$$\text{total calories required} = 2870 + 36T$$

$$\text{heat lost by 200 g of } H_2O \atop \text{in cooling from } 25° \text{ to } T° = (200 \text{ g})\left(1 \dfrac{\text{cal}}{\text{g deg}}\right)[(25 - T)\text{deg}]$$

$$\text{heat lost by calorimeter walls} \atop \text{in cooling from } 25° \text{ to } T° = \left(31 \dfrac{\text{cal}}{\text{deg}}\right)[(25 - T)\text{deg}]$$

$$\text{total calories lost} = (231)(25 - T)$$

If we go on the principle that

$$\text{total calories required} = \text{total calories lost}$$

then

$$2{,}870 + 36T = (231)(25 - T)$$

$$267T = 2{,}905$$

$$T = \frac{2{,}905}{267} = 10.9°C, \text{ resultant temperature}$$

The enthalpy of transition divided by the absolute temperature at which it occurs is a common way to describe the change in molecular order that occurs during the transition. We refer to this change in molecular order as the "change in entropy" (ΔS_T), or the "entropy of transition." If ΔS_T is positive, the change results in an increase in molecular *dis*order. Changes in state offer some of the simplest examples from which one can obtain a feeling for the relationship between changes in entropy and changes in molecular order. Crystals have a very high degree of order; in them, the movement of atoms, ions, or molecules is restricted primarily to vibration about their locations in the crystalline lattice. When crystals melt, the component atoms, ions, or molecules can move fairly independently of each other in the liquid, and slowly change their neighbors by diffusion; the molecular order represented by the lattice disappears. When liquids vaporize, the component atoms or molecules, now in the gaseous phase, move about very independently in a very chaotic, random manner. Each stage, melting and vaporization, represents an increase in molecular chaos, and is described in terms of an increase in entropy.

Just as ΔH_T represents the difference in enthalpies between "state 2" and "state 1," so does ΔS_T represent the difference between the entropies in "state 2" and "state 1";

$$\Delta S_T = S_2 - S_1$$

$$\Delta S_T = \frac{\Delta H_T}{T} = \frac{cal}{mole\ deg}$$

For water,

$$\Delta S_f = \frac{1435\ \dfrac{cal}{mole}}{273\ deg} = 5.26\ \frac{cal}{mole\ deg} = \text{entropy of fusion at } 0°C$$

$$\Delta S_v = \frac{9713\ \dfrac{cal}{mole}}{373\ deg} = 26.0\ \frac{cal}{mole\ deg} = \text{entropy of vaporization at } 100°C$$

For very many liquids, the entropy of vaporization at the normal boiling point is approximately 21 cal/mole deg; water is not typical. The units for changes in entropy are the same as for molar specific heats, but they should not be confused with each other. In order to avoid such confusion later, note now that this method of calculating ΔS from $\Delta H/T$ is valid only under equilibrium conditions. For transitions, for example, this method can be used only at temperatures where the two phases in question can coexist in equilibrium with each other.

Enthalpies of Reaction

Most reactions either liberate or absorb heat. To say that heat is liberated means that the atoms, in the molecular arrangement they have as products, must possess less energy than they did in their arrangement as reactants, and that this difference in energy is evolved as heat; the reaction is exothermic. When it is important to show this heat change, one way to do so is to include it as part of the chemical equation, as illustrated by the burning of methane gas:

$$CH_{4(g)} + 2O_{2(g)} \rightarrow CO_{2(g)} + 2H_2O_{(l)} + 212,800\ cal$$

Another, more useful, way is to say that the enthalpy of the reactants ("state 1") is H_1, that the enthalpy of the products ("state 2") is H_2, and that the "heat of reaction" is simply the "change in enthalpy" (ΔH):

$$\Delta H = H_2 - H_1 = \text{heat of reaction} = \text{enthalpy of reaction}$$

The actual amount of heat we measure experimentally for a given reaction depends somewhat on (a) the temperature of the experiment and (b) whether it is run at constant volume or constant pressure. The basic reasons for this are that: (a) each reactant and product has a characteristic specific heat that varies individualistically with temperature; and (b) at constant pressure some of the heat of reaction may expand or compress gases if they are not confined to a fixed volume. We shall avoid these complications here by saying that $\Delta H°$ refers only to changes occurring at 25°C and at a constant pressure of 1 atm; that is, the reactants and the products are in their *standard states*.

Also, ΔH will always be negative for an exothermic reaction, because the products collectively have a smaller enthalpy than the reactants. For the burning of CH_4,

$$CH_{4(g)} + 2O_{2(g)} \rightarrow CO_{2(g)} + 2H_2O_{(l)} \qquad \Delta H° = -212{,}800 \text{ cal}$$

For an endothermic reaction, ΔH is +. For example,

$$2HCl_{(g)} \rightarrow H_{2(g)} + Cl_{2(g)} \qquad \Delta H° = +44{,}120 \text{ cal}$$

If we had written the previous equation in the reverse order, the sign of ΔH would have been negative. The positive sign means that it takes energy to decompose HCl, and the negative sign means that energy is liberated when HCl is formed from the elements, H_2 and Cl_2:

$$\tfrac{1}{2}H_{2(g)} + \tfrac{1}{2}Cl_{2(g)} \rightarrow HCl_{(g)} \qquad \Delta H° = -22{,}060 \text{ cal}$$

When the reaction produces a compound from elements in their common physical state at 25°C and 1 atm, the value of $\Delta H°$ is called the standard *enthalpy of formation*. The standard enthalpy of formation of 1 mole of HCl is $\Delta H° = -22.06$ kcal/mole. The standard enthalpies of formation of hundreds of compounds have been determined and are listed in tables in chemistry handbooks. A few are listed here in Table 23-1. The standard enthalpy of formation of all *elements*, in their common form at 25°C and 1 atm pressure, is assumed to be zero.

One real value of tables of standard enthalpies of formation is that they permit the *calculation* of the standard enthalpy of any reaction for which all the reactants and products are listed; it is not necessary to do an experimental measurement. The basic premise of the use of the tables is that the enthalpy of reaction is the difference between the sum of the enthalpies of formation of the products and the sum of the enthalpies of formation of the reactants. That is,

$\Delta H° = \Sigma$ (standard enthalpies of formation of products)
$\qquad\qquad - \Sigma$ (standard enthalpies of formation of reactants)

$\qquad = \Sigma(\Delta H°_f)_{\text{products}} - \Sigma(\Delta H°_f)_{\text{reactants}}$

TABLE 23-1.
Standard Enthalpies of Formation at 25°C and 1 atm (ΔH_f° in kcal/mole)

All elements in normal state		0.00	C_6H_6	l	11.72	N (atom)	g	85.56
Ag$^+$	aq[a]	25.31	C_6H_6	g	19.82	Na (atom)	g	25.98
AgCl	s	−30.36	Cu^{++}	aq	15.49	Na$^+$	aq	−57.30
Br (atom)	g	26.71	F (atom)	g	18.30	NH_3	g	−11.04
Br$^-$	aq	−28.90	F$^-$	aq	−78.66	NH_3	aq	−19.32
BrCl	g	3.51	FeO	s	−63.70	NH_4^+	aq	−31.74
C (atom)	g	171.70	Fe_2O_3	s	−196.50	NO	g	21.60
C (diamond)	s	0.45	H (atom)	g	52.09	NO_2	g	8.09
Ca^{++}	aq	−129.77	H$^+$	aq	0.00	NO_3^-	aq	−49.37
Cd^{++}	aq	−17.30	HBr	g	−8.66	N_2O	g	19.49
Cl (atom)	g	29.01	HCl	g	−22.06	O (atom)	g	59.16
Cl$^-$	aq	−40.02	HI	g	6.20	OH$^-$	aq	−54.96
ClO_4^-	aq	−31.41	H_2O	l	−68.32	P (atom)	g	75.18
CO	g	−26.42	H_2O	g	−57.80	PCl_3	g	−73.22
CO_2	g	−94.05	H_2S	g	−4.82	PCl_5	g	−95.35
CH_3OH	l	−57.02	I (atom)	g	25.48	S (atom)	g	53.25
CH_3OH	g	−48.08	I$^-$	aq	−13.37	S$^=$	aq	10.00
C_2H_5OH	l	−66.36	ICl	g	4.20	SO_2	g	−70.96
C_2H_5OH	g	−56.24	K$^+$	aq	−60.04	$SO_4^=$	aq	−216.90
C_4H_{10}	g	−29.81	Li$^+$	aq	−66.54	Zn^{++}	aq	−36.43

[a] The symbol (aq) indicates a very dilute aqueous solution.

In the following problems we first write the chemical equation. Then, below each substance we write its standard enthalpy of formation, multiplied by the number of moles of the substance used in the balanced equation. The standard enthalpy of reaction is the difference between the sum of the enthalpies of formation of the products and the sum of the enthalpies of formation of the reactants.

PROBLEM:
Compute the standard enthalpy of reaction for the gaseous dissociation of PCl_5 into PCl_3 and Cl_2.

SOLUTION:
We write the balanced chemical equation and take the needed values of ΔH_f° from Table 23-1.

$$PCl_{5(g)} \rightleftarrows PCl_{3(g)} + Cl_{2(g)}$$
$$(1)(-95.35) \quad (1)(-73.22) \quad (1)(0.00)$$

$$\Delta H_f^\circ \text{ of products} = (1)(-73.22) = -73.22 \text{ kcal}$$
$$\text{plus } (1)(0.00) \quad = \underline{\quad 0.00 \text{ kcal}}$$
$$= -73.22 \text{ kcal}$$

$$\Delta H_f^\circ \text{ of reactants} = (1)(-95.35) = -95.35 \text{ kcal}$$

$$\Delta H^\circ \text{ of reaction} = (-73.22) - (-95.35) = +22.13 \text{ kcal}$$

The positive sign of the answer indicates that the reaction is endothermic and that the dissociation at 25°C requires 22,130 calories/mole.

PROBLEM:
What is the standard enthalpy of combustion of ethyl alcohol, C_2H_5OH?

SOLUTION:
Take the needed standard enthalpies of formation from Table 23-1, and use them with the balanced equation

$$C_2H_5OH_{(l)} + 3\ O_{2(g)} \rightarrow 2\ CO_{2(g)} + 3\ H_2O_{(l)}$$
$$(1)(-66.36) \quad (3)(0) \qquad (2)(-94.05) \quad (3)(-68.32)$$

$$\Delta H_f^\circ \text{ of products} = (2)(-94.05) = -188.10 \text{ kcal}$$
$$\text{plus } (3)(-68.32) = \underline{-204.96 \text{ kcal}}$$
$$= -393.06 \text{ kcal}$$

$$\Delta H_f^\circ \text{ of reactants} = (1)(-66.36) = -66.36 \text{ kcal}$$
$$\text{plus } (3)(0.00) \quad = \underline{\quad 0.00 \text{ kcal}}$$
$$= -66.36 \text{ kcal}$$

$$\Delta H^\circ \text{ of reaction} = (-393.06) - (-66.36) = -326.70 \text{ kcal}$$

The reaction is exothermic.

Enthalpies of combustion are relatively simple to determine, and they are often used to find other energy values that are very difficult or impossible to determine directly. For example, the enthalpy of formation of methyl alcohol corresponds to the reaction

$$2\ H_{2(g)} + C_{(s)} + \tfrac{1}{2}O_{2(g)} \rightarrow CH_3OH_{(l)},$$

yet it is impossible to make CH_3OH directly from the elements. However, we can burn CH_3OH in an excess of O_2 in a bomb calorimeter and measure the heat produced ($\Delta H^\circ = -173.67$ kcal/mole) in the reaction

$$CH_3OH_{(l)} \quad + \quad \tfrac{3}{2}O_{2(g)} \rightarrow \quad CO_{2(g)} + \quad 2H_2O_{(l)}$$
$$(1)(\Delta H_{CH_3OH}) \quad (1.5)(0.00) \quad (1)(-94.05) \quad (2)(-68.32)$$

$$\Delta H_f^\circ \text{ of products} = (1)(-94.05) + (2)(-68.32) = -230.69 \text{ kcal}$$

$$\Delta H_f^\circ \text{ of reactants} = (1)(\Delta H_{CH_3OH}^\circ) + (1.5)(0.00) = \Delta H_{CH_3OH}^\circ$$

$$\Delta H° \text{ of reaction} = -173.67 = (-230.69) - (\Delta H°_{CH_3OH})$$

$$\Delta H°_{CH_3OH} = (-230.69) + (+173.67) = -57.02 \text{ kcal/mole}$$

$$= \text{standard enthalpy of formation of } CH_3OH_{(l)}$$

You will note that Table 23-1 also contains standard enthalpies of formation for ions in aqueous solution. It is worth noting here that calorimetry was a strong argument favoring the view that all strong acids and bases exist in dilute water solution only as ions and not as molecules. No matter which combination of strong acid and base is used in a neutralization reaction, the heat value obtained is always very close to the value $\Delta H° = -13.36$ kcal per mole of water formed. The implication, of course, is that although the *chemicals are different* in each case, the *reaction is the same;* i.e., it must be

$$H^+_{(aq)} + OH^-_{(aq)} \rightarrow H_2O_{(l)} \qquad \Delta H° = -13.36 \text{ kcal}$$

If weak acids or bases are used, the observed heat values are always less and quite variable. In essence, the reaction is also the same for the weak ones, except that some of the enthalpy of the reaction (the -13.36 kcal) must be used to remove the H^+ or OH^- from the weak acids or bases. The subscript *aq* refers to the fact that the substance in question is in dilute aqueous solution; $\Delta H°_f$ for $H^+_{(aq)}$ in Table 23-1 has also been set arbitrarily equal to zero, just as were the elements in their standard states.

Bond Energies

The term *bond energy* is defined as the ΔH required to form a bond between two atoms in an isolated gaseous molecule. At first sight one might think that the value of $\Delta H° = -22.06$ kcal/mole would be the H-Cl bond energy, but it is not. The value -22.06 represents the difference between the energy required to dissociate the H_2 and the Cl_2 molecules and the energy liberated when the H atoms and the Cl atoms combine to form HCl. The ΔH bond energy corresponds to the reaction

$$H_{(g)} + Cl_{(g)} \rightarrow HCl_{(g)}$$

We could calculate ΔH for this if we knew for the elements the heat of formation of molecules from their atoms. Some crystalline elements (especially metals) vaporize as monatomic gases and it is not too difficult to determine their heats of sublimation. Some elements, such as H_2, O_2, and Br_2 are diatomic gases that dissociate into atoms at high temperatures; these dissociation energies may also be determined. Table 23-1 also includes the standard enthalpies of formation of a number of atoms; these are based on the normal physical form of the element at 25°C. For HCl we find

$$\Delta H_f^\circ \text{ for } HCl_{(g)} = -22.06 \text{ kcal (product)}$$

$$\Delta H_f^\circ \text{ for } H_{(g)} \quad = +52.09 \text{ kcal}$$
$$\Delta H_f^\circ \text{ for } Cl_{(g)} \quad = \underline{+29.01 \text{ kcal}}$$
$$\text{sum} = +81.10 \text{ kcal (reactants)}$$

$$\Delta H^\circ \text{ or bond energy} = (-22.06) - (81.10) = -103.16 \text{ kcal}$$

For further discussion of bond energies, see Chapter 9.

Free Energies of Reaction

In Chapter 16 you learned two methods by which to calculate the maximum amount of work, called the change in free energy (ΔG°), that is available from a given electron-transfer reaction when the reactants and products are in their standard states. By one method the standard half-cell potentials from Table 16-1 are used to calculate the standard cell potential (E_{cell}°) for the whole reaction, and then ΔG° is calculated from the equation (see p. 191):

$$\Delta G^\circ = -(23,060)(n)(E_{cell}^\circ)$$

By the other method, one not limited to electron-transfer reactions, ΔG° is calculated from the equilibrium constant for the reaction at temperature T, by the equation (see p. 192)

$$\Delta G^\circ = -2.3RT \log K_e$$

There is a third way by which ΔG° may be calculated. Just as for enthalpy (H), it is not possible to know the actual free-energy content (G) of any substance; only changes can be measured. Also, as for enthalpy, it is possible to construct a table of standard free energies of formation for each substance, based on the arbitrary assumption that the elements in their standard states (physical forms stable at 25°C and 1 atm) have zero free energy of formation. You recall that these "formation" reactions correspond to the formation of compounds directly from the elements. Table 23-2 shows a few selected values of ΔG_f°. Once such a table has been constructed, it is possible to *calculate* the ΔG° or K_e for any reaction for which data are available — even though the reaction might not actually occur because the activation energy is too high or because other reactions occur instead. Just as with enthalpy, we can say

$$\Delta G^\circ = \Sigma(\Delta G_f^\circ)_{products} - \Sigma(\Delta G_f^\circ)_{reactants}$$

and apply it in the following typical manner.

PROBLEM:

What is the change in standard free energy, and what is the value of K_e, at 25°C for the gaseous dissociation of PCl_5 into PCl_3 and Cl_2?

SOLUTION:

Take the needed standard free energies of formation from Table 23-2 and use them with the balanced chemical equation

$$PCl_{5(g)} \rightleftharpoons PCl_{3(g)} + Cl_{2(g)}$$

$$(-77.59) \quad (-68.42) \quad (0.00)$$

$$\begin{aligned}\Delta G_f^\circ \text{ of products} &= (1)(-68.42) = -68.42 \text{ kcal}\\ \text{plus } (1)(0.00) &= \underline{0.00 \text{ kcal}}\\ &= -68.42 \text{ kcal}\end{aligned}$$

$$\Delta G_f^\circ \text{ of reactants} = (1)(-77.59) = -77.59 \text{ kcal}$$

$$\Delta G^\circ \text{ of reaction} = (-68.42) - (-77.59) = +9.17 \text{ kcal}$$

$$\log K_e = \frac{-\Delta G^\circ}{2.3RT} = \frac{-(9,170)}{(2.3)(1.987)(298)} = -6.73$$

$$K_e = 10^{-6.73} = 10^{.27} \times 10^{-7} = 1.86 \times 10^{-7}$$

Since K_e is relatively small, you would conclude that PCl_5 is only slightly dissociated at room temperature.

Entropies of Reaction

We are now in a position to see the relationship between H, G, and S. We say that the energy content or enthalpy (H) of a substance is composed of two parts; one that can do work and is called free energy (G), and another that cannot do work and is the product of temperature (T) and entropy (S). The product of T and S is necessary in order to obtain the units of energy, since S has the units of cal/deg. In equation form we write

$$H = G + TS$$

In chemical and physical changes, these properties of H, G, and S change for each substance, so that for a *change* (Δ) we would write

$$\Delta H = \Delta G + \Delta(TS)$$

and if the change takes place at constant temperature and pressure

$$\Delta H = \Delta G + T(\Delta S)$$

When changes occur with all materials in their standard states, the superscript $^\circ$ would be used. For example, for ΔS° we would have

$$\Delta S^\circ = \frac{\Delta H^\circ - \Delta G^\circ}{T}$$

TABLE 23-2.
Standard Free Energies of Formation at 25°C (ΔG_f° in kcal/mole)

Substance	State	ΔG_f°	Substance	State	ΔG_f°	Substance	State	ΔG_f°
All elements in								
normal state		0.00	C_6H_6	l	29.76	N (atom)	g	81.47
Ag^+	aq	18.43	C_6H_6	g	30.99	Na (atom)	g	18.67
AgCl	s	−26.22	Cu^{++}	aq	15.53	Na^+	aq	−62.59
Br (atom)	g	19.69	F (atom)	g	14.20	NH_3	g	−3.98
Br^-	aq	−24.57	F^-	aq	−66.08	NH_3	aq	−6.37
BrCl	g	−0.21	FeO	s	−58.40	NH_4^+	aq	−19.00
C (atom)	g	160.84	Fe_2O_3	s	−177.10	NO	g	20.72
C (diamond)	s	0.68	H (atom)	g	48.58	NO_2	g	12.39
Ca^{++}	aq	−132.18	H^+	aq	0.00	NO_3^-	aq	−26.41
Cd^{++}	aq	−18.58	HBr	g	−12.72	N_2O	g	24.76
Cl (atom)	g	25.19	HCl	g	−22.77	O (atom)	g	54.99
Cl^-	aq	−31.35	HI	g	0.31	OH^-	aq	−37.60
ClO_4^-	aq	−2.57	H_2O	l	−56.69	P (atom)	g	66.71
CO	g	−32.81	H_2O	g	−54.64	PCl_3	g	−68.42
CO_2	g	−94.26	H_2S	g	−7.89	PCl_5	g	−77.59
CH_3OH	l	−39.73	I (atom)	g	16.77	S (atom)	g	43.57
CH_3OH	g	−38.69	I^-	aq	−12.35	$S^=$	aq	20.00
C_2H_5OH	l	−41.77	ICl	g	−1.32	SO_2	g	−71.79
C_2H_5OH	g	−40.30	K^+	aq	−67.47	$SO_4^=$	aq	−177.34
C_4H_{10}	g	−3.75	Li^+	aq	−70.22	Zn^{++}	aq	−35.18

PROBLEM:
Calculate the change in standard entropy for the dissociation of PCl_5 into PCl_3 and Cl_2 at 25°C.

SOLUTION:
Take the values of $\Delta H° = +22,130$ cal and $\Delta G° = +9,170$ cal as obtained in problems solved earlier in this chapter and substitute them into the equation:

$$\Delta S° = \frac{\Delta H° - \Delta G°}{T} = \frac{22,130 - 9,170}{298} = \frac{12,960}{298}$$

$$= +43.5 \text{ cal/deg}$$

The positive sign of the 43.5 cal/deg indicates that the reaction goes with an increase in entropy, that the products represent a state of greater molecular disorder than do the reactants.

When a substance changes from one physical state to another at a temperature where the two states can coexist at 1 atm (such as at the melting point or the normal boiling point), the two phases are at equilibrium, and $\Delta G_T^\circ = 0$ for the transition. Under these conditions,

$$\Delta S_T^\circ = \frac{\Delta H_T^\circ - \Delta G_T^\circ}{T} = \frac{\Delta H_T^\circ}{T}.$$

It is for this reason (that $\Delta G_T^\circ = 0$) that we were able to calculate standard entropies of transition by simply dividing the enthalpies of transition by the absolute temperature as we did on p. 294. Entropies of reaction can *not* be calculated this way; the more general expression must be used.

PROBLEMS A   Answers on page 353

1. Calculate the approximate specific heat of each of the following elements: (a) S; (b) Zn; (c) La; (d) U; (e) Pb.

2. Calculate the resultant temperature when 150 g of water at 75°C is mixed with 75 g of water at 20°C. (Assume no heat loss to container or surroundings.)

3. Calculate the resultant temperature when 50 g of silver metal at 150°C is mixed with 50 g of water at 20°C. (Assume no heat loss to container or surroundings.)

4. Suppose 150 ml of water at 50°C, 25 g of ice at 0°C, and 100 g of Cu at 100°C are mixed together. Calculate the resultant temperature, assuming no heat loss, and using the approximate specific heat of Cu.

5. Suppose 150 ml of water at 20°C, 50 g of ice at 0°C, and 70 g of Cu at 100°C are mixed together. Calculate the resultant temperature, assuming no heat loss, and using the approximate specific heat of Cu.

6. A Dewar flask (vacuum-jacketed bottle) was used as a calorimeter, and the following data were obtained. Measurements in parts (a) and (b) were made to obtain the heat capacity of the calorimeter, and parts (c) and (d) were performed on an unknown metal.
   (a) Calorimeter with 150 ml of H_2O has a temperature of 21.3°C.
   (b) When 35 ml of H_2O at 99.5°C was added to the calorimeter and water of (a), the resultant temperature was 35.6°C.
   (c) Calorimeter with 150 ml of H_2O has a temperature of 22.7°C.
   (d) A 50.3-g sample of metal at 99.5°C added to the calorimeter and water of (c) gave a resultant temperature of 24.3°C.
   Calculate the approximate atomic weight of the metal. (Assume that the specific heat of water is 1.0 cal/g deg.)

7. A 100-g sample of glycerol was put into the calorimeter calibrated in Problem 6, and its temperature observed to be 20.5°C. Then 45.7 g of iron

at 165°C was added to the glycerol, giving a resultant temperature of 37.4°C. Calculate the specific heat of the glycerol. (Use the approximate specific heat of iron.)

8. A metal, X, whose specific heat is 0.1193 cal/g degree forms an oxide whose composition is 32.00% O. (a) What is the empirical formula of the oxide? (b) What is the exact atomic weight of the metal?

9. A metal, Y, whose specific heat is 0.0504 cal/g degree forms two chlorides, whose compositions are 46.71% Cl and 59.42% Cl. (a) Find the formulas of the chlorides. (b) What is the exact atomic weight of the metal?

10. The enthalpy of combustion of rhombic sulfur is −70.96 kcal/mole. The enthalpy of combustion of monoclinic sulfur is −70.88 kcal/mole. What are the standard enthalpy, free energy, and entropy of transition from rhombic to monoclinic sulfur?

11. Using the tables of standard enthalpies and free energies of formation, determine the standard enthalpy, free energy, and entropy of reaction for each of the following reactions:

(a) $FeO_{(s)} + H_{2(g)} \rightleftarrows Fe_{(s)} + H_2O_{(l)}$
(b) $4NH_{3(g)} + 5O_{2(g)} \rightleftarrows 6H_2O_{(g)} + 4NO_{(g)}$
(c) $Zn_{(s)} + 2HCl_{(aq)} \rightleftarrows H_{2(g)} + ZnCl_{2(aq)}$
(d) $2FeO_{(s)} + \frac{1}{2}O_{2(g)} \rightleftarrows Fe_2O_{3(s)}$
(e) $3NO_{2(g)} + H_2O_{(l)} \rightleftarrows 2HNO_{3(aq)} + NO_{(g)}$
(f) $2N_2O_{(g)} \rightleftarrows 2N_{2(g)} + O_{2(g)}$

12. What is the enthalpy of combustion of $C_6H_{6(l)}$? (Give the thermochemical equation.)

13. Calculate the bond energies of the following gaseous molecules: (a) NO; (b) H_2O; (c) NH_3; (d) PCl_5.

14. Calculate the standard enthalpy, free energy, and entropy of vaporization for C_2H_5OH at 25°C.

15. The equilibrium constant at 25°C for the reaction $H_{2(g)} + \frac{1}{2}S_{(s)} \rightleftarrows H_2S_{(g)}$ is 6.32×10^5 (with gas concentrations expressed in atm). Calculate $\Delta H°$, $\Delta G°$, and $\Delta S°$ for this reaction at 25°C.

PROBLEMS B   No answers given

16. Calculate the approximate specific heat of each of the following elements: (a) Pt; (b) P; (c) Sr; (d) As; (e) Au.

17. Calculate the resultant temperature when 250 g of water at 25°C is mixed with 100 g of water at 80°C. (Assume no heat loss to container or surroundings.)

18. Calculate the resultant temperature when 100 g of lead metal at 200°C is mixed with 200 g of water at 20°C. (Assume no heat loss to container or surroundings.)

19. Suppose 200 ml of water at 55°C, 35 g of ice at 0°C, and 120 g of Zn at 100°C are mixed, Calculate the resultant temperature, assuming no heat loss, and using the approximate specific heat of Zn.

20. Suppose 200 ml of water at 15°C, 60 g of ice at 0°C, and 90 g of Cd at 125°C are mixed. Calculate the resultant temperature, assuming no heat loss, and using the approximate specific heat of Cd.

21. A metal, X, whose specific heat is 0.1121 cal/g degree forms an oxide whose composition is 27.90% O. (a) What is the exact atomic weight of the metal? (b) What is the formula of the oxide?

22. A metal, Y, whose specific heat is 0.0312 cal/g degree forms two chlorides whose compositions are 35.10% Cl and 15.25% Cl. (a) What is the exact atomic weight of the metal? (b) What are the formulas of the chlorides?

23. Calculate the heat capacity of a calorimeter (Figure 23-1) that when containing 250 ml of water, requires a current of 0.650 amp passing for 5 min 25 sec through the 8.35 ohm immersed resistance in order to raise the water temperature from 25.357°C to 26.213°C.

24. Compute the following quantities, using the calorimeter calibrated in Problem 23.
    (a) The specific heat of an unknown metal when 40.0 g of a finely divided sample are stirred with 220 ml of water and a current of 0.720 amp for 12 min 24 sec raises the temperature from 25.265°C to 27.880°C.
    (b) The specific heat of an unknown liquid when 250 g of it are placed in the calorimeter in place of the water and a current of 0.546 amp is passed for 4 min 45 sec in order to raise the temperature from 26.405°C to 27.033°C.
    (c) The enthalpy change per mole of precipitated $BaSO_4$ when 125 ml of 0.500M $BaCl_2$ solution are mixed in the calorimeter with 125 ml of 0.500M Na_2SO_4 solution. This is the so-called "heat of reaction." After the reaction mixture has cooled to the same initial temperature possessed by the reactants, a current of 0.800 amp passing for 3 min 47 sec was required to regain the final temperature reached by the reaction. Write the equation for the reaction also.
    (d) The enthalpy change per mole of NH_4NO_3 when 16.0 g of solid NH_4NO_3 are dissolved in 250 ml of water in the calorimeter. This is the so-called "heat of solution." After the salt is dissolved, it was found that a current of 1.555 amp passing for 4 min 16 sec was required in order to regain the initial temperature of the water. Write the equation for this "reaction."
    (e) The enthalpy of fusion per mole of diiodobenzene, $C_6H_4I_2$. When 20.0 g of diiodobenzene are added to 250 ml of water in the calorimeter, the crystals, being immiscible with water and also more dense, sink to the

bottom. As current is passed through the heater, the temperature of the mixture rises until the diiodobenzene starts to melt; at this point all of the electrical energy is used for fusion and none for raising the temperature. A current of 0.500 amp passing for 6 min 49 sec is required for the period in which the temperature stays constant and before the temperature again begins to rise, as both the water and *liquid* diiodobenzene rise above the melting point of 27°C. Write the equation for the "reaction."

25. The enthalpy of combustion of diamond is -94.50 kcal/mole. The enthalpy of combustion of graphite is -94.05 kcal/mole. What is the standard enthalpy, free energy, and entropy of transition from diamond to graphite?

26. Using the tables of standard enthalpies and free energies of formation, determine the standard enthalpy, free energy, and entropy of reaction for each of the following reactions:
    (a) $C_{(s)} + H_2O_{(g)} \rightleftarrows H_{2(g)} + CO_{(g)}$
    (b) $C_6H_{6(l)} + 7\frac{1}{2}O_{2(g)} \rightleftarrows 6CO_{2(g)} + 3H_2O_{(l)}$
    (c) $2K_{(s)} + H_2SO_{4(aq)} \rightleftarrows K_2SO_{4(aq)} + H_{2(g)}$
    (d) $N_2O_{(g)} + H_{2(g)} \rightleftarrows H_2O_{(l)} + N_{2(g)}$
    (e) $NH_{3(g)} \rightleftarrows \frac{1}{2}N_{2(g)} + 1\frac{1}{2}H_{2(g)}$
    (f) $2H_2S_{(g)} + 3O_{2(g)} \rightleftarrows 2H_2O_{(l)} + 2SO_{2(g)}$

27. What is the standard enthalpy of combustion of butane, $C_4H_{10(g)}$? (Give the thermochemical equation.)

28. Calculate the bond energies in the following gaseous molecules: (a) CO; (b) CO_2; (c) SO_2; (d) NaCl. The enthalpy of sublimation of NaCl is $\Delta H = +54.7$ kcal/mole, and the standard enthalpy of formation of $NaCl_{(s)}$ is $\Delta H_f^\circ = -98.3$ kcal/mole.

29. Calculate the standard enthalpy, free energy, and entropy of vaporization of C_6H_6 at 25°C.

30. The equilibrium constant at 25°C for the reaction $NO_{(g)} + O_{2(g)} \rightleftarrows NO_{2(g)}$ is 1.26×10^6 (with gas concentrations expressed in atm). Calculate ΔH°, ΔG°, and ΔS° for this reaction at 25°C.

Nuclear Chemistry

All the chemical changes and many of the physical changes that we have studied so far have involved alterations in the electronic structures of atoms. Oxidation-reduction reactions, emission and absorption spectra, and X-rays result from the movement of electrons from one energy level to another. In all of these the nuclei of the atoms remain unchanged, and different isotopes of the same element have the same chemical activity. Nuclear chemistry, or radioactivity, differs from other branches of chemistry in that the important changes occur in the nucleus. These nuclear changes are also represented by chemical equations; but since the isotopes of the same element may, from a *nuclear* standpoint, be very different in reactivity, it is necessary that the equations show which isotopes are involved.

Nuclear symbolism disregards valence and electrons and always includes (1) the mass number of the isotope (this is the whole number nearest the atomic weight of the isotope), (2) the atomic number of the element, and (3) the symbol of the element. The mass number is used as a superscript preceding the symbol, and the atomic number is used as a subscript preceding the symbol. This is illustrated by the following examples:

$_1^1H$ = a proton, or a hydrogen atom with mass number 1 and atomic number 1;

$_1^2H$ = a deuteron, or a hydrogen atom with mass number 2 and atomic number 1;

$_0^1n$ = a neutron, or a particle with mass number 1 and no electrical charge (atomic number 0).

PbI_2

... of .01 M Pb(NO_3)_2

$$273$$
$$78.4$$
$$35 \sqrt{} = $$

| $K_{sp}(PbI_2)$ | 9×10^{-9} |

$$273$$
$$34.9$$
$$307 \; g$$

.01
$Pb(NO_3)$
Hint work for 1 liter of .01M Pb(NO_3)_2

$$Pb(NO_3)_2 \longrightarrow Pb^{++} + 2NO_3^{-}$$

x .01 + x $2x$

$$PbI_2 \longrightarrow Pb^{++} + 2I^{-}$$

$K_{sp} = 9 \times 10^{-9} = (.01 + x)(2x)^2$

$x = 4.8 \times 10^{-4} \frac{m}{\ell}$ of PbI_2

$4.8 \times 10^{-4} \frac{m \; PbI_2}{\ell} \times \frac{461 \; g \; PbI_2}{1 \; mole \; PbI_2} \times \frac{.25 \; \ell}{}$

$$5.4 \times 10^{-2} \; g$$

P.211

3.(a) K_{sp} for $Ag_2SO_4 = 1.2 \times 10^{-5}$

$Ag_2SO_4 \rightleftharpoons 2Ag^+ + SO_4^=$

$K_{sp} = [Ag^+]^2 [SO_4]^1$

$(1 \times 10^{-4})^2$ $.01 = 1 \times 10^{-10}$

$.01\ M\ SO_4 = x\ \frac{1000}{1001}$ ppt will not form

$.1\ M\ Ag^+ \times \frac{1}{1001} = 1 \times 10^{-4}$

if product is = or > than K_{sp} ppt will form

"   "   " < " K_{sp} ppt will not form

$\times 10^{-18}$

1.14
(.14)/6
4 ? 2
1 4
1.3 7 9 6

Three other particles that occur in nuclear reactions, but that are not nuclear particles, must also be described by similar symbols if we are to write nuclear reactions. They are:

$_{-1}^{0}e$ = an electron, with mass number 0 and atomic number -1;

$_{1}^{0}e$ = a positron, with mass number 0 and atomic number $+1$;

$_{0}^{0}\nu$ = a neutrino, with mass number 0 and atomic number 0.

The actual mass of $_{1}^{0}e$ and $_{-1}^{0}e$ is about $\frac{1}{1,845}$ that of a proton, and the mass of $_{0}^{0}\nu$ is virtually negligible.

In the early days of nuclear chemistry, before an established system of symbols was used, three other terms were invented: (1) alpha particle, α; (2) beta particle, β; and (3) gamma ray, γ. We still use these terms, but for writing nuclear equations we must know that:

$\alpha = _{2}^{4}He$ (a He^{++} ion);

$\beta = _{-1}^{0}e$ (an electron);

γ = radiation similar to X-rays, which have no mass.

Two rules must be followed in balancing nuclear equations: (1) the sum of the mass numbers on the left side of the equation must equal the sum of the mass numbers on the right side; and (2) the sum of the atomic numbers on the left side must equal the sum of the atomic numbers on the right side. These rules are illustrated in the following nuclear reactions:

$$_{88}^{226}Ra \rightarrow \,_{86}^{222}Rn + \,_{2}^{4}He$$

$$_{90}^{232}Th \rightarrow \,_{88}^{228}Ra + \,_{2}^{4}He$$

$$_{93}^{239}Np \rightarrow \,_{94}^{239}Pu + \,_{-1}^{0}e$$

$$_{6}^{14}C \rightarrow \,_{7}^{14}N + \,_{-1}^{0}e$$

These equations also illustrate the following general statements. (1) Whenever a radioactive element emits an α particle, its product (daughter) has a mass number that is 4 less than that of the parent and an atomic number that is 2 less than that of the parent. (2) Whenever a radioactive element emits a β particle, its daughter has a mass number that is the same as that of the parent and an atomic number that is 1 greater than that of the parent. Since electrons as such are not present in the nuclei of atoms, it is not obvious at first why the loss of a β particle should cause an increase in atomic number. What actually happens is that a neutron disintegrates:

$$_{0}^{0}n \rightarrow \,_{1}^{1}H + \,_{-1}^{0}e + \,_{0}^{0}\nu$$

The $_{-1}^{0}e$ is ejected as a β particle, and the proton, $_{1}^{1}H$, stays in the nucleus in the place of the neutron. The additional positive charge causes the atomic number to be increased by 1.

Alpha particles from radioactive samples, or He^{++} ions accelerated in a cyclotron, may be used to bring about other nuclear reactions. For example, they may bombard a beryllium metal target:

$$_{4}^{9}Be + _{2}^{4}He \longrightarrow \ _{6}^{12}C + _{0}^{1}n$$

The neutrons produced by such bombardments may be used to cause additional nuclear reactions. For example, they may bombard lithium:

$$_{3}^{6}Li + _{0}^{1}n \longrightarrow \ _{2}^{4}He + _{1}^{3}H$$

Protons ($_{1}^{1}H$) and deuterons ($_{1}^{2}H$) are the other common particles that may be accelerated and used to bring about nuclear reactions:

$$_{7}^{14}N + _{1}^{1}H \longrightarrow \ _{6}^{11}C + _{2}^{4}He$$

$$_{26}^{54}Fe + _{1}^{2}H \longrightarrow \ _{27}^{55}Co + _{0}^{1}n$$

If a product of one of these man-made reactions is unstable and spontaneously disintegrates further, it is said to be "artificially radioactive." In the preceding equation, for example, $_{27}^{55}Co$ is artificially radioactive, disintegrating as

$$_{27}^{55}Co \longrightarrow \ _{26}^{55}Fe + _{1}^{0}e$$

In all the nuclear equations we have just written, we have shown that there is no change in the sum of the mass numbers. However, there actually is a small change in mass, and this change is one of the most important properties of nuclear reactions. In the nuclear fusion of hydrogen and tritium,

$$_{1}^{1}H + _{1}^{3}H \longrightarrow \ _{2}^{4}He$$

the sum of the weight (per mole) of the products is 0.0246 g less than the sum of the weights of the reactants. In the Bethe cycle (the series of reactions in the sun by which solar energy is produced), 4 moles of protons weighing 4.03228 g are converted to 1 mole of helium weighing 4.00336 g, a loss of 0.02892 g. In the nuclear fission of $_{92}^{235}U$,

$$_{92}^{235}U + _{0}^{1}n \longrightarrow \ \text{fission products} + 2 \text{ or } 3 \ _{0}^{1}n$$

there is a loss of 0.205 g per mole. The weight that is lost is converted to energy, according to the Einstein equation

$$E = mc^{2}$$

where E is the energy, in ergs, liberated by converting mass to energy, m is the mass, in grams, converted to energy, and c is the velocity of light, 3×10^{10} cm/sec.

The ergs of energy may be expressed as calories if we remember that 1 cal $= 4.186 \times 10^7$ ergs.

(1) In nuclear fusion, 0.0246 g gives 5.3×10^{11} cal.
(2) In the Bethe cycle, 0.02892 g gives 5.88×10^{11} cal.
(3) In the nuclear fission, 0.205 g gives 4.4×10^{12} cal.

Radioactive substances vary in their activity, or in the rate at which they disintegrate. Each radioactive isotope disintegrates at a rate that is unaffected by temperature, pressure, or other external conditions. The rate of disintegration $\left(-\dfrac{dN}{dt}\right)$ of a given isotope has been found to be proportional to the number of atoms (N) of the isotope that are present at any given instant:

$$-\frac{dN}{dt} = kN = \frac{\text{atoms}}{\text{unit time}} \qquad (24\text{-}1)$$

The proportionality constant (k) is different for each isotope, and has the units of $(\text{time})^{-1}$. We can rearrange this fundamental rate equation to

$$-\frac{dN}{N} = k \, dt \qquad (24\text{-}2)$$

The calculus student will recognize that a simple integration will make it possible for us to find the total number of atoms that remain after the lapse of a given time, t. If we have N_1 atoms at $t = 0$ and N_2 atoms at $t = t$, then

$$-\int_{N_1}^{N_2} \frac{dN}{N} = k \int_0^t dt$$

and

$$\ln \frac{N_1}{N_2} = kt$$

or, expressed in terms of common logarithms (base 10) instead of natural logarithms (base e),

$$2.3 \log \frac{N_1}{N_2} = kt \qquad (24\text{-}3)$$

Thus, as disintegration proceeds, fewer and fewer atoms remain, and the rate of disintegration slows.

When half of the atoms have disintegrated, $N_2 = \tfrac{1}{2}N_1$ and the elapsed time is the *half-life*, $t_{\frac{1}{2}}$. Thus

$$k = \frac{2.3 \log 2}{t_{\frac{1}{2}}} = \frac{0.693}{t_{\frac{1}{2}}} \qquad (24\text{-}4)$$

TABLE 24-1.
Half-Lives of Selected Radioactive Elements

Isotope	Half-life, $t_{\frac{1}{2}}$	Particle emitted	Isotope	Half-life, $t_{\frac{1}{2}}$	Particle emitted
$^{237}_{93}Np$	2.20×10^6 yr	β	$^{226}_{88}Ra$	1.62×10^3 yr	α
$^{239}_{94}Pu$	2.40×10^4 yr	α	$^{238}_{92}U$	4.51×10^9 yr	α
$^{227}_{89}Ac$	21.7 yr	α	$^{14}_{6}C$	5.57×10^3 yr	β
$^{210}_{84}Po$	138 days	α	$^{35}_{16}S$	87.1 days	β
$^{211}_{82}Pb$	36.1 min	β	$^{211}_{85}At$	7.2 hr	α

The most common way to characterize a radioactive isotope is to give its half-life rather than its k value. A list of characteristic half-lives is given in Table 24-1, along with the particles emitted.

In a problem involving the half-life of an isotope, the given time and the particular isotope determine whether you use Equation 24-1 or 24-3. If the elapsed time given in the problem is insignificant compared to the half-life, Equation 24-1 is appropriate, but if the time is an appreciable fraction (or multiple) of the half-life, Equation 24-3 should be used, as is illustrated in the following problems.

PROBLEM:
How much of a 1-g sample of $^{238}_{92}U$ will disintegrate in a period of ten years?

SOLUTION:
Ten years is an infinitesimal period of time compared to the half-life of 4.51×10^9 years. If we tried using Equation 24-3 we would have

$$\log \frac{N_1}{N_2} = \frac{kt}{2.3} = \frac{0.693\,t}{2.3\,t_{\frac{1}{2}}} = \frac{(0.693)(10 \text{ yr})}{(2.3)(4.51 \times 10^9 \text{ yr})} = 0.00000000067$$

It is impractical to evaluate N_2 in this way because the log is far smaller than that shown in any normal log table. Instead we use Equation 24-1 and

$$-dN = k \cdot N \cdot dt$$

$-dN =$ the number of atoms (or grams) that disintegrate, $N =$ the number of atoms (or grams) present, an amount that stays virtually constant in ten years, and $dt = 10$ years, an infinitesimal period of time in this problem. Consequently,

$$-dN = \left(\frac{0.693}{4.51 \times 10^9 \text{ yr}}\right)(1.0 \text{ g})(10 \text{ yr})$$

$$= 1.54 \times 10^{-9} \text{ grams of } ^{238}_{92}U \text{ that disintegrate}$$

PROBLEM:
You have 0.200 g of $^{210}_{84}Po$, whose half-life is 138 days (Table 24-1). How much of it will remain 21 days from now? Write the nuclear equation for the reaction.

SOLUTION:

Table 24-1 shows that $^{210}_{84}Po$ emits α particles; therefore the nuclear equation is

$$^{210}_{84}Po \rightarrow {}^{206}_{82}Pb + {}^{4}_{2}He$$

In this problem, 21 days is an appreciable fraction of the half-life, 138 days, and it is appropriate to use Equation 24-3. Since the number of atoms is proportional to the number of grams, we have

$$\log \frac{N_1}{N_2} = \log \frac{0.200}{N_2} = \frac{kt}{2.3} = \frac{0.693\,t}{2.3\,t_{\frac{1}{2}}} = \frac{(0.693)(21\ \text{days})}{(2.3)(138\ \text{days})} = 0.0459$$

This logarithm is large enough to be found in normal log tables. Taking the antilog of both sides, we get

$$\frac{0.200}{N_2} = 1.11$$

$$N_2 = \frac{0.200}{1.11} = 0.180\ g$$

PROBLEM:

How much time must pass for $\frac{12}{13}$ of a sample of $^{227}_{89}Ac$ to disintegrate? Its half-life (Table 24-1) is 21.7 years.

SOLUTION:

Here we find that

$$k = \frac{0.693}{t_{\frac{1}{2}}} = \frac{0.693}{21.7} = 3.20 \times 10^{-2}\ \text{yr}^{-1}$$

Therefore

$$t = \frac{2.3}{k} \log \frac{N_1}{N_2} = \frac{2.3}{3.2 \times 10^{-2}} \log \frac{1}{\frac{1}{13}} = \frac{2.3}{3.2 \times 10^{-2}} \log 13$$

$$= \frac{2.3 \times 1.114}{3.2 \times 10^{-2}} = 80.1\ \text{yr}$$

PROBLEM:

What is the half-life of an isotope if a sample of it gives 10,000 counts per minute (cpm), and $3\frac{1}{2}$ hr later it gives 8,335 cpm?

SOLUTION:

Since the cpm are proportional to the number of atoms of the isotope present, we may write

$$\log \frac{10,000}{8,335} = \frac{k \times 3.5}{2.3}$$

$$k = \frac{2.3}{3.5} \log \frac{10,000}{8,335} = \frac{2.3 \times 0.079}{3.5} = 0.0519\ \text{hr}^{-1}$$

$$t_{\frac{1}{2}} = \frac{0.693}{k} = \frac{0.693}{0.0519} = 13.35\ \text{hr}$$

PROBLEMS A   Answers on page 354

1. Write nuclear equations to show the disintegration of the first five radio-active isotopes listed in Table 24-1.

2. The following isotopes are artificially radioactive. They emit the particles shown in parentheses. Write nuclear equations to show the disintegration of these isotopes: (a) $^{27}_{14}Si$ ($^{0}_{1}e$); (b) $^{28}_{13}Al$ ($_{-1}^{0}e$); (c) $^{30}_{15}P$ ($^{0}_{1}e$); (d) $^{24}_{11}Na$ ($_{-1}^{0}e$); (e) $^{17}_{9}F$ ($^{0}_{1}e$).

3. Bombardment reactions are often summarized in a terse form, such as $^{9}_{4}Be(\alpha,n)$. This means that the target, $^{9}_{4}Be$, is bombarded by α particles, $^{4}_{2}He$, and that neutrons, $^{1}_{0}n$, are produced. By the rules for balancing nuclear equations, we know that $^{12}_{6}C$ is also produced. Give the complete balanced nuclear equation for each of these transmutation bombardments (p stands for proton, and d stands for deuteron in the abbreviated method): (a) $^{14}_{7}N(n,p)$; (b) $^{26}_{12}Mg(n,\alpha)$; (c) $^{59}_{27}Co(d,p)$; (d) $^{14}_{7}N(\alpha,p)$; (e) $^{63}_{29}Cu(p,n)$.

4. (a) Calculate the energy liberated per gram-atom of Li when the following reaction (which might occur in one type of H-bomb) takes place:

$$^{7}_{3}Li + {}^{1}_{1}H \rightarrow 2\,{}^{4}_{2}He$$

The actual atomic masses involved are: $^{7}_{3}Li = 7.01818$ g; $^{1}_{1}H = 1.00813$ g; $^{4}_{2}He = 4.00386$ g. (b) How many tons of carbon would have to be burned to give as much energy as 100 g of $^{7}_{3}Li$ consumed by this reaction? (Carbon gives 94.1 kcal/mole when burned to CO_2.)

5. The halogen astatine can be obtained only by artificial methods through bombardment. It has been found useful for the treatment of certain types of cancer of the thyroid gland, for it migrates to this gland just as iodine does. If a sample containing 0.1 mg of $^{211}_{85}At$ is given to a person at 9 A.M. one morning, how much will remain in his body at 9 A.M. the following morning?

6. A purified sample of a radioactive compound is found to have 1,365 cpm at 10 A.M. but only 832 cpm at 1 P.M. the same day. What is the half-life of this sample?

7. If in the explosion of one atom bomb 50 g of plutonium were scattered about in the atmosphere before it had a chance to undergo nuclear fission, how much would be left in the atmosphere after 1,000 yr? (Assume that none of it settled out.)

8. If you sealed a 100-g sample of very pure uranium in a container for safe-keeping, and if in the distant future a scientist dissolved it and found it to contain 0.8 g $^{206}_{82}Pb$, how many years would have elapsed since you sealed the sample? (Lead is the stable endproduct of the radioactive decay of uranium.)

9. Libby and his coworkers have found that a very small amount of the CO_2 in the air is radioactive as a result of continuous bombardment of nitrogen in the upper atmosphere by neutrons of cosmic origin. This radioactive $^{14}_{6}C$ ($t_{\frac{1}{2}} = 5,570$ yr) is uniformly distributed among all forms of carbon that are in equilibrium with the atmosphere, and it is found to show 15.3 cpm per gram of carbon. All plants use CO_2 for growth, and while living are in equilibrium with the atmosphere. When plants die, they no longer remain in equilibrium with the air, and the carbon slowly loses its activity. Cyprus wood from the ancient Egyptian tomb of Sneferu at Meydum had an activity of 6.88 cpm/g of carbon. Estimate the age of this wood (and presumably the age of the tomb).

10. A 0.5-g sample of radium is sealed into a very thin-walled tube so that the α particles emitted from the radium and its decay products can penetrate and be collected in an evacuated volume of 25 ml. What helium pressure at 20°C will build up in 100 yr? (Five α particles are produced as radium disintegrates to a stable isotope of lead.)

11. Determine how many milligrams of a 100-milligram sample of Ra will disintegrate in 30 days.

PROBLEMS B   No answers given

12. Write nuclear equations to show the disintegration of the last five radioactive isotopes listed in Table 24-1.

13. The following isotopes are artificially radioactive. They emit the particles shown in parentheses. Write nuclear equations to show the disintegration of these isotopes: (a) $^{23}_{10}Ne$ $(_{-1}^{0}e)$; (b) $^{41}_{18}Ar$ $(_{-1}^{0}e)$; (c) $^{38}_{19}K$ $(^{0}_{1}e)$; (d) $^{84}_{37}Rb$ $(_{-1}^{0}e)$; (e) $^{13}_{7}N$ $(^{0}_{1}e)$.

14. Bombardment reactions are often summarized in a terse form, as is explained in Problem 3. Give the complete balanced nuclear equation for each of these transmutation bombardments: (a) $^{9}_{4}Be(p,d)$; (b) $^{16}_{8}O(d,\alpha)$; (c) $^{27}_{13}Al(n,\alpha)$; (d) $^{40}_{18}Ar(n,\gamma)$; (e) $^{25}_{12}Mg(\alpha,p)$.

15. (a) Calculate the energy liberated per gram-atom of He when the following reaction takes place:

$$^{3}_{1}H + ^{2}_{1}H \rightarrow \, ^{4}_{2}He + ^{1}_{0}n$$

The actual atomic masses involved are: $^{3}_{1}H = 3.01710$ g; $^{2}_{1}H = 2.01470$ g; $^{4}_{2}He = 4.00386$ g; $^{1}_{0}n = 1.00897$ g. (b) How many tons of carbon would have to be burned to give as much energy as 1 lb of hydrogen consumed by this reaction? (Carbon gives 94.1 kcal/mole when burned to CO_2.)

16. Radioactive sulfur has been used as a tracer in the study of the action of sulfur as an insecticide for red spider (a citrus-fruit pest). If 10 mg of $^{35}_{16}S$ is absorbed by an orange, how much will remain after 6 months' storage?

17. A pure sample of a radioactive compound is found to have 1,555 cpm at 3:30 P.M. and 960 cpm at 4:00 P.M. the same day. What is the half-life of this isotope?

18. If you sealed a 100-g sample of very pure thorium ($t_{\frac{1}{2}} = 1.39 \times 10^{10}$ yr) in a container for safekeeping, and if in the distant future a scientist dissolved it and found it to contain 0.6 g of $^{208}_{82}Pb$, how many years would have elapsed since you sealed the sample? (Lead is the stable endproduct of the radioactive decay of thorium.)

19. Libby and his coworkers have shown that a very small amount of the CO_2 in the air is radioactive, as is explained in Problem 9. Acacia wood from the ancient Egyptian tomb of Zoser at Sakkara had an activity of 7.62 cpm per gram of carbon. Estimate the age of this wood (and presumably the age of the tomb).

20. If a sample of rock contains 2 mg of helium for each 100 mg of thorium ($t_{\frac{1}{2}} = 1.39 \times 10^{10}$ yr), calculate the minimum age of this thorium deposit. (Six α particles are produced as thorium disintegrates to a stable isotope of lead. Assume that none of the helium escaped from the rock.)

21. How many grams of Th will disintegrate in 1.0 yr if, at the outset, you have a 1.00-g sample?

22. The positrons produced in the Bethe cycle (the production of solar energy) react with electrons to produce γ radiation. This "annihilation" reaction is

$$_{-1}^{0}e + {}_{1}^{0}e \longrightarrow \gamma$$

(a) Assuming that the mass of both $_{-1}^{0}e$ and $_{1}^{0}e$ is $\frac{1}{1,845}$ that of a proton, calculate the energy produced in the annihilation of a single electron by a single positron. (The gram-atomic weight of $_{1}^{1}H$ is 1.00813.) (b) If a single γ ray is produced by this annihilation, calculate its wavelength from the fundamental equation $E = h\nu$, where

E = energy of the quantum of γ radiation, in ergs
h = Planck's constant = 6.624×10^{-27} erg $\times$ sec
ν = frequency of the radiation, in sec^{-1}

[The fundamental relation between wavelength (λ), velocity (c), and frequency (ν) is $\lambda = c/\nu$.]

Reactions:
Prediction and Synthesis

Two important questions asked in the study of chemistry are: (1) "Will a reaction take place when you mix reactants A and B?" and (2) "How do you prepare a given compound, X?"

The first question concerns *prediction* and the second concerns *synthesis*. They are related. In the first you are given the reactants (the lefthand side of the equation) and are asked to predict the products. In the second you are given a product (the righthand side of the equation) and are asked to suggest suitable reactants. For synthesis you must use a reaction that is predicted to "go;" the products of a reaction that is predicted to "go" actually correspond to a chemical synthesis. This chapter gives an organized approach to answering these questions, including writing the chemical equations for the reactions involved.

PREDICTION

Fundamentally, a reaction "goes" if it is accompanied by a decrease in free energy (ΔG is $-$), as discussed in Chapter 23. The experimental conditions that lead to the fulfillment of this principle are summarized in the following statement.

A REACTION WILL "GO" FOR ANY ONE OF THE FOLLOWING REASONS:

1. For electron-transfer reactions, if the electrode potential (E) for the reducing half-reaction is more negative than that for the oxidizing half-reaction;
2. For non-electron-transfer reactions, if
   (a) an insoluble product is formed ($K_e > 1$ because $K_{sp} < 1$), or
   (b) a weak (or non-) electrolyte is formed ($K_e > 1$ because $K_i < 1$), or
   (c) a gas is formed in an open vessel (loss of gas product continually shifts equilibrium to right).

Guidelines

Some simple, general guides for using these principles are listed below.

1. *Electrode potentials* for half-reactions are most easily obtained from the Table of Standard Electrode Potentials (p. 185); they may be modified for marked changes in concentration as needed (p. 208).
2. *General solubility rules* for water as a solvent must be stated in two parts, one emphasizing the negative ions of the salts, and the other emphasizing the positive ions, as follows:
   (a) All nitrates, acetates, perchlorates, halides (Cl^-, Br^-, and I^-), and sulfates *are soluble,* with the following important exceptions:
       (1) the halides of Ag^+, Hg_2^{++}, and Pb^{++} are insoluble;
       (2) the sulfates of Ba^{++} and Pb^{++} are insoluble.
   (b) All alkali metal (Li^+, Na^+, K^+, Rb^+, and Cs^+) and ammonium (NH_4^+) salts and bases *are soluble.* $Ba(OH)_2$, a fairly important base, is also soluble.
   (c) In general, all other common types of inorganic compounds are insoluble.
3. *Strong electrolytes* (those that ionize in water) comprise, with minor exception:
   (a) all soluble salts (this does not include acids and bases);
   (b) all strong acids (there are seven common ones: HCl, HBr, HI, HNO_3, $HClO_4$, H_2SO_4, and $H_3PO_4 - H_3PO_4$ is here only because it loses its first H^+ readily);
   (c) all strong bases (these are the alkali metal and Ba^{++} hydroxides).
4. *Weak electrolytes* include all those substances that are not classed as strong electrolytes. A particularly important weak electrolyte is water.
5. *Common gases* at room temperature are: H_2, O_2, N_2, CO_2, CO, NO, N_2O, NO_2, Cl_2, NH_3, HCl, HCN, H_2S. Other substances, particularly H_2O, may be driven off as a gas at elevated temperatures.

Balancing Ionic Equations

A properly balanced chemical equation shows all the information we have just discussed. Soluble salts and strong electrolytes in aqueous solution are always written in ionic form, for example, $Na^+ + Cl^-$, not $NaCl$, or H^+ (or H_3O^+) $+ ClO_4^-$, not $HClO_4$. Insoluble salts and weak electrolytes are not written in ionic form, even though a minor fraction of each may actually exist in solution. For example, we indicate the formula for calcium carbonate as $CaCO_3$, not $Ca^{++} + CO_3^=$. Similarly, for slightly dissociated acetic acid, we use the molecular formula $HC_2H_3O_2$ instead of $H^+ + C_2H_3O_2^-$.

Insoluble products are usually indicated by ↓, and evolved gases by ↑. The following equations illustrate correct procedures:

$$BaCO_3 + 2\ HC_2H_3O_2 \rightarrow Ba^{++} + 2\ C_2H_3O_2^- + H_2O + CO_2\uparrow$$

$$BaCO_3 + 2H^+ + SO_4^= \rightarrow BaSO_4\downarrow + H_2O + CO_2\uparrow$$

$$BaCO_3 + 2H^+ + 2Cl^- \rightarrow Ba^{++} + 2Cl^- + H_2O + CO_2\uparrow$$

Whenever there are chemical species in solution that are *identical* on both sides of the equation, like Cl^- in the last example, this species does not participate in any way in the reaction, and it may be omitted from the final balanced equation:

$$BaCO_3 + 2\ H^+ \rightarrow Ba^{++} + H_2O + CO_2\uparrow$$

Such equations often emphasize the generality of a reaction. This last one makes it obvious that *any* strong acid will dissolve $BaCO_3$ to produce CO_2 and H_2O.

There are times when the conditions of an experiment determine whether a reaction goes. In the reaction

$$CdS + 2\ H^+ \rightleftharpoons Cd^{++} + H_2S\uparrow$$

CdS dissolves in an excess of strong acid because the equilibrium is shifted to the right and the H_2S is evolved to the atmosphere. However, if we should saturate an aqueous solution of Cd^{++} (with no added acid) with H_2S, we should actually obtain a precipitate of CdS. The reduced acidity shifts the equilibrium to the left, as does maintaining a constant high pressure of H_2S.

Remember that a weak electrolyte does not have to be a molecule; it may also be a weakly dissociated ion. For example, $Ca_3(PO_4)_2$ dissolves in an excess of strong acid, because the weak acid ion $HPO_4^=$ (as well as some $H_2PO_4^-$) is formed by the reaction

$$Ca_3(PO_4)_2 + 2\ H^+ \rightarrow 3\ Ca^{++} + 2\ HPO_4^=$$

In general, you could expect the insoluble salt of a weak acid to dissolve in an excess of strong acid for the reasons just given (extremely insoluble substances, such as some of the metal sulfides, are exceptions).

The last two examples illustrate how precipitation or solution of an insoluble compound may be controlled by controlling the *p*H (see also pp. 266-270).

SUMMARY OF PREDICTION PROCEDURE

1. First, decide whether an electron-transfer reaction is possible, using approximate half-reaction potentials (see p. 206). If it is, follow the procedure outlined on pp. 202-208. A reducing agent *and* an oxidizing agent must be present for electron transfer; one alone is not enough. If one of the reactants is a metal, it *must* be a reducing agent if anything.

2. If the reaction is non-electron-transfer, then treat it as a metathesis reaction.

   (a) Write a preliminary balanced molecular equation in which the reaction partners have been interchanged:

   $$A_2B + 2\,CD \rightarrow 2\,AC + BD_2$$

   (b) Study each compound in the molecular equation and decide whether it is an insoluble solid, a weak (or non-) electrolyte, or a gas. Rewrite the equation as is appropriate. If A_2B is a strong electrolyte, CD a weak electrolyte, AC a gas, and BD_2 an insoluble solid, then

   $$2A^+ + B^- + 2\,CD \rightleftharpoons 2\,AC\uparrow + BD_2\downarrow$$

   Eliminate all entities that are identical in form on both sides to get the final balanced ionic equation.

   (c) The reaction will go if any *one* of the products is a gas, a solid, or a weak electrolyte. If there is competition (insoluble solids or weak electrolytes on *both* sides), you must consider (1) whether one of the reactants is used in excess or high concentration to favor a shift to the right, and (2) the relative insolubility of solids or weakness of electrolytes.

SYNTHESIS

To prepare compounds you must use reactions that, by the principles just outlined, are predicted to "go." Not all reactions that take place, however, are suitable for synthesis, and attention must be paid to purity of product, cost of reactants, and ease of preparation. Comments on these factors accompany the answers to the A problems and the suggestions that follow for the preparation of X. If your laboratory work includes qualitative analysis, you might well consider each evolved gas and each precipitate to be a chemical preparation, even though the prime purpose of the project before you is separation or identification.

The general methods of synthesis outlined below are merely a collection of simple, common-sense, practical techniques presented in an organized manner. Some of the methods will already be familiar to you from previous

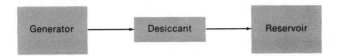

FIGURE 25-1
Schematic arrangement of apparatus needed for preparing a pure gas.

experience, and you will probably gain experience in additional methods as you progress through your present chemistry course.

1. *X* is a gas. For equipment, connect together a generator, a drying agent (probably), and a reservoir for collection, as shown schematically in Figure 25-1.

If possible, add one reagent to the other in a controlled manner through a funnel whose tip lies below the liquid level of the reaction mixture, as in Figure 25-2a. Warm if necessary. If the reaction involves heating a solid or a mixture of solids, then generator (b) can be used. Discard the first volume of gas generated as it sweeps the air out of the generator and connecting tubes. If possible, use relatively nonvolatile reagents.

If water is produced, or if water solutions are used in the generator, then an appropriate drying agent (desiccant) should be used. Some common desiccants are listed in Table 25-1, along with their principal limitations. Appropriate containers for the different types of desiccants are shown in Figure 25-3. Always pay attention to the *capacity* of the drying apparatus and chemicals, to avoid plugging up the desiccant tube as water accumulates, or else some part of the generator system will blow apart.

TABLE 25-1.
Desiccants.

Formula	Physical state	Type of container in Figure 25-3	Cannot be used to dry	Removes H_2O by forming
$Mg(ClO_4)_2$	Solid	a		$Mg(ClO_4)_2 \cdot 6\ H_2O$
P_4O_{10}	Solid	a	Basic gases	H_3PO_4
$CaCl_2$[a]	Solid	a		$CaCl_2 \cdot 6\ H_2O$
CaO[a]	Solid	a	Acidic gases	$Ca(OH)_2$
NaOH	Solid	a	Acidic gases	NaOH solution
SiO_2 (silica gel)[a]	Solid	a		Adsorbed water
H_2SO_4 (concentrated)	Liquid	b	Basic gases	Dilute H_2SO_4
Dry Ice (CO_2)[a]	Solid	c	Gases with b.p. $< -78°C$	Solid H_2O
Liquid N_2	Liquid	c	Gases with b.p. $< -195°C$	Solid H_2O

[a]Relatively cheap, safe, and simple to use.

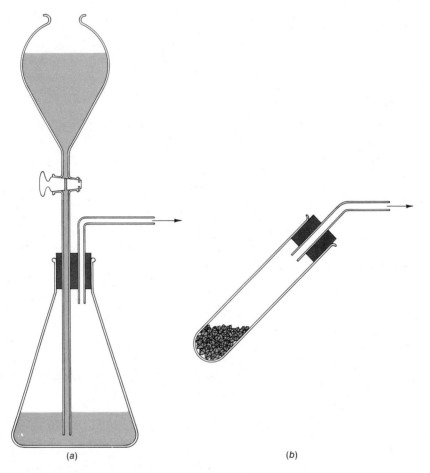

(a)   (b)

FIGURE 25-2
Simple gas generators.

The most satisfactory way to collect a gas is by upward displacement over mercury, if mercury is available and if the gas is unreactive toward it. If a gas is collected over water, don't dry it first. And, of course, collection over water is impossible if the gas is very soluble in water. If the gas is nontoxic and abundant, it may be collected by the upward displacement of air. For whatever upward displacement method you employ, collection bottle (a) in Figure 25-4 may be used, filled at the outset with mercury, water, or air. Bottle (b) may be used for collecting gases more dense than air by downward displacement. Bottles (a) and (b) can be capped with a glass plate and are simple for use in qualitative experiments. Reservoir (c) illustrates one way in which a gas may be collected and then used as desired at a later time.

Do not try to prepare gases such as CO_2 and SO_2 by burning C or S in an

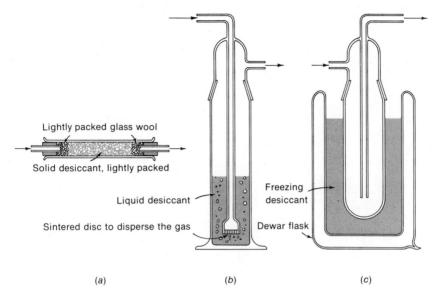

Lightly packed glass wool

Solid desiccant, lightly packed

Liquid desiccant

Freezing desiccant

Sintered disc to disperse the gas

Dewar flask

(a)   (b)   (c)

FIGURE 25-3
Simple methods for drying gases.

excess of air (or even pure O_2); the product will be contaminated by the huge excess of N_2, excess O_2, and other components in the air. A better method is to add an excess of acid to a carbonate or a sulfite contained in generator (a) in Figure 25-2.

2. X is a salt.

(a) X is insoluble. Choose two soluble reactants that by exchange of partners will give a precipitate of X; the other product must be soluble. No care is needed in measuring quantities, because X can be filtered from the excesses in solution, washed with distilled water, and dried.

(b) X is soluble.

(1) Exactly neutralize the appropriate acid with the appropriate base if both are soluble, and evaporate the water from the resulting salt. An indicator or a pH meter may be used to determine the endpoint.

(2) Add insufficient acid to an insoluble base (or vice versa, if appropriate) for complete reaction with the base, filter off and discard the excess base, and evaporate the water from the filtrate to get X. In this context, a base is defined as an oxide, a hydroxide, or a carbonate. The acid is completely used up, and therefore does not contaminate the filtrate.

(3) Add insufficient acid to a metal whose half-reaction (Table 15-1) lies above the H_2-H^+ half-reaction, filter off and discard the ex-

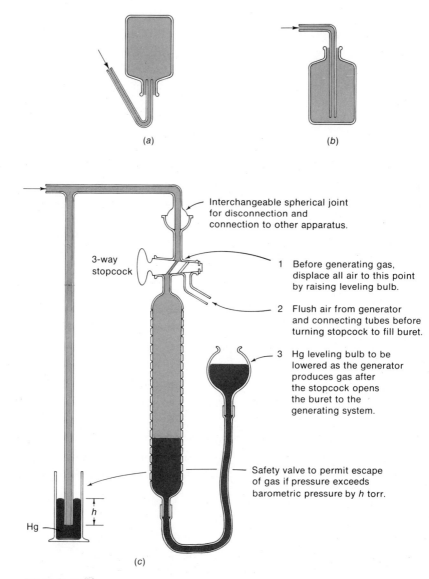

(a)

(b)

Interchangeable spherical joint for disconnection and connection to other apparatus.

3-way stopcock

1   Before generating gas, displace all air to this point by raising leveling bulb.

2   Flush air from generator and connecting tubes before turning stopcock to fill buret.

3   Hg leveling bulb to be lowered as the generator produces gas after the stopcock opens the buret to the generating system.

Safety valve to permit escape of gas if pressure exceeds barometric pressure by *h* torr.

Hg

(c)

FIGURE 25-4
Simple methods for collecting gases.

cess metal, and evaporate the water from the filtrate to get X.
(4) Add an excess of HNO_3 to a metal (almost any metal except Al, Pt, Sn, or Au) to produce the soluble metal nitrate. To this acidic solution add an excess of appropriate base to precipitate the metal hydroxide. (If the metal forms a soluble NH_3 complex, do *not* use NH_3; if it is amphoteric and forms a soluble OH^- com-

plex, do *not* use NaOH.) Treat this hydroxide as in (2) above to get the desired salt.

3. X is an acid. Many of the common inorganic acids can be easily prepared by heating the appropriate salt with a concentrated, nonvolatile acid, such as H_2SO_4 or with the more expensive H_3PO_4. In the following general scheme, the most commonly available or practical salt has been selected as a starting material, but Na^+ or almost any other metal ion could be used as well.

$$
\left.\begin{array}{l} CaF_2 \\ NaCl \\ NaBr \\ NaI \\ NaNO_3 \\ FeS \\ NH_4CN \\ CaCO_3 \\ Na_2SO_3 \end{array}\right\} \; + \; \begin{array}{c} \text{concentrated } H_2SO_4 \\ \text{or} \\ \text{concentrated } H_3PO_4 \end{array} \; \xrightarrow{\Delta} \; \left.\begin{array}{l} HF \\ HCl \\ HBr \\ HI \\ HNO_3 \\ H_2S \\ HCN \\ CO_2 \\ SO_2 \end{array}\right\} \uparrow \; + \; \begin{array}{c} \text{metal } SO_4^=, \\ HSO_4^- \\ \text{or} \\ \text{metal } PO_4^{\equiv}, \\ HPO_4^=, \\ H_2PO_4^- \end{array}
$$

A few specific comments on this preparative method are in order.

(a) All the acids prepared are gases at room temperature except HF and HNO_3.

(b) Only HCl, HBr, HI, and HNO_3 require *vigorous* heating. The others require little or none, depending on the salt used.

(c) H_2S and HCN are HIGHLY TOXIC gases, and must be produced with proper ventilation. With the exception of CO_2, all the gases are extremely irritating, corrosive, and obnoxious.

(d) An all-glass apparatus must be used for HNO_3 because it is corrosive to rubber, cork, etc.

(e) Hydrogen fluoride etches glass,

$$4 \, HF + SiO_2 \rightarrow SiF_4 \uparrow + 2 \, H_2O$$

Therefore all equipment for generation and collection of HF must be made of polyethylene, paraffin-coated glass, metallic lead, or something equally unreactive to HF.

(f) The preparation of HBr and HI requires H_3PO_4 in order to avoid the strong oxidizing action of hot concentrated H_2SO_4, which will also yield such products as the halogens, SO_2 and H_2S.

(g) In order to get H_2CO_3 and H_2SO_3, CO_2 and SO_2 must be passed into H_2O.

Many organic acids are insoluble in water, and so can be prepared from their soluble salts by simply adding a dilute solution of a strong inorganic acid to their aqueous solutions. Hot concentrated H_2SO_4 and H_3PO_4 should never be used with organic acids. All you would probably get is a mess of carbon.

4. X is a hydroxide. As noted by the solubility rules, most hydroxides are only slightly soluble and would be prepared as in 2(a). NaOH, KOH, and NH_3 are commonly available, but the choice of base must take into account the possibility that soluble OH^- complexes and NH_3 complexes may form. NH_3 is easily prepared, as a gas, by warming any NH_4^+ salt with an excess of NaOH.

5. Miscellaneous compounds.

(a) Metal oxides. With the exception of the alkali metals, prepare by vigorously heating the metal hydroxide to eliminate water. Do *not* try to burn the metal in air.

(b) Sulfides of alkaline earth metals, Al, and Cr. Prepare by heating the pulverized metals with an excess of sulfur and with protection from the air. They cannot be precipitated by H_2S from solution; they decompose when they come in contact with water.

(c) Acid salts, such as $NaHCO_3$, KH_2PO_4, or K_2HPO_4. Usually the acid solution (of unknown strength) to be used is divided into appropriate parts, one part completely and exactly neutralized, then the neutralized and unneutralized portions recombined. Water is evaporated. For example, (1) divide a given H_2SO_4 solution into two equal parts, (2) exactly neutralize one part with NaOH (its strength need not be known — simply add it until an indicator endpoint is reached) to give a Na_2SO_4 solution, (3) recombine the two parts and evaporate the water to get the residue of $NaHSO_4$. With H_3PO_4, you would exactly neutralize either one-third or two-thirds, depending on which salt you wanted, recombine, and evaporate the water. For $NaHCO_3$, $NaHSO_3$, and NaHS, simply saturate an NaOH solution with the appropriate gas (CO_2, SO_2, or H_2S), and the resulting solution will contain the desired acid salt.

(d) Metasalts and pyrosalts. Many solid acid salts serve as starting materials for the preparation of metasalts and pyrosalts because they lose water when heated vigorously. For example, dihydrogen salts will yield metasalts, such as

$$NaH_2AsO_4 \xrightarrow{\Delta} NaAsO_3 + H_2O \uparrow$$

and monohydrogen salts will yield pyrosalts:

$$2\,NaHSO_4 \xrightarrow{\Delta} Na_2S_2O_7 + H_2O \uparrow$$

(e) Metal hydrides. These compounds can usually be prepared by heating the metal in an excess of H_2 and cooling it in H_2. The temperature of treatment varies from about 150°C for some alkali metals to 700°C for others. Extreme care must be used when heating with H_2; the system should be absolutely leakproof against air, flames should not be used for heating or be in the vicinity, and the H_2-flow should be safely vented. Further, some hydrides are spontaneously inflammable in air; they should all be handled with extreme care, and never be thoughtlessly added to water, with which they react to form H_2. The alkali and alkaline earth metals form straightforward hydrides (MH and MH_2) composed of M^+ (or M^{++}) and H^-. The hydrides of other metals are usually nonstoichiometric (e.g., $CeH_{2.69}$, $TaH_{0.76}$, etc.), with the H atoms occupying interstitial positions in the metal's crystal lattice.

FRACTIONAL CRYSTALLIZATION

Mixtures of soluble crystalline substances are often separated by a process known as fractional crystallization. This method is based on differences in the rates of change of solubility, with temperature, of the constituents of the mixture. The following problem illustrates the method.

PROBLEM:
At 100°C the solubility of KNO_3 is 246 g per 100 ml of water, and that of NaCl is 40 g per 100 ml of water. At 10°C the solubility of KNO_3 is 21 g per 100 ml and that of NaCl is 36 g per 100 ml. A mixture contains 500 g each of KNO_3 and NaCl. What weights of the pure substances can be obtained in a single-step fractional crystallization?

SOLUTION:
The mixture is treated with just enough water to dissolve the more soluble constituent (KNO_3) at 100°C. Part of the NaCl is also dissolved, but a large fraction of it remains as a residue. This residue of pure NaCl is separated by filtration. When the solution is cooled to 10°C, much of the dissolved KNO_3 precipitates. Before cooling, enough water is added to keep all the dissolved NaCl in the solution, so the precipitate obtained is pure KNO_3.

Step 1. Treat the original mixture with enough water to dissolve all the KNO_3 at 100°C.

vol of water to dissolve 500 g KNO_3

$$= 500 \text{ g } KNO_3 \times \frac{100 \text{ ml water}}{246 \text{ g } KNO_3} = 203 \text{ ml}$$

wt of NaCl dissolved at 100°C in 203 ml of water

$$= 203 \text{ ml of water} \times \frac{40 \text{ g NaCl}}{100 \text{ ml of water}} = 82 \text{ g}$$

The undissolved NaCl, 500 g $-$ 82 g, or 418 g, is removed by filtration.

Step 2. Dilute the solution to the volume required to hold 82 g of NaCl in solution at 10°C (if this is not done the precipitate obtained on cooling will contain some NaCl):

vol of water to hold 82 g NaCl at 10°C

$$= 82 \text{ g NaCl} \times \frac{100 \text{ ml of water}}{36 \text{ g NaCl}} = 227 \text{ ml}$$

vol water added $= 227 \text{ ml} - 203 \text{ ml} = 24 \text{ ml}$

Step 3. Cool the solution to 10°C. Part of the KNO_3 precipitates, but none of the NaCl precipitates:

wt of KNO_3 in solution in 227 ml of water at 10°C

$$= 227 \text{ ml of water} \times \frac{21 \text{ g } KNO_3}{100 \text{ ml water}} = 48 \text{ g}$$

The remainder of the original KNO_3 precipitates:

wt of KNO_3 precipitate $= 500 \text{ g} - 48 \text{ g} = 452 \text{ g}$

This calculation shows that 452 g of pure KNO_3 and 418 g of pure NaCl can be obtained by a single crystallization. The remainder, 48 g of KNO_3 and 82 g of NaCl, remains in the solution. A further recovery can be made by concentrating the solution and repeating the process as described in the following problem.

PROBLEM:

What weights of pure salts can be recovered by further crystallization from the solution of the preceding problem?

SOLUTION:

The solution is evaporated at 100°C to the volume at which it is saturated with KNO_3:

vol of water to hold 48 g of KNO_3 at 100°C

$$= 48 \text{ g } KNO_3 \times \frac{100 \text{ ml of water}}{246 \text{ g } KNO_3} = 20 \text{ ml}$$

At 100°C the weight of NaCl dissolved in 20 ml of water is

wt of NaCl in solution $= 20 \text{ ml of water} \times \dfrac{40 \text{ g NaCl}}{100 \text{ ml of water}} = 8 \text{ g}$

wt NaCl precipitate $= 82 \text{ g} - 8 \text{ g} = 74 \text{ g}$

The precipitate is filtered off, as before. The solution is now diluted to the volume required to keep 8 g NaCl in solution at 10°C:

vol of water to hold 8 g of NaCl at 10°C

$$= 8 \text{ g NaCl} \times \frac{100 \text{ ml of water}}{36 \text{ g NaCl}} = 22 \text{ ml}$$

On cooling to 10°C, a second crop of KNO_3 crystals is obtained:

$$\text{wt of } KNO_3 \text{ remaining in solution} = 22 \text{ ml water} \times \frac{21 \text{ g } KNO_3}{100 \text{ ml of water}} = 5 \text{ g}$$

$$\text{wt of } KNO_3 \text{ precipitate} = 48 \text{ g} - 5 \text{ g} = 43 \text{ g}.$$

Of the original 500 g of each constituent the loss in the remaining solution is now only 8 g of NaCl and 5 g of KNO_3. In practice, the recovery is not as high as the computed values, because some solution is held up by the precipitate at each stage of the operation.

In the preceding problems the separation of KNO_3 from NaCl was described. If KCl and $NaNO_3$ are mixed in equimolar proportions, they will react in the saturated solutions at 100°C:

$$Na^+ + NO_3^- + K^+ + Cl^- = NaCl_{(s)} + K^+ + NO_3^-$$

because NaCl is the least soluble substance. The process of fractional crystallization will recover pure NaCl and KNO_3.

PROBLEMS A   Answers on page 354

1. None of the reactions listed below takes place. Explain why.
   (a) $2KNO_3 + ZnBr_2 \rightarrow\!\!\!\!/\ \ 2KBr + Zn(NO_3)_2$
   (b) $NaCl + H_2O \rightarrow\!\!\!\!/\ \ NaOH + HCl$
   (c) $Br_2 + 2NaCl \rightarrow\!\!\!\!/\ \ 2NaBr + Cl_2$
   (d) $CaCO_3 + 2NaCl \rightarrow\!\!\!\!/\ \ CaCl_2 + Na_2CO_3$
   (e) $2Au + 2H_3PO_4 \rightarrow\!\!\!\!/\ \ 2AuPO_4 + 3H_2$
   (f) $KMnO_4 + 5Fe(NO_3)_3 + 8HNO_3 \rightarrow\!\!\!\!/\ \ Mn(NO_3)_2 + 5Fe(NO_3)_4$
   $ + KNO_3 + 4H_2O$

   (g) $H_2S + MgCl_2 \rightarrow\!\!\!\!/\ \ MgS + 2HCl$

2. The following reactions go to the right. Rewrite each in ionic form, and tell why it goes to the right.
   (a) $Mg(OH)_2 + 2\ NH_4Cl \rightarrow MgCl_2 + 2\ NH_3 + 2\ H_2O$
   (b) $AgCl + 2NH_3 \rightarrow Ag(NH_3)_2Cl$
   (c) $Ag_2S + 4KCN \rightarrow K_2S + 2KAg(CN)_2$
   (d) $AgCl + KI \rightarrow AgI + KCl$
   (e) $SrSO_4 + Na_2CO_3 \rightarrow SrCO_3 + Na_2SO_4$
   (f) $FeSO_4 + H_2S + 2NaOH \rightarrow FeS + Na_2SO_4 + 2H_2O$
   (g) $NH_4Cl + NaOH \rightarrow NH_3 + H_2O + NaCl$
   (h) $KBr + H_3PO_4 \text{ (conc.)} \rightarrow HBr + KH_2PO_4$
   (i) $3CuS + 8HNO_3 \rightarrow 3Cu(NO_3)_2 + 3S + 2NO + 4H_2O$
   (j) $Ag(NH_3)_2Cl + KI \rightarrow AgI + KCl + 2NH_3$

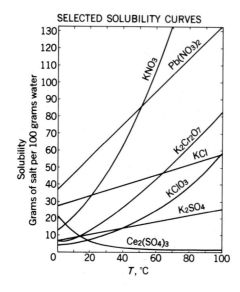

SELECTED SOLUBILITY CURVES

FIGURE 25-5
Selected solubility curves.

3. Give balanced ionic equations for the following reactions in aqueous solution. If a reaction does not occur, write NR. Indicate precipitates by ↓ and gases by ↑.

(a) $Pb(NO_3)_2 + NaCl \rightarrow$

(b) $CuSO_4 + NaCl \rightarrow$

(c) $AgNO_3 + H_2S \rightarrow$

(d) $Cd + H_3PO_4 \rightarrow$

(e) $BaCO_3 + HNO_3 \rightarrow$

(f) $Au + HNO_3 \rightarrow$

(g) $Ba_3(PO_4)_2 + HCl \rightarrow$

(h) $AlCl_3 + NH_3 \rightarrow$

(i) $Fe(OH)_3 + HCl \rightarrow$

(j) $FeS + NaCl \rightarrow$

(k) $Ca_3(PO_4)_2 + NaNO_3 \rightarrow$

(l) $CuO + H_2SO_4 \rightarrow$

(m) $NH_3 + HCl \rightarrow$

(n) $KMnO_4 + K_2SO_4 \rightarrow$

(o) $HC_2H_3O_2 + HCl \rightarrow$

(p) $AgC_2H_3O_2 + HCl \rightarrow$

(q) $Fe + Cu(NO_3)_2 \rightarrow$

(r) $CuO + NaOH \rightarrow$

(s) $KMnO_4 + H_3AsO_3 \rightarrow$

(t) $CuSO_4 + HCl \rightarrow$

(u) $Ba(NO_3)_2 + H_2SO_4 \rightarrow$

(v) $ZnS + H_2SO_4 \rightarrow$

(w) $ZnSO_4 + Na_2S \rightarrow$

(x) $Cl_2 + KBr \rightarrow$

(y) $KClO_3 + KI + HCl \rightarrow$

(z) $Br_2 + KCl \rightarrow$

4. Write balanced ionic equations for reactions that would be suitable for the practical *laboratory* preparation of each of the following compounds in pure form and good yield. You may have to supplement some equations with a few words of explanation. For gases, give a suitable desiccant if it is needed, and state what kind of a generator is used.

(a) $PbSO_4$

(b) $KC_2H_3O_2$

(c) CdS

(d) $KHCO_3$

(e) NH_3

(f) $Cu(OH)_2$

(g) $NaHSO_4$

(h) Al_2S_3

(i) Fe_2O_3

(j) Hg_2Cl_2

(k) CO_2

(l) Br_2

(m) Ag

(n) $KAsO_3$

5. Devise methods for the laboratory preparation and isolation of each of the following compounds in pure form and in good yield. Use pure metals as starting materials.
   (a) $Al(NO_3)_3$ (d) $HgCl_2$
   (b) $Cu_3(PO_4)_2$ (e) ZnS
   (c) $Ni(OH)_2$

6. Describe the stepwise procedure, stating volumes and temperatures, that you would use for the separation by fractional crystallization of each of the solid mixtures below. Calculate the number of grams of pure salts obtained in each case. Use the curves shown in Figure 25-5 for solubility data required (KNO_3 has a solubility of 246 grams per 100 grams of water at 100°C). The mixtures are
   (a) 1 mole each of KCl and $KClO_3$
   (b) 100 g of $K_2Cr_2O_7$ and 50 g of KCl
   (c) 50 g each of KNO_3 and K_2SO_4

PROBLEMS B   No answers given

7. None of the reactions listed below takes place. Explain why.
   (a) $Na_2SO_4 + CuCl_2 \rightarrow CuSO_4 + 2NaCl$
   (b) $Cl_2 + 2KF \rightarrow F_2 + 2KCl$
   (c) $H_2S + 2NaCl \rightarrow Na_2S + 2HCl$
   (d) $2H_2O + K_2SO_4 \rightarrow 2KOH + H_2SO_4$
   (e) $H_2SO_4 + 2Ag \rightarrow Ag_2SO_4 + H_2$
   (f) $Na_2Cr_2O_7 + 6Mg(NO_3)_2 + 14HNO_3 \rightarrow 2NaNO_3 + 6Mg(NO_3)_3$
   $$+ 7H_2O + 2Cr(NO_3)_3$$
   (g) $Cu + ZnSO_4 \rightarrow CuSO_4 + Zn$

8. The following reactions go to the right. Rewrite each in ionic form, and tell why it goes to the right.
   (a) $Fe(OH)_2 + H_2S \rightarrow FeS + 2H_2O$
   (b) $Al(OH)_3 + NaOH \rightarrow NaAlO_2 + 2H_2O$
   (c) $Al(OH)_3 + 3HCl \rightarrow AlCl_3 + 3H_2O$
   (d) $BiCl_3 + 3NH_3 + 3H_2O \rightarrow Bi(OH)_3 + 3NH_4Cl$
   (e) $K_2Cr_2O_7 + Pb(C_2H_3O_2)_2 + H_2O \rightarrow 2PbCrO_4 + 2HC_2H_3O_2$
   $$+ 2KC_2H_3O_2$$
   (f) $FeCl_3 + 6KSCN \rightarrow K_3Fe(SCN)_6 + 3KCl$
   (g) $H_2S + Br_2 \rightarrow S + 2HBr$
   (h) $PbSO_4 + 2NH_4C_2H_3O_2 \rightarrow Pb(C_2H_3O_2)_2 + (NH_4)_2SO_4$
   (i) $3HgS + 2HNO_3 + 6HCl \rightarrow 3HgCl_2 + 2NO + 3S + 4H_2O$
   (j) $SnS_2 + 2NH_4HS \rightarrow (NH_4)_2SnS_3 + H_2S$

9. Give balanced ionic equations for the following reactions in aqueous solution. If a reaction does not occur, write NR. Indicate precipitates by $\downarrow$ and gases by $\uparrow$.

(a) $NaOH + HNO_3 \rightarrow$

(b) $Cd + AgNO_3 \rightarrow$

(c) $Cu + HNO_3 \rightarrow$

(d) $Cu + HCl \rightarrow$

(e) $CuSO_4 + NaCl \rightarrow$

(f) $CaCl_2 + Na_3PO_4 \rightarrow$

(g) $Ca_3(PO_4)_2 + HCl \rightarrow$

(h) $Na_2CO_3 + KCl \rightarrow$

(i) $Na_2CO_3 + BaCl_2 \rightarrow$

(j) $H_2S + FeSO_4 \rightarrow$

(k) $KMnO_4 + HCl \rightarrow$

(l) $FeS + H_2SO_4 \rightarrow$

(m) $CaCl_2 + Na_2C_2O_4 \rightarrow$

(n) $CaC_2O_4 + HCl \rightarrow$

(o) $KNO_3 + NaCl \rightarrow$

(p) $AgNO_3 + NaCl \rightarrow$

(q) $Na_2Cr_2O_7 + SnSO_4 + H_2SO_4 \rightarrow$

(r) $PbO_2 + HI \rightarrow$

(s) $AgNO_3 + KBr \rightarrow$

(t) $AgCl + NH_3 \rightarrow$

(u) $Na_2CrO_4 + CuSO_4 \rightarrow$

(v) $CuSO_4 + NaOH \rightarrow$

(w) $CuSO_4 + H_3PO_4 \rightarrow$

(x) $Zn + H_2SO_4 \rightarrow$

(y) $ZnO + H_2SO_4 \rightarrow$

(z) $CuCrO_4 + HNO_3 \rightarrow$

10. Write balanced ionic equations for reactions that would be suitable for the practical *laboratory* preparation of each of the following compounds in pure form and good yield. You may have to supplement some equations with a few words of explanation. For gases, give a suitable desiccant if it is needed and state what kind of generator is used.

(a) SO_2

(b) $AgCl$

(c) MgO

(d) BaS

(e) $KHSO_4$

(f) $Cd(OH)_2$

(g) N_2O

(h) $NaHCO_3$

(i) Bi_2S_3

(j) $NaC_2H_3O_2$

(k) $BaSO_4$

(l) Cu

(m) Cl_2

(n) $K_4P_2O_7$

11. Devise methods for the laboratory preparation and isolation of each of the following compounds in pure form and in good yield. Use pure metals as starting materials.

(a) $Pb(C_2H_3O_2)_2$

(b) $MgSO_4$

(c) Ag_2SO_4

(d) CdS

(e) $SnCl_2$

12. Describe the stepwise procedure, stating volumes and temperatures, that you would use for the separation by fractional crystallization of each of the solid mixtures below. Calculate the number of grams of pure salts obtained in each case. Use the solubility curves of Problem 6 for solubility data required (KNO_3 has a solubility of 246 grams per 100 grams of water at $100°C$). The mixtures are:

(a) 100 g each of $KClO_3$ and $Ce_2(SO_4)_3$

(b) 1 mole each of $Pb(NO_3)_2$ and KNO_3

(c) 100 g of $K_2Cr_2O_7$ and 50 g of K_2SO_4

CHAPTER 2

1. (a) 10^5
   (b) $10^1 = 10$
   (c) 10^{-1}
   (d) $10^0 = 1$
   (e) 10^{-3}
   (f) $10^1 = 10$

2. (a) 6.225×10^6
   (b) 7.213×10^8
   (c) 3.24×10^2
   (d) 1.2×10^{-1}
   (e) 2.57×10^{-3}
   (f) 8.6×10^{-10}

3. (a) 1.2×10^6
   (b) 6×10^{10}
   (c) 2.2
   (d) 4.5×10^2
   (e) 2×10^{-22}
   (f) 8×10^3
   (g) 4×10^{-8}
   (h) 2×10^{-13}
   (i) 1.5×10^{-2}

4. (a) 6.00
   (b) 24.0
   (c) 0.500
   (d) 3.00
   (e) 8.50
   (f) 5.64
   (g) 3.52
   (h) 25.15
   (i) 2,740
   (j) 124.8
   (k) 9.00
   (l) 900
   (m) 22.1
   (n) 8.01×10^5
   (o) 2.25×10^{-4}
   (p) 2.215
   (q) 7.00
   (r) 22.15
   (s) 4.10
   (t) 7.35×10^{-2}
   (u) 27.0
   (v) 2.74×10^4
   (w) 104
   (x) 7.19×10^8
   (y) 1.16×10^{-6}
   (z) 1.391

5. (a) 1.21×10^6
   (b) 5.42×10^{10}
   (c) 2.16
   (d) 4.18×10^2
   (e) 2.23×10^{-22}
   (f) 8.14×10^3
   (g) 4.29×10^{-8}
   (h) 1.975×10^{-13}
   (i) 1.132×10^{-2}

6. (a) 0.4048
   (b) 2.9304
   (c) $\overline{2}.6628$ or $8.6628 - 10$
   (d) $\overline{12}.9777$ or $8.9777 - 20$
   (e) 3.5781
   (f) 6.2103

7. (a) 9,860
   (b) 0.0108
   (c) 3.00×10^{-5}
   (d) 3.267
   (e) 70.07
   (f) 4.999×10^{-11}

8. (a) 1.21×10^6
   (b) 5.40×10^{10}
   (c) 2.16
   (d) 4.18×10^2
   (e) 2.23×10^{-22}
   (f) 8.14×10^3
   (g) 4.29×10^{-8}
   (h) 1.98×10^{-13}
   (i) 1.13×10^{-2}

9. (a) 1.209×10^6
   (b) 5.421×10^{10}
   (c) 2.162
   (d) 4.180×10^2
   (e) 2.231×10^{-22}
   (f) 8.140×10^3
   (g) 4.289×10^{-8}
   (h) 1.975×10^{-13}
   (i) 1.132×10^{-2}

10. (a) $x = 5.25$
    (b) $v = 31$
    (c) $x = 9$
    (d) $F = 113$
    (e) $R = 5.56$
    (f) $P = 2.56$
    (g) $T_2 = 23.6$
    (h) $M = 12$
    (i) $x = 0.472$
    (j) $x = -0.425$
        and 1.175
    (k) $x = -2$ and $+1$

11. (a) 2^9 (b) 2^{-3} (c) 3^{11}

CHAPTER 3

1. 2.6784×10^6 sec   3. 20 sec   5. 1.20×10^3 lbs

2. $F = 8.64 \times 10^4$ sec/day   4. 720 mph

6. (a) $F = \left(\dfrac{1 \text{ lb}}{16 \text{ oz}}\right)\left(\dfrac{1 \text{ ton}}{2000 \text{ lb}}\right) = 3.125 \times 10^{-5}$ tons/oz

   (b) $F = \left(\dfrac{1 \text{ ft}}{12 \text{ in.}}\right)^3\left(\dfrac{1 \text{ yd}}{3 \text{ ft}}\right)^3 = 2.14 \times 10^{-4}$ yd^3/in.^{3}

   (c) $F = \left(\dfrac{1 \text{ mi}}{5280 \text{ ft}}\right)\left(60 \dfrac{\text{sec}}{\text{min}}\right)\left(60 \dfrac{\text{min}}{\text{hr}}\right) = 0.682 \dfrac{\text{mi/hr}}{\text{ft/sec}}$

   (d) $F = \left(2000 \dfrac{\text{lb}}{\text{ton}}\right)\left(\dfrac{1 \text{ yd}}{3 \text{ ft}}\right)^2\left(\dfrac{1 \text{ ft}}{12 \text{ in.}}\right)^2 = 1.54 \dfrac{\text{lb/in.}^2}{\text{ton/yd}^2}$

   (e) $F = \left(\dfrac{1\$}{100\cent}\right)\left(2000 \dfrac{\text{lb}}{\text{ton}}\right) = 20 \dfrac{\$/\text{ton}}{\cent/\text{lb}}$

   (f) $F = \left(\dfrac{1 \text{ min}}{60 \text{ sec}}\right)\left(\dfrac{1 \text{ hr}}{60 \text{ min}}\right)\left(\dfrac{1 \text{ day}}{24 \text{ hr}}\right)\left(\dfrac{1 \text{ wk}}{7 \text{ days}}\right) = 1.65 \times 10^{-6} \dfrac{\text{weeks}}{\text{sec}}$

   (g) $F = \left(60 \dfrac{\text{sec}}{\text{min}}\right)\left(12 \dfrac{\text{in.}}{\text{ft}}\right)^3\left(2.54 \dfrac{\text{cm}}{\text{in.}}\right)^3\left(\dfrac{1 \text{ liter}}{1000 \text{ cm}^3}\right)\left(\dfrac{1 \text{ qt}}{.94 \text{ liter}}\right) = 1{,}794 \dfrac{\text{qt/min}}{\text{ft}^3/\text{sec}}$

(h) $F = \left(5280 \dfrac{\text{ft}}{\text{mi}}\right)\left(\dfrac{1 \text{ fathom}}{6 \text{ ft}}\right) = 880 \dfrac{\text{fathom}}{\text{mi}}$

(i) $F = \left(3 \dfrac{\text{ft}}{\text{yd}}\right)\left(12 \dfrac{\text{in.}}{\text{ft}}\right)\left(1000 \dfrac{\text{mil}}{\text{in.}}\right) = 3.6 \times 10^4 \dfrac{\text{mil}}{\text{yd}}$

CHAPTER 4

1. $A = 72 \text{ cm}^2$, $V = 36 \text{ cm}^3$

2. $A = 197.8 \text{ cm}^2$, $V = 95.8 \text{ cm}^3$

3. 24¢

4. 44.1 lb

5. 76.5 miles

6. 23.65 cm

7. (a) 6,700 Å, 3,700 Å
   (b) 0.67μ, 0.37μ

8. (a) 1.537395×10^{-8} cm
   (b) 2.28503×10^{-8} cm
   (c) 7.0783×10^{-9} cm
   (d) 2.0862×10^{-9} cm

9. 69.8°F

10. −45.6°C, 227.4°K

11. −38.0°F, 234.13°K

12. −459.4°F

13. −40°C

14. (a) 6 cm²
    (b) 60 cm²
    (c) 6×10^6 cm², 0.134 football field

15. $3.33 \times 10^3 \text{m}^2$

16. 0.1 ml

17. $A = 8.89 \text{ cm}^2$, diam $= 1.682$ cm

18. (a) 2.99×10^{-23} ml/molecule
    (b) 1.93Å

19. 0.44 mm

20. 4.41×10^5 tons

21. 0.25 g

22. 200 ml

23. $\dfrac{\text{g}}{\text{cm} \times \text{sec}}$

24. erg sec

25. force $=$ mass $\times$ acceleration $= \text{g} \times \dfrac{\text{cm}}{\text{sec}^2} = $ dynes

pressure $= \dfrac{\text{force}}{\text{area}} = \dfrac{\text{g} \times \text{cm}}{\text{sec}^2 \times \text{cm}^2}$

volume $= \text{cm}^3$

pressure $\times$ volume $= \dfrac{\text{g} \times \text{cm} \times \text{cm}^3}{\text{sec}^2 \times \text{cm}^2} = \dfrac{\text{g} \times \text{cm}}{\text{sec}^2} \times \text{cm} = \text{dyne} \times \text{cm} = \text{ergs} = \text{work}$

26. (a) $35.1°M$ (b) $-59.2°M$

27. 1.581×10^{-5} light-year

28. 4×10^{-4} mm

CHAPTER 5

1. (a) 3   (c) 4   (e) 4   (g) 2   (i) 1
   (b) 3   (d) 5   (f) 3   (h) 2   (j) 1

2. (a) 15.60   (e) 1.60×10^3
   (b) 721.17   (f) 4.00×10^{17}
   (c) 1,308.4   (g) 3.401
   (d) 0.099   (h) 2.472×10^{-42}

3. 0.16 p.p.m.

4. (a) 5.3 p.p.t. $= 0.53\%$ (e) 0.032 p.p.t. $= 0.0032\%$
   (b) 5.3 p.p.t. $= 0.53\%$ (f) 13.2 p.p.t. $= 1.32\%$
   (c) 0.65 p.p.t. $= 0.065\%$ (g) 5.3 p.p.t. $= 0.53\%$
   (d) 9.7 p.p.t. $= 0.97\%$

5. av. dev. $= 4.6$ and $S = 6.0$
   90% confidence interval for a single value $= 2643 \pm 11.4$
   90% confidence interval for the mean $= 2643 \pm 4.0$

6. av. dev. $= 0.0034$
   95% confidence interval for the mean $= 5.017 \pm 0.0042$

7. av. dev. $= 0.10\%$
   70% confidence limits for a single value $= \pm 0.15\%$

8. av. dev. $= 3.2$ p.p.t.
   99% confidence limits for the mean $= \pm 4.2$ p.p.t.

9. av. dev. $= 2.9$ p.p.t.
   80% confidence limits for the mean $= \pm 2.3$ p.p.t.

10. $d = 0.997$ g/ml
    $\gamma = 72.0$ dynes/cm

11. 4 g

CHAPTER 6

1.

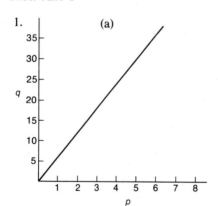

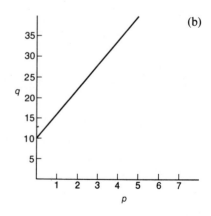

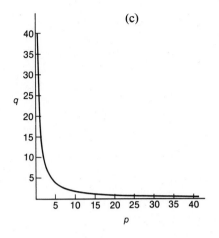

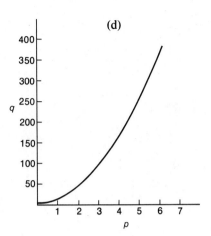

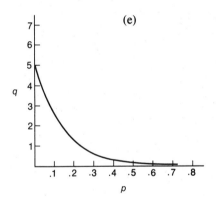

2.

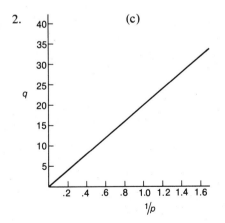

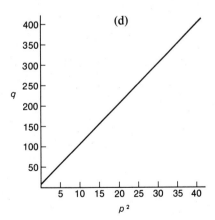

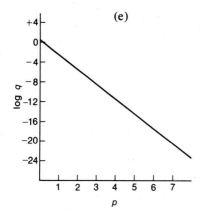

(e)

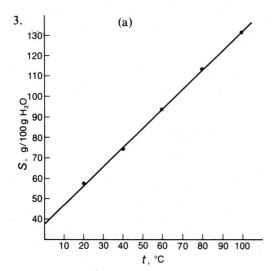

3. (a)

(b) $S = 0.95\,t + 37.0$

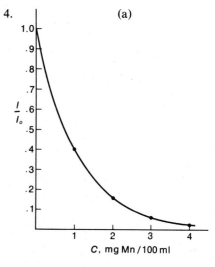

4. (a)

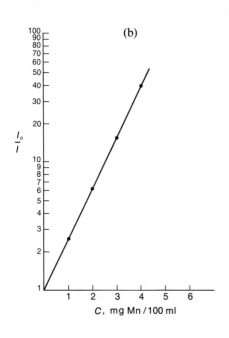

(b)

I_o/I

C, mg Mn /100 ml

(c) $\log\left(\dfrac{I^\circ}{I}\right) = 0.4C$

5. (a) $k = 6.12 \times 10^{-3}$
    $K = 3.986$

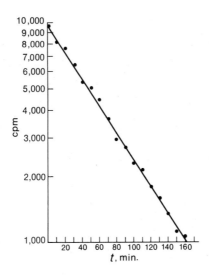

cpm

t, min.

(b) K is the logarithm of the cpm at $t = 0$.

(c) Let $(cpm)'$ = counts per min at the beginning

$(cpm)''$ = counts per min when material is half gone at $t_{\frac{1}{2}}$

$(cpm)'' = \frac{1}{2}(cpm)'$

$\log \dfrac{(cpm)''}{(cpm)'} = \log \frac{1}{2} = -\log 2 = -0.301 = -kt_{\frac{1}{2}}$

$$t_{\frac{1}{2}} = \frac{0.301}{k}$$

(d) 49.5 min

6. (a) $\log k = -5.27 \times 10^3 \times \dfrac{1}{T} + 23.28$

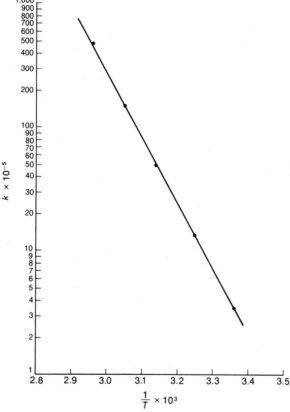

(b) 24.1 kcal

(c) Q corresponds to the logarithm of the rate constant at infinitely high temperature.

7. (a) $\log K_e = 665 \times \dfrac{1}{T} + 0.77$

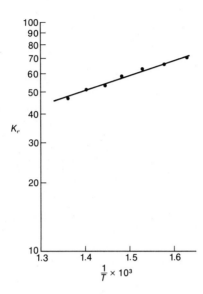

(b) 3,040 cal

(c) Z corresponds to the logarithm of the equilibrium constant at infinitely high temperature.

CHAPTER 7

1. 33.5 ml
2. 1,540 g
3. 7.81 ml
4. gold
5. 3.165 g/ml
6. 9.97 g/ml
7. 25.4032 g
8. 56.3668 g
9. 16.3428 g
10. 1.0041 ml
11. The correction for bouyancy is 0.1 mg, which is less than the 0.2 mg sensitivity of the balance.

12. 24.9250 g
13. 0.8018 g/ml
14. 1.6093
15. 34.0 g
16. (a) 5.869
    (b) 4.658
    (c) 5.869 and 4.640
17. 11 ¢/ml
18. 0.166 g/liter
19. (a) 497 g
    (b) 1,988 g
20. 0.90 pints
21. 0.15 cm

22. 3.16×10^4 lb   24. (a) 1.158 ml
23. 0.8021 g/ml   (b) 2.12 mm per small division

CHAPTER 8

1. calcium hydroxide
2. silver phosphate
3. silver thiocyanate
4. magnesium phthalate
5. ammonium sulfate
6. zinc sulfide
7. cadmium cyanide
8. barium iodate
9. cupric sulfite, copper(II) sulfite
10. cuprous iodide, copper(I) iodide
11. ferric nitrate, iron(III) nitrate
12. ferrous oxalate, iron(II) oxalate

13. mercurous chloride, mercury(I) chloride
14. manganous carbonate, manganese(II) carbonate
15. manganic hydroxide, manganese(III) hydroxide
16. nickelous hypochlorite, nickel(II) hypochlorite
17. chromic arsenate, chromium(III) arsenate
18. stannic bromide, tin(IV) bromide
19. chromous fluoride, chromium(II) fluoride
20. plumbous permanganate, lead(II) permanganate

21. sodium silicate
22. bismuthic oxide, bismuth(V) oxide
23. aluminum perchlorate
24. mercuric acetate, mercury(II) acetate
25. cesium chlorate
26. strontium hypoiodite
27. rubidium arsenite
28. beryllium nitride

29. calcium bicarbonate, calcium monohydrogen carbonate
30. antimonous nitrate, antimony(III) nitrate
31. phosphorus trichloride
32. bismuthous cyanate, bismuth(III) cyanate
33. aluminum thiosulfate

34. $Al(BrO_3)_3$
35. Hg_3PO_4
36. Bi_2O_3
37. $Sr(HCO_3)_2$
38. AuI
39. $Cr(IO_3)_3$
40. $Mn(OH)_2$
41. Li_3As
42. $As_2(SO_4)_3$
43. $SnCl_4$
44. $Ni(IO_4)_2$
45. Cl_2O_7

46. $Ag_2C_2O_4$
47. $Cr_3(BO_3)_2$
48. Sb_2S_3
49. $Al(C_2H_3O_2)_3$
50. CaC_2O_4
51. $NaClO_2$
52. $Sn(N_3)_2$
53. $Hg(CN)_2$
54. $(NH_4)_2SO_3$
55. $Co(MnO_4)_2$
56. $PbCO_3$
57. Zn_3P_2

58. Cu_2SiO_4
59. $Ba(IO)_2$
60. (a) $ZnSe$
    (b) Fr_3PO_4
    (c) $CoSeO_3$
    (d) SeO_2
    (e) Fr_2SeO_4
    (f) SeF_6
    (g) FrH

CHAPTER 9

1. Only major resonance forms are considered in providing these approximate answers. The electron-pair geometries on which the sketches are based are referred to by numbers in parentheses.

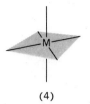 

(1)   (2)   (3)   (4)   (5)

FIGURE A9-1
Electron-pair geometries.

Shape	Bond Angles	Bond Distance, Å	Bond energies, kcal/bond
(a) O=C=O	180°	1.21	
(b) tetrahedral (2)	109° 28′	1.83	104.3
(c) Δ pyramid (2)	< 109° 28′	1.70	65.1
(d) F—Xe—F	180°		
(e) angular (1)	< 120°	~1.25	
(f) angular (1)	> 120°	~1.25	
(g) O=N=O⁺	180°	1.15	
(h) angular (1)	< 120°	N—F = 1.34	60.6
		N=O = 1.15	
(i) seesaw (3)	4 angles, 90°	2.16	58.3
	1 angle, < 120°		
	1 angle, < 180°		
(j) octahedral (4)	90°	1.81	150.2
(k) Δ coplanar (1)	~120°	N—Cl = 1.69	48.2
		N—O = ~1.25	
(l) dodecahedral			
(m) square coplanar (4)	90°	1.78	74.6
(n) octahedral (4)	90°		
(o) O—C≡N⁻	180°	C—O = 1.43	81.2
		C≡N = 1.15	
(p) tetrahedral (2)	~109° 28′	S—S = 2.08	50.9
		S—O = 1.70	65.1
(q) pentagonal bypyramid (5)			
(r) Δ coplanar about each N (1)	O—N—O = ~120°	N—N = 1.40	38.4
	O—N—N = ~120°	N—O = 1.36	41.6
(s) Δ coplanar (1)	H—C—H = < 120°	C—H = 1.07	97.3
		C=O = 1.43	
(t) square-base pyramid (4)	90°	1.78	74.6
(u) T-shape (3)	Cl—I—Cl = ~90°	2.32	52.8
(v) Δ pyramid (2)	< 109° 28′	2.25	47.3
(w) tetrahedral (2)	O—S—O = < 109° 28′	S—O = 1.70	65.1
	Cl—S—Cl = > 109° 28′	S—Cl = 2.03	60.2
(x) Δ coplanar about each C; all atoms in same plane	C—C—C = 120°	C—C = 1.43	
	H—C—C = 120°	C—H = 1.07	97.3
(y) [structure: H₂C=CHCl coplanar]	~120°	C—H = 1.07	97.3
		C=C = 1.21	
		C—Cl = 1.76	76.3
(z) [structure: —O—Si—O—Si—O—Si— tetrahedral about each Si]	109° 28′	1.83	104.3

(y)
H   H
   C=C
H   Cl
coplanar

(z)
  O   O   O
  Si   Si   Si
O O O O
  O   O   O
tetrahedral about each Si

CHAPTER 10

1. (a) 63.6% N, 36.4% O
   (b) 46.7% N, 53.3% O
   (c) 30.4% N, 69.6% O
   (d) 32.4% Na, 22.5% S, 45.1% O
   (e) 29.0% Na, 40.4% S, 30.6% O
   (f) 14.3% Na, 9.9% S, 6.2% H, 69.6% O
   (g) 18.5% Na, 25.7% S, 4.0% H, 51.8% O
   (h) 43.5% Ca, 26.1% C, 30.4% N
   (i) 29.2% N, 8.3% H, 12.5% C, 50.0% O
   (j) 47.5% U, 5.6% N, 2.4% H, 44.5% O
   (k) 56.1% C, 7.6% H, 8.2% N, 9.4% S, 18.7% O

2. Units are grams:

(a) 44	(d) 142.1	(g) 248.2	(j) 502.1
(b) 30	(e) 158.2	(h) 92	(k) 342.1
(c) 46	(f) 322.1	(i) 96	

3. Units are moles:

(a) 10.3	(d) 3.19	(g) 1.83	(j) 0.905
(b) 15.1	(e) 2.87	(h) 4.94	(k) 1.33
(c) 9.88	(f) 1.41	(i) 4.73	

4. (a) FeO (f) CH
   (b) Fe_2O_3 (g) CH_4
   (c) Fe_3O_4 (h) $Mg_2P_2O_7$
   (d) K_2CrO_4 (i) $Ag_4V_2O_7$
   (e) $K_2Cr_2O_7$ (j) $Sn_2Fe(CN)_6$

5. (a) $CuSO_4 \cdot 5H_2O$ (d) $CoCl_2 \cdot 6H_2O$
   (b) $Hg(NO_3)_2 \cdot \frac{1}{2}H_2O$ (e) $CaSO_4 \cdot 2H_2O$
   (c) $Pb(C_2H_3O_2)_2 \cdot 3H_2O$

6. (a) Cu_2S (d) $AlCl_3$
   (b) Mg_3N_2 (e) La_2O_3
   (c) LiH

7. (a) 9, 18, 27, 36   (b) MCl_3

8. 58.9 or some multiple of this

9. 334 g/mole

10. C_6H_6OS

CHAPTER 11

1. 240 ml
2. 840 ml
3. 203 ml
4. (a) 402 ml
   (b) 389 ml
5. 0.279 g
6. 830 ml
7. 133 g/mole
8. 2.73 g/liter
9. (a) 1.965 g/liter
   (b) 1.730 g/liter
10. 68.2 g/mole
11. 276 ml
12. 498 ml
13. 22.0 ml
14. 619 cm
15. 13.9 lb/in.²
16. 475 torr
17. 106.8 torr
18. 35.6 g/mole
19. 0.35 atm
20. (a) 0.075 atm N_2,
    0.25 atm N_2O,
    0.175 atm CO_2
    (b) total $p = 0.325$ atm,
    0.075 atm N_2,
    0.25 atm N_2O

21. 3.3 atm
22. 9.54 g/mole
23. 87.9 g/mole
24. $C_4H_8O_2$
25. 36.8 lb/in.²
26. 48.9 sec
27. 1.66 ml/min
28. 3.54×10^{10} molecules/ml
29. 30.7 kg
30. C_2H_4NO
31. 21.1% CO_2, 14.5% H_2,
    24.4% CO,
    40.0% N_2
32. (a) $P_1 = \dfrac{\pi r^2 lh}{V - \pi r^2 lh} \cong \dfrac{\pi r^2 lh}{V}$
    (b) 9.42×10^{-2} torr
33. (a) 7.94×10^3 liters/day
    (b) 10.24 kg/day
34. (a) 314 liters/day
    (b) 617 g/day
35. 74.8%
36. 11.7 g
37. (a) Balloon would rise
    (b) 2.9 g
38. 28.4% H_2 and 71.6% O_2

CHAPTER 12

1. The answers to these problems are given as the numerical coefficients needed to balance the equations, and in the order in which the substances appear in the unbalanced equations:

   (a) 2,2,1
   (b) 2,2,4,1
   (c) 2,2,2,1
   (d) 3,4,1,4

   (e) 1,3,2,3
   (f) 1,4,3,4
   (g) 1,1,1
   (h) 1,4,2,1,1

   (i) 1,2,1,1
   (j) 1,2,2,1,1
   (k) 1,3,3,1,3
   (l) 1,4,1,3,2

2. (a) (i) 395 g
     (ii) 1,170 g

  (b) 16.2 tons

  (c) 5.35 g NH_4Cl, 2.81 g CaO

  (d) 1.53×10^3 g

  (e) 7.8 liters HCl, 5.05 kg $CaCO_3$

  (f) (i) 1,031 g
      (ii) 561 ml

  (g) (i) 4.45 kg
      (ii) 3.06 liters

  (h) 157 g

  (i) $36.45 for NH_4Cl, $22.91 for $NaNO_3$

  (j) 86.0%

  (k) 76.1%

3. 8.75 g

4. (a) 7.62 g
  (b) 0.1165 mole

5. 1.5 liters

6. 1×10^6 ft³

7. 3.87 liters

8. (a) 17.4 g
  (b) 0.8 mole

9. 241 ml

10. (a) 1,345 liters
   (b) 6,410 liters

11. (a) 7.05 g
   (b) 0.357 mole
   (c) 29.8 ml
   (d) 5.61 g

12. (a) 1.81 g
   (b) 7.6×10^{-2} mole
   (c) 7.6 ml

13. 2.41×10^{-4} ml

14. 266 g

15. 1.74×10^4 ft³

16. 2.5×10^7 liters

17. 1.93×10^7 ft³

18. 8.33×10^5 ft³

19. 14.6%

20. 0.247 g

21. (a) 11.45%
   (b) $C_6H_6N_2O_3$

CHAPTER 13

1. Dissolve each of the following weights in distilled water, then dilute to the volumes requested in the problems:
  (a) 382 g   (b) 27.4 g   (c) 4.71 g   (d) 4,040 g   (e) 20.6 g

2. Dilute each of the following volumes with distilled water to the volumes requested in the problems:
  (a) 0.5 ml   (b) 56 ml   (c) 68.4 ml   (d) 0.262 ml   (e) 8.14 ml

3.

	molarity	molality						
(a)	8.40	9.6	(d)	3.41	4.28	(g)	0.535	0.534
(b)	11.45	18.6	(e)	0.871	0.954	(h)	0.805	0.825
(c)	17.74	193.6	(f)	0.834	0.861			

4. (a) 1.59 g   (e) 2.02 g
   (b) 0.703 g   (f) 484 g
   (c) 0.182 g   (g) 37.2 g
   (d) 0.0911 g

5. (a) 9.26 ml diluted to 500 ml
   (b) 0.787 g diluted to 500 ml
   (c) 3.93 g diluted to 1 liter
   (d) 315 ml diluted to 1 liter
   (e) dissolve 0.25 g of Al in 4.63 ml
       of $6M$ HCl and dilute to 250 ml
   (f) 4.61 g diluted to 2 liters

6. 22.9%

7. $4.2M$

8. 767 ml

9. (a) $0.1688M$
   (b) 6.161 mg

10. $0.4812M$

11. $0.2151M$

12. 13.76 ml

13. $0.0656M$

14. 50.40 ml

15. 4.931%

16. 44.0%

17. 71.26%

18. (a) 192.1 ml
    (b) dilute to 390 ml

19. (a) 0.8604 g
    (b) 0.3510 g
    (c) 25.00 ml

20. 51.3 ml

21. 54.0 ml

22. 162 g
    333 ml
    13.35 liters

23. 210

24. 32.35%

25. 35.8% oxalic acid,
    64.2% benzoic acid

CHAPTER 14

1.

A. Effect on		B. Effect on		C. Effect on	
position	K_e	position	K_e	position	K_e
R	O	O	O	L	O
R	+	R	+	L	−
	O	R	O	R	O
O	O	O	O	O	O
R	O	O	O	L	O

2. 1.06

3. 1.1×10^{-13}

4. (a) $4.2 \times 10^{-7}\%$
   (b) 0.28%

5. (a) 30.6% dissociation
   (b) 36.4% dissociation

6. (a) 66.0% conversion
   (b) 83.0% conversion

7. (a) $CO = H_2O = 23\%$ (b) $CO_2 = H_2 = 14.5\%$
   $CO_2 = H_2 = 27\%$ $CO = 2.2\%$
                               $H_2O = 68.8\%$

CHAPTER 15

1. (i) (a) 3.16 v   (ii) (a) 2.98 v
      (b) 0.45 v   (b) 0.51 v
      (c) 1.21 v   (c) 1.27 v
      (d) 0.92 v   (d) 1.16 v

2. Arrange beakers and salt bridge as in Fig. 15-1. Use solutions and electrodes as follows:
   (a) Mg electrode in Mg^{++} solution in left beaker. Ag electrode in Ag^+ solution in right beaker.
   (b) Cu electrode in Cu^{++} solution in left beaker. Hg electrode in Hg_2^{++} solution in right beaker.
   (c) Fe electrode in Fe^{++} solution in left beaker. Pt electrode in $Fe^{++} + Fe^{+++}$ solution in right beaker.
   (d) Pt electrode in $Sn^{++} + Sn^{++++}$ solution in left beaker. Pt electrode in liquid Br_2 in contact with Br^- solution in right beaker.

3. (a) 1×10^{107}
   (b) 2×10^{19}
   (c) 6.5×10^9
   (d) 1.26×10^{31}

4. (a) -146 kcal/mole Mg
   (b) -26.3 kcal/mole Cu
   (c) -55.7 kcal/mole Fe metal
   (d) -42.5 kcal/mole Sn

5. Half-cell reactions are:
   $2 e^- + Fe(OH)_2 \rightleftarrows 2 OH^- + Fe$
   $e^- + Ni(OH)_3 \rightleftarrows OH^- + Ni(OH)_2$
   Fe is the negative electrode.

6. + electrode   − electrode
   (a) O_2 H_2
   (b) I_2 Cd
   (c) O_2 Zn
   (d) Cl_2 Au
   (e) O_2 H_2

7. (a) 28.0 liters
   (b) 56.0 liters
   (c) 56.0 liters

8. (a) 8 g
   (b) 35.5 g
   (c) 12.15 g
   (d) 29.45 g
   (e) 119.6 g

9. 2.98 amp

10. 3.22 hr

11. $0.041M$

12. $2.3M$. As much Ni goes into solution as is removed.

13. 32.7 hr

14. 1.128×10^5 amp hr

15. 265 amp hr

CHAPTER 16

1. The oxidation numbers are given for the atoms in the order in which they occur in the given molecule or ion.

(a) +4, −2   (f) +1, −1
(b) +1, +5, −2   (g) +3, −2
(c) +2, −1   (h) +4, −1
(d) +3, +6, −2   (i) +3, −2
(e) +2, +5, −2   (j) +6, −2

2. (a) $HNO_2 + H_2O \rightleftarrows 3H^+ + NO_3^- + 2e^-$
   (b) $H_3AsO_3 + H_2O \rightleftarrows H_3AsO_4 + 2H^+ + 2e^-$
   (c) $Al \rightleftarrows Al^{+++} + 3e^-$
   (d) $Ni \rightleftarrows Ni^{++} + 2e^-$
   (e) $Hg_2^{++} \rightleftarrows 2Hg^{++} + 2e^-$
   (f) $H_2O_2 \rightleftarrows O_2 + 2H^+ + 2e^-$
   (g) $2I^- \rightleftarrows I_2 + 2e^-$

3. (a) $PbO_2 + 4H^+ + 2e^- \rightleftarrows Pb^{++} + 2H_2O$
   (b) $NO_3^- + 2H^+ + e^- \rightleftarrows NO_2 + H_2O$ (conc'd acid)
       $NO_3^- + 4H^+ + 3e^- \rightleftarrows NO + 2H_2O$ (dilute acid)
   (c) $Co^{+++} + e^- \rightleftarrows Co^{++}$
   (d) $ClO_4^- + 8H^+ + 8e^- \rightleftarrows Cl^- + 4H_2O$
   (e) $BrO^- + 2H^+ + 2e^- \rightleftarrows Br^- + H_2O$
   (f) $Ag^+ + e^- \rightleftarrows Ag$
   (g) $F_2 + 2e^- \rightleftarrows 2F^-$
   (h) $Sn^{++} + 2e^- \rightleftarrows Sn$

4. (a) $Zn + Cu^{++} \rightarrow Zn^{++} + Cu$
   (b) $Zn + 2H^+ + SO_4^= \rightarrow Zn^{++} + SO_4^= + H_2$
   (c) $Cr_2O_7^= + 6I^- + 14H^+ \rightarrow 2Cr^{+++} + 3I_2 + 7H_2O$
   (d) $2MnO_4^- + 10\,Cl^- + 16H^+ \rightarrow 2Mn^{++} + 5Cl_2 + 8H_2O$
   (e) $ClO_3^- + 6Br^- + 6H^+ \rightarrow Cl^- + 3Br_2 + 3H_2O$
   (f) $2MnO_4^- + 5H_2O_2 + 6H^+ \rightarrow 2Mn^{++} + 5O_2 + 8H_2O$
   (g) $MnO_2 + 4H^+ + 2Cl^- \rightarrow Mn^{++} + Cl_2 + 2H_2O$
   (h) $3Ag + 4H^+ + NO_3^- \rightarrow 3Ag^+ + NO + 2H_2O$
   (i) $PbO_2 + Sn^{++} + 4H^+ \rightarrow Pb^{++} + Sn^{++++} + 2H_2O$
   (j) $2MnO_4^- + 5H_2C_2O_4 + 6H^+ \rightarrow 2Mn^{++} + 10CO_2 + 8H_2O$
   (k) $H_2O_2 + HNO_2 \rightarrow H^+ + NO_3^- + H_2O$
   (l) $Fe^{++} + Zn \rightarrow Fe + Zn^{++}$
   (m) $2Fe^{+++} + 2I^- \rightarrow 2Fe^{++} + I_2$
   (n) $ClO_4^- + 4H_3AsO_3 \rightarrow Cl^- + 4H_3AsO_4$
   (o) $H_2S + ClO^- \rightarrow S + Cl^- + H_2O$

5. (a) $Zn + CuSO_4 \rightarrow ZnSO_4 + Cu$
   (b) $Zn + H_2SO_4 \rightarrow ZnSO_4 + H_2$
   (c) $K_2Cr_2O_7 + 6KI + 7H_2SO_4 \rightarrow Cr_2(SO_4)_3 + 3I_2 + 7H_2O + 4K_2SO_4$
   (d) $2KMnO_4 + 10KCl + 8H_2SO_4 \rightarrow 2MnSO_4 + 5Cl_2 + 8H_2O + 6K_2SO_4$
   (e) $KClO_3 + 6KBr + 3H_2SO_4 \rightarrow KCl + 3Br_2 + 3H_2O + 3K_2SO_4$
   (f) $2KMnO_4 + 5H_2O_2 + 3H_2SO_4 \rightarrow 2MnSO_4 + 5O_2 + 8H_2O + K_2SO_4$
   (g) $MnO_2 + 2H_2SO_4 + 2KCl \rightarrow MnSO_4 + Cl_2 + 2H_2O + K_2SO_4$
   (h) $3\,Ag + 4HNO_3 \rightarrow 3\,AgNO_3 + NO + 2H_2O$
   (i) $PbO_2 + SnSO_4 + 2H_2SO_4 \rightarrow PbSO_4 + Sn(SO_4)_2 + 2H_2O$

(j) $2KMnO_4 + 5H_2C_2O_4 + 3H_2SO_4 \rightarrow 2MnSO_4 + 10CO_2 + 8H_2O + K_2SO_4$

(k) $H_2O_2 + HNO_2 \rightarrow HNO_3 + H_2O$

(l) $FeSO_4 + Zn \rightarrow Fe + ZnSO_4$

(m) $2Fe_2(SO_4)_3 + 2KI \rightarrow 2FeSO_4 + I_2 + K_2SO_4$

(n) $KClO_4 + 4H_3AsO_3 \rightarrow KCl + 4H_3AsO_4$

(o) $H_2S + KClO \rightarrow S + KCl + H_2O$

6. (a) $CH_2O + Ag_2O + OH^- \rightarrow 2Ag + HCO_2^- + H_2O$

(b) $C_2H_2 + 2MnO_4^- + 6H^+ \rightarrow 2CO_2 + 2Mn^{++} + 4H_2O$

(c) $2C_2H_3OCl + 3Cr_2O_7^= + 24H^+ \rightarrow 4CO_2 + Cl_2 + 6Cr^{+++} + 15H_2O$

(d) $6Ag^+ + AsH_3 + 3H_2O \rightarrow 6Ag + H_3AsO_3 + 6H^+$

(e) $2OH^- + CN^- + 2Fe(CN)_6^\equiv \rightarrow CNO^- + 2Fe(CN)_6^\equiv + H_2O$

(f) $3C_2H_4O + 2NO_3^- + 2H^+ \rightarrow 2NO + 3C_2H_4O_2 + H_2O$

7. (a) $4Zn + 10H^+ + NO_3^- \rightarrow 4Zn^{++} + NH_4^+ + 3H_2O$

(b) $2S_2O_3^= + I_2 \rightarrow S_4O_6^= + 2I^-$

(c) $IO_3^- + 5I^- + 6H^+ \rightarrow 3I_2 + 3H_2O$

(d) $Cu + 2H_2SO_4 \rightarrow CuSO_4 + SO_2 + 2H_2O$

(e) $Zn + 2Fe^{+++} \rightarrow 2Fe^{++} + Zn^{++}$

(f) $8I^- + 5H_2SO_4 \rightarrow 4I_2 + H_2S + 4SO_4^= + 4H_2O$

(g) $3ClO^- \rightarrow 2Cl^- + ClO_3^-$

CHAPTER 17

1. (a) 2.43 g
   (b) 14.2 g
   (c) 234 g

2. (a) 32.3 ml
   (b) 1.74 ml
   (c) 96.0 ml

3. 8.33 liters

4. (a) 26.67 ml
   (b) 224.4 mg
   (c) 212 mg
   (d) 112 mg
   (e) 130.8 mg

5. 266.7 ml

6. $0.3229N$

7. $0.4650N$

8. $0.07540N$

9. 574.4 mg

10. $0.1642N$

11. (a) $0.2020N$
    (b) $0.2104N$

12. 97.2 ml

13. (a) 130.3 g
    (b) 500 ml HCl
    (c) 37.1 liters

14. 20.10 ml

15. (a) $0.3104N$
    (b) 0.62 ml

16. 45.96%

CHAPTER 18

1. (a) 5.12°C
   (b) 0.36°C

(c) 0.928
(d) 92.8 torr

(e) 82.68°C

2. 98.4 g/mole

3. −0.604°C

4. 84.65°C

5. 169 g/mole

6. S_8

7. $C_8H_{10}O_2$

8. $C_{10}H_6N_2O_4$

9. 48.7

10. (a) 1.21 molal, mole fraction 0.0213
    (b) 2.92 molal, mole fraction 0.1496
    (c) 1.296 molal, mole fraction 0.0562
    (d) 0.333 molal, mole fraction 0.00596
    (e) 4.39 molal, mole fraction 0.255

11. 13.9 qt

12. 45.5 g in one and 204.5 g in the other

13. 6.1 torr

14. The apparent molecular weight of 121.9 in water indicates a very slight dissociation in water into benzoate and H^+ ions. The apparent molecular weight of 244 in benzene indicates a polymerization in benzene into a compound of the formula $(C_7H_6O_2)_2$.

15. 41.0 atm

16. 3.50×10^4 g/mole

17. 51 units

18. 20.9 torr

19. 2.1%

20. 79.0%

21. −0.0194°C

22. 78.3%

23. 91.0%

CHAPTER 19

1. (a) 4
   (b) 6
   (c) 8
   (d) −1
   (e) 1.92
   (f) 1.05
   (g) 4.43
   (h) 7.19
   (i) −0.54
   (j) 0.30

2. (a) 10
   (b) 8
   (c) 6
   (d) 15
   (e) 12.40
   (f) 12.90
   (g) 9.67
   (h) 6.41
   (i) 14.81
   (j) 13.86

3. (a) 2.45×10^{-4} mole/liter
   (b) 3.02×10^{-8} mole/liter
   (c) 3.72×10^{-14} mole/liter
   (d) 0.17 mole/liter
   (e) 3.55×10^{-7} mole/liter
   (f) 1.10×10^{-9} mole/liter
   (g) 1.00 mole/liter
   (h) 1.58×10^{-3} mole/liter
   (i) 3.98 moles/liter
   (j) 1.58×10^{-15} mole/liter

4. (a) 10
   (b) 8
   (c) 6
   (d) 15
   (e) 12.08
   (f) 12.95
   (g) 9.57
   (h) 6.81
   (i) 14.54
   (j) 13.70

5. (a) 1.00
   (b) 1.18
   (c) 1.48
   (d) 3.00
   (e) 7.00
   (f) 11.00
   (g) 11.68

6. (a) 13.00
   (b) 12.82
   (c) 12.52
   (d) 11.00
   (e) 7.00
   (f) 3.00
   (g) 2.32

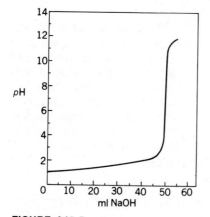

FIGURE A19-5.
Titration of HCl with NaOH; pH against ml of NaOH.

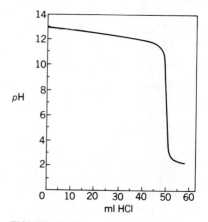

FIGURE A19-6.
Titration of NaOH with HCl; pH against ml of HCl.

CHAPTER 20

1. (a) 4.00×10^{-10}
   (b) 1.82×10^{-5}
   (c) 1.82×10^{-5}
   (d) 1.82×10^{-5}
   (e) 2.12×10^{-4}

2. (a) 1.82×10^{-5}
   (b) 4×10^{-10}
   (c) 1.81×10^{-5}
   (d) 2.22×10^{-4}
   (e) 6.15×10^{-4}

3. (a) 2.0
   (b) 2.86
   (c) 5.62
   (d) 2.67
   (e) 1.74
   (f) 11.72
   (g) 12.22
   (h) 9.22
   (i) 10.48
   (j) 11.98

4. (a) $3.66 \times 10^{-2}\%$
   (b) 1.83%
   (c) 2.44%
   (d) 0.15%
   (e) 0.12%

5. (a) 2.76
   (b) 4.61

6. 117 g

7. 4.92

8. 8.46

9. 305 g

10. 9.16

11. 4.74

12. 9.08

13. (a) 6.61×10^{-10}
    (b) 2.5×10^{-5}
    (c) 1.9×10^{-11}
    (d) 4.78×10^{-11}
    (e) 5.5×10^{-10}

14. (a) 9.16
    (b) 4.54
    (c) 8.07
    (d) 7
    (e) 5.74
    (f) 7
    (g) 8.70
    (h) 7. The [H$^+$] from the water is more important than that from HCl.

15. (a) 7
    (b) 5.37
    (c) 11.2
    (d) 8.81

16. 5.14

17. Prepare the solution so that the molar ratio of NH_4NO_3 to NH_3 is 18:1.

18. (a) 9.26 (b) 9.34 (c) 9.17 (d) 14 (e) 0
    (f) Small amounts of strong acid and base, when added to water ($pH = 7$), normally cause a tremendous change in pH, as in d and e. In a buffer solution, small amounts of acid and base cause only a very small change in pH.

19.
ml NaOH added	pH
0	2.88
5	4.14
10	4.57
20	5.35
24	6.13
24.9	7.14
25.0	8.72
25.1	10.30
26	11.29
30	11.96

20.
ml NaOH added	pH
0	1.00
5	1.18
10	1.37
24	2.69
24.9	3.70
25.0	7.00
25.1	10.30
26	11.29
30	11.96

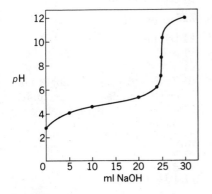

FIGURE A20-19.
Titration of $HC_2H_3O_2$ with NaOH; pH against ml of NaOH.

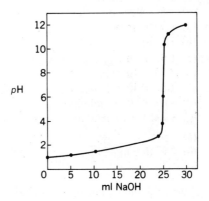

FIGURE A20-20.
Titration of HCl with NaOH; pH against ml of NaOH.

CHAPTER 21

1. (a) 1.50×10^{-16} (c) 1.11×10^{-14} (e) 4.54×10^{-18} (g) 4.41×10^{-27}
   (b) 1.06×10^{-10} (d) 3.66×10^{-9} (f) 1.29×10^{-29} (h) 5.01×10^{-9}

2. Solids (b), (c), (d), and (f) will dissolve.

3. (a) no ppt
   (b) no ppt
   (c) ppt
   (d) no ppt
   (e) no ppt
   (f) no ppt
   (g) no ppt

4. (a) 1×10^{-27} mole/liter
   (b) 6×10^{-3} mole/liter
   (c) 4.47×10^{-5} mole/liter
   (d) 6.3×10^{-18} mole/liter
   (e) 1.31×10^{-3} mole/liter
   (f) 1.36×10^{-6} mole/liter

5. (a) 5.72×10^{-3} g
   (b) 2.56×10^{-6} g
   (c) 2.56×10^{-5} g
   (d) 5.72×10^{-3} g

6. (a) 0.151 g
   (b) 5.46×10^{-2} g
   (c) 2.59×10^{-3} g

7. (a) 1×10^{-39} mole/liter
   (b) 1×10^{-43} mole/liter
   (c) 2.5×10^{-19} mole/liter
   (d) 1×10^{-22} mole/liter
   (e) 1×10^{-48} mole/liter

8. (a) 1.95×10^{-9} mole/liter
   (b) 3.73×10^{-10} mole/liter
   (c) 2.56×10^{-8} mole/liter
   (d) 1.09×10^{-12} mole/liter
   (e) 3×10^{-11} mole/liter

9. (a) Pb^{++}
   (b) 2×10^{-6} mole/liter

10. 32.2 mg

11. 4.08×10^{-7} mole/liter

12. $8 \times 10^{-5}\%$

13. 2×10^{-14} mole/liter

14. 0.124 mole/liter

15. 18.0 g

16. 0.0293 mole

17. The $pH = 7$ because water will contribute $[OH^-] = 10^{-7}$ mole/liter, while $Fe(OH)_3$ will contribute less than 4×10^{-10} mole/liter.

18. (a) -0.21 (b) $2 \times 10^{-7}M$

19. -0.26

20. (a) -1.55
    (b) -0.22
    (c) 1.94
    (d) 4.44

21. AgI is converted to Ag_2S because the $[Ag^+]$ available from AgI and the $[S^=]$ possible in $5M$ HCl exceeds the K_{sp} for Ag_2S.

22. 1,660 liters

23. (a) $3.08 \times 10^{-21}M$
    (b) $3.08 \times 10^{-9}M$
    (c) $3.07 \times 10^{-4}M$

24. $6.25 \times 10^{-13}M$

25. (a) $[H^+]^2[CO_3^=] = 9.18 \times 10^{-19}$
    (b) 5.72
    (c) $4 \times 10^{-7}M$
    (d) The method can't be applied because there is not enough difference between the K_{sp} values for the various carbonates.

26. MnS

27. (a) $8.8 \times 10^{-12}M$
    (b) $1.18 \times 10^{-6}M$

CHAPTER 22

1. $9.6 \times 10^{-17}M$

2. (a) $0.1M$ (c) $3.1 \times 10^{-10}M$ (e) $1.76 \times 10^{-12}M$
   (b) $1.8M$ (d) $0.1M$ (f) $5.7 \times 10^{-3}M$

3. 3.34 g

4. 0.485 g

5. 1.08M

6. 4.35M

7. 1.16 × 10^{-5} g

8. 4.33M

9. 2.35 × 10^{-11}M

10. 2.0 × 10^{-3}M

11. 6.27M

12. (a) 4.50
    (b) 1.74

13. (a) $[Pb^{++}] = 2.93 \times 10^{-6}M$
       $[OH^-] = 5.86 \times 10^{-6}M$
       $[Pb(OH)_3^-] = 1.17 \times 10^{-7}M$
    (b) $[Pb^{++}] = 1 \times 10^{-16}M$
       $[Pb(OH)_3^-] = 0.02M$
    (c) $[Sn^{++}] = 4 \times 10^{-26}M$
       $[Sn(OH)_3^-] = 2.5 \times 10^{-4}M$
    (d) 10.2 liters

CHAPTER 23

1. (a) 0.193 cal/g-deg
   (b) 0.0949 cal/g-deg
   (c) 0.0447 cal/g-deg
   (d) 0.026 cal/g-deg
   (e) 0.0299 cal/g-deg

2. 56.7°C

3. 27.1°C

4. 35.0°C

5. 0°C

6. 93.4

7. 0.314 cal/g-deg

8. (a) X_2O_3
   (b) 51.0

9. (a) YCl_3 and YCl_5
   (b) 121.3

10. $\Delta H° = -80$ cal/mole
    $\Delta S° = -0.27$ cal/mole deg
    $\Delta G° = 0.0$

11.

	$\Delta H°$	$\Delta G°$	$\Delta S°$
(a)	−4.62 kcal	1.71 kcal	−21.2 cal/mole deg
(b)	−216.24 kcal	−229.04 kcal	+43.0 cal/mole deg
(c)	−36.43 kcal	−35.18 kcal	−4.2 cal/mole deg
(d)	−69.10 kcal	−60.30 kcal	−29.5 cal/mole deg
(e)	−33.09 kcal	−12.58 kcal	−68.7 cal/mole deg
(f)	−38.98 kcal	−49.52 kcal	+35.4 cal/mole deg

12. $C_6H_{6(l)} + 7.5\ O_{2(g)} \rightarrow 6\ CO_{2(g)} + 3\ H_2O_{(l)}$
    $\Delta H° = -780.98$ kcal

13. (a) $\Delta H° = -123.12$ kcal/bond
    (b) $\Delta H° = -110.57$ kcal/bond
    (c) $\Delta H° = -84.29$ kcal/bond
    (d) $\Delta H° = -63.12$ kcal/bond

14. $\Delta H = 10.12$ kcal/mole
    $\Delta S = 34.0$ cal/mole deg
    $\Delta G = 0.00$

15. $\Delta G° = -7.89$ kcal
    $\Delta H° = -4.82$ kcal
    $\Delta S° = +10.3$ cal/mole deg

CHAPTER 24

1. (a) $^{237}_{93}Np \rightarrow\ ^{237}_{94}Pu + ^{0}_{-1}e$
   (b) $^{239}_{94}Pu \rightarrow\ ^{235}_{92}U + ^{4}_{2}He$
   (c) $^{227}_{89}Ac \rightarrow\ ^{223}_{87}Fr + ^{4}_{2}He$
   (d) $^{210}_{84}Po \rightarrow\ ^{206}_{82}Pb + ^{4}_{2}He$
   (e) $^{211}_{82}Pb \rightarrow\ ^{211}_{83}Bi + ^{0}_{-1}e$

2. (a) $^{27}_{14}Si \rightarrow\ ^{27}_{13}Al + ^{0}_{1}e$
   (b) $^{28}_{13}Al \rightarrow\ ^{28}_{14}Si + ^{0}_{-1}e$
   (c) $^{30}_{15}P \rightarrow\ ^{30}_{14}Si + ^{0}_{1}e$
   (d) $^{24}_{11}Na \rightarrow\ ^{24}_{12}Mg + ^{0}_{-1}e$
   (e) $^{17}_{9}F \rightarrow\ ^{17}_{8}O + ^{0}_{1}e$

3. (a) $^{14}_{7}N + ^{1}_{0}n \rightarrow\ ^{1}_{1}H + ^{14}_{6}C$
   (b) $^{26}_{12}Mg + ^{1}_{0}n \rightarrow\ ^{4}_{2}He + ^{23}_{10}Ne$
   (c) $^{59}_{27}Co + ^{2}_{1}H \rightarrow\ ^{1}_{1}H + ^{60}_{27}Co$
   (d) $^{14}_{7}N + ^{4}_{2}He \rightarrow\ ^{1}_{1}H + ^{17}_{8}O$
   (e) $^{63}_{29}Cu + ^{1}_{1}H \rightarrow\ ^{1}_{0}n + ^{63}_{30}Zn$

4. (a) 4×10^8 kcal
   (b) 802 tons

5. 0.01 mg

6. 4.2 hr

7. 48.5 g

8. 6×10^7 yr

9. 6.42×10^3 yr

10. 345 torr

11. 3.52×10^{-3} mg

CHAPTER 25

1. In (a), (b), (d), and (g) the reactions do not take place because none of the products is insoluble, un-ionized, or goes off as a gas. In (c) and (e) the cell potentials are negative and the reaction would take place in the reverse direction. In (f) all reactants are oxidizing agents and there is no reducing agent to be oxidized; the apparent oxidation of Fe^{+++} is false, since the higher oxidation state of $+4$ does not exist.

2. Many of the statements made below are true because an excess of reagent is used; for example, NH_4^+ in (a). This excess shifts the equilibrium to the right.
   (a) $Mg(OH)_2 + 2NH_4^+ \rightarrow Mg^{++} + 2NH_3 + 2H_2O$
   OH^- removed more completely by NH_4^+ than by Mg^{++}
   (b) $AgCl + 2NH_3 \rightarrow Ag(NH_3)_2^+ + Cl^-$
   Ag^+ removed more completely by NH_3 than by Cl^-
   (c) $Ag_2S + 4CN^- \rightarrow 2Ag(CN)_2^- + S^=$
   Ag^+ removed more completely by CN^- than by $S^=$
   (d) $AgCl + I^- \rightarrow AgI + Cl^-$
   AgI more insoluble than $AgCl$
   (e) $SrSO_4 + CO_3^= \rightarrow SrCO_3 + SO_4^=$
   Sr^{++} removed more completely by $CO_3^=$ than by $SO_4^=$
   (f) $Fe^{++} + H_2S + 2OH^- \rightarrow FeS \downarrow + 2H_2O$
   FeS insoluble (hence reaction goes to right) if acidity is kept low
   (g) $NH_4^+ + OH^- \rightarrow NH_3 \uparrow + H_2O$
   NH_3 is a gas and a weak base
   (h) $Br^- + H_3PO_4$ (conc.) $\rightarrow HBr \uparrow + H_2PO_4^-$
   HBr goes off as a gas, especially if warmed

(i) $3CuS + 8H^+ + 2NO_3^- \rightarrow 3Cu^{++} + 3S\downarrow + 2NO\uparrow + 4H_2O$
   $S^=$ is irreversibly oxidized to S, and NO goes off as a gas

(j) $Ag(NH_3)_2^+ + I^- \rightarrow AgI\downarrow + 2NH_3$
   I^- removes Ag^+ from solution more completely than does NH_3

3. (a) $Pb^{++} + 2Cl^- \rightarrow PbCl_2\downarrow$
   (b) NR
   (c) $2Ag^+ + H_2S \rightarrow Ag_2S\downarrow + 2H^+$
   (d) $3Cd + 2H^+ + 2H_2PO_4^- \rightarrow Cd_3(PO_4)_2 + 3H_2\uparrow$
   (e) $BaCO_3 + 2H^+ \rightarrow Ba^{++} + H_2O + CO_2$
   (f) NR
   (g) $Ba_3(PO_4)_2 + 4H^+ \rightarrow 3Ba^{++} + 2H_2PO_4^-$
   (h) $Al^{+++} + 3NH_3 + 3H_2O \rightarrow Al(OH)_3\downarrow + 3NH_4^+$
   (i) $Fe(OH)_3 + 3H^+ \rightarrow Fe^{+++} + 3H_2O$
   (j) NR
   (k) NR
   (l) $CuO + 2H^+ \rightarrow Cu^{++} + H_2O$
   (m) $NH_3 + H^+ \rightarrow NH_4^+$
   (n) NR
   (o) NR
   (p) $AgC_2H_3O_2 + H^+ + Cl^- \rightarrow AgCl\downarrow + HC_2H_3O_2$
   (q) $Fe + Cu^{++} \rightarrow Fe^{++} + Cu$
   (r) NR
   (s) $2MnO_4^- + 5H_3AsO_3 + 6H^+ \rightarrow 2Mn^{++} + 5H_3AsO_4 + 3H_2O$
   (t) NR
   (u) $Ba^{++} + SO_4^= \rightarrow BaSO_4\downarrow$
   (v) $ZnS + 2H^+ \rightarrow Zn^{++} + H_2S\uparrow$
   (w) $Zn^{++} + S^= \rightarrow ZnS\downarrow$
   (x) $Cl_2 + 2Br^- \rightarrow Br_2 + 2Cl^-$
   (y) $ClO_3^- + 6I^- + 6H^+ \rightarrow Cl^- + 3I_2 + 3H_2O$
   (z) NR

4. (a) Mix solutions of $Pb(NO_3)_2$ and Na_2SO_4, and filter off ppt

$$Pb^{++} + SO_4^= \rightarrow PbSO_4\downarrow$$

   (b) Neutralize KOH with $HC_2H_3O_2$, then evaporate off water

$$K^+ + OH^- + HC_2H_3O_2 \rightarrow K^+ + C_2H_3O_2^- + H_2O$$

   (c) Pass H_2S into a dilute $Cd(NO_3)_2$ solution, and filter off CdS

$$Cd^{++} + H_2S \rightarrow CdS + 2H^+$$

   (d) Saturate a KOH solution with CO_2, and evaporate off water

$$K^+ + OH^- + CO_2 \rightarrow K^+ + HCO_3^-$$

(e) Mix slaked lime with NH_4Cl, and warm. Pass the gas through a drying tube filled with solid NaOH or $CaCl_2$.

$$2NH_4^+ + Ca(OH)_2 \xrightarrow{\Delta} 2NH_3 \uparrow + 2H_2O$$

(f) Add NaOH to $CuSO_4$ solution, and filter off the ppt. NH_3 cannot be used.

$$Cu^{++} + 2OH^- \rightarrow Cu(OH)_2 \downarrow$$

(g) Divide a given volume of H_2SO_4 in half. Neutralize one of the halves with NaOH, then add the remaining half of H_2SO_4. Evaporate off the water.

$$2Na^+ + 2OH^- + 2H^+ + SO_4^= \rightarrow 2Na^+ + SO_4^= + 2H_2O$$
$$2Na^+ + SO_4^= + 2H^+ + SO_4^= \rightarrow 2Na^+ + 2HSO_4^-$$

(h) Heat Al metal with an excess of S and distill off the excess sulfur. Al_2S_3 will not precipitate from solution saturated with H_2S.

$$2Al + 3S \xrightarrow{\Delta} Al_2S_3$$

(i) Add excess NH_3 to a solution of $FeCl_3$. Filter off the ppt of $Fe(OH)_3$, and ignite.

$$Fe^{+++} + 3OH^- \rightarrow Fe(OH)_3 \downarrow$$
$$2Fe(OH)_3 \xrightarrow{\Delta} Fe_2O_3 + 3H_2O \uparrow$$

(j) Add HCl to a solution of $Hg_2(NO_3)_2$. Filter off the ppt.

$$2Cl^- + Hg_2^{++} \rightarrow Hg_2Cl_2 \downarrow$$

(k) Add HCl to $CaCO_3$, and pass the gas through conc. H_2SO_4 or over P_2O_5 to dry it.

$$CaCO_3 + 2H^+ \rightarrow Ca^{++} + CO_2 \uparrow + H_2O$$

(l) Add HBr to MnO_2 or $KMnO_4$. Warm. Condense Br_2 in very cold container. Caution with eyes and skin!

$$MnO_2 + 4H^+ + 2Br^- \rightarrow Mn^{++} + 2H_2O + Br_2$$

(m) Add excess metallic Zn to $AgNO_3$ solution to reduce Ag^+, then add excess dilute HNO_3 to dissolve the excess Zn (Ag will not dissolve). Filter off Ag and wash.

$$Zn + 2Ag^+ \rightarrow Zn^{++} + 2Ag \downarrow$$
$$3Zn + 8H^+ + 2NO_3^- \rightarrow 3Zn^{++} + 2NO \uparrow + 4H_2O$$

(n) Vigorously heat solid KH_2AsO_4:

$$KH_2AsO_4 \xrightarrow{\Delta} KAsO_3 + H_2O \uparrow$$

5. (a) Al will not react directly with HNO_3; therefore dissolve it in HCl. Neutralize the acid with NH_3, and add an excess to precipitate $Al(OH)_3$. NaOH cannot be used because of the formation of $Al(OH)_4^-$, which is soluble. Filter off the ppt, and just neutralize with HNO_3. Evaporate off the water.

$$2Al + 6H^+ \rightarrow 2Al^{+++} + 3H_2 \uparrow$$

$$Al^{+++} + 3\ NH_3 + 3\ H_2O \rightarrow Al(OH)_3 \downarrow + 3NH_4^+$$

$$Al(OH)_3 + 3H^+ + 3NO_3^- \rightarrow Al^{+++} + 3NO_3^- + 3H_2O$$

(b) Dissolve Cu in HNO_3, then make basic with NaOH to precipitate $Cu(OH)_2$. Filter and dissolve the ppt with H_3PO_4 (just neutralize).

$$3Cu + 2NO_3^- + 8H^+ \rightarrow 3Cu^{++} + 2NO \uparrow + 4H_2O$$

$$Cu^{++} + 2OH^- \rightarrow Cu(OH)_2 \downarrow$$

$$3Cu(OH)_2 + 2H_3PO_4 \rightarrow Cu_3(PO_4)_2 \downarrow + 6H_2O$$

(c) Dissolve Ni in HCl, then make basic with NaOH to precipitate $Ni(OH)_2$. NH_3 cannot be used.

$$Ni + 2H^+ \rightarrow Ni^{++} + H_2 \uparrow$$

$$Ni^{++} + 2OH^- \rightarrow Ni(OH)_2 \downarrow$$

(d) Dissolve Hg in conc. HNO_3, then make basic with NaOH to precipitate $Hg(OH)_2$. NH_3 cannot be used. Just dissolve ppt with HCl, and evaporate off the water.

$$Hg + 4H^+ + 2NO_3^- \rightarrow Hg^{++} + 2NO_2 \uparrow + 2H_2O$$

$$Hg^{++} + 2OH^- \rightarrow Hg(OH)_2 \downarrow$$

$$Hg(OH)_2 + 2H^+ + 2Cl^- \rightarrow Hg^{++} + 2Cl^- + 2H_2O$$

(e) Dissolve Zn in HCl, then make solution slightly basic with NH_3, and saturate with H_2S. Filter off ZnS.

$$Zn + 2H^+ \rightarrow Zn^{++} + H_2 \uparrow$$

$$Zn^{++} + H_2S \rightarrow ZnS \downarrow + 2H^+$$

$$H^+ + NH_3 \rightarrow NH_4^+$$

6. The volumes and weights used in the following procedures are based on the ideal experiment performed with flawless technique. Certain losses will occur because of inability to control the temperature perfectly during filtration, and because crystals and solutions cling to containers.

(a) Assume from the curves that the solubility of KCl is 58 g at 100°C and 28 g at 0°C, and that of $KClO_3$ is 58 g at 100°C and 4 g at 0°C. The original mixture contains 74.6 g of KCl and 122.6 g of $KClO_3$.

(1) Dissolve the mixture in $\frac{74.6}{28} \times 100 = 266$ ml of hot water. Cool to 0°C, and filter off $122.6 - (4 \times 2.66) = 112$ g of $KClO_3$.

(2) Evaporate cold filtrate from step 1 to $\frac{10.6}{58.0} \times 100 = 18.3$ ml, and filter at 100°C. This volume will just dissolve all remaining $KClO_3$, and you will obtain $74.6 - \frac{18.3}{100}(58) = 64$ g of KCl.

(3) Add 19.7 ml of H_2O to the filtrate from step 2, and cool to 0°C. You need 38.0 ml to dissolve the remaining 10.6 g of KCl at 0°C. You will obtain $10.6 - \frac{38}{100} \times 4 = 9.1$ g of $KClO_3$ crystals.

(4) Evaporate the filtrate from step 3 to $\frac{1.5}{58} \times 100 = 2.6$ ml, and filter at 100°C. You will then obtain $10.6 - \frac{2.6}{100}(58) = 9.1$ g of KCl.

(5) Since it is not practical to work with smaller volumes, stop and combine crystals. Your yield will be:

$$64.1 + 9.1 = 73.2 \text{ g of pure KCl}$$
$$112.0 + 9.1 = 121.1 \text{ g of pure KClO}_3$$

(b) Assume from the curves that the solubility of $K_2Cr_2O_7$ is 81 g at 100°C and 7 g at 0°C, and that the solubility of KCl is 58 g at 100°C and 28 g at 0°C.

(1) Add $\frac{50}{58} \times 100 = 86.1$ ml of H_2O. Heat to 100°C, and filter off $100 - 0.861 \times 81 = 30.2$ g of $K_2Cr_2O_7$.

(2) Add 92 ml of H_2O, giving a total volume of 178 ml, the volume needed to keep all the KCl in solution at 0°C. Cool to 0°C, and filter off $69.8 - 7(1.78) = 57.3$ g of $K_2Cr_2O_7$.

(3) Evaporate filtrate from step 2 to $\frac{12.5}{81} \times 100 = 15.4$ ml, the minimum volume needed to dissolve all $K_2Cr_2O_7$ at 100°C. Heat to 100°C, and filter off $50 - (0.154)(58) = 41$ g of KCl.

(4) Add 16.8 ml of H_2O to the filtrate from step 3 to give $\frac{9}{28} \times 100 = 32.2$ ml, the minimum volume needed to keep all KCl in solution at 0°C. Cool to 0°C, and filter off $12.5 - (0.32)(7) = 10.3$ g of $K_2Cr_2O_7$.

(5) Evaporate the filtrate from step 4 to $\frac{2.2}{81} \times 100 = 2.8$ ml, the minimum volume needed to keep all $K_2Cr_2O_7$ in solution at 100°C. Heat to 100°C, and filter off $9 - (0.028)(58) = 7.4$ g of KCl.

(6) It is not practical to continue further. Your yield is:

$$30.2 + 57.3 + 10.3 = 97.8 \text{ g of pure K}_2\text{Cr}_2\text{O}_7$$
$$41.0 + 7.4 = 48.4 \text{ g of pure KCl}$$

(c) Assume from the curves that the solubility of KNO_3 is 246 g at 100°C and 12 g at 0°C, and that the solubility of K_2SO_4 is 25 g at 100°C and 7 g at 0°C.

(1) Add $\frac{50}{246} \times 100 = 20.3$ ml, the minimum volume needed to dissolve all KNO_3 at 100°C. Heat to 100°C, and filter off $50 - (25)(0.203) = 44.9$ g of K_2SO_4.

(2) Dilute the filtrate from step 1 to $\frac{5.1}{7} \times 100 = 73$ ml, the minimum volume needed to dissolve the K_2SO_4 at 0°C. Cool to 0°C, and filter off $50 - (0.73)(12) = 41.2$ g of KNO_3.

(3) Evaporate the filtrate from step 2 to $\frac{8.8}{246} \times 100 = 3.6$ ml, the minimum volume needed to dissolve the KNO_3 at 100°C. Heat to 100°C, and filter off $5.1 - (0.036)(25) = 4.2$ g of K_2SO_4.

(4) Dilute the filtrate from step 3 to $\frac{0.9}{7} \times 100 = 12.9$ ml, and cool to 0°C. No K_2SO_4 precipitates, and you obtain $8.8 - (0.129)(12) = 7.2$ g of KNO_3.

(5) It is not practical to go further. Your yield will be:

$$41.2 + 7.2 = 48.4 \text{ g of pure } KNO_3$$

$$44.9 + 4.2 = 49.1 \text{ g of pure } K_2SO_4$$

Four-Place Logarithms

N	0	1	2	3	4	5	6	7	8	9	Proportional parts								
											1	2	3	4	5	6	7	8	9
10	0000	0043	0086	0128	0170	0212	0253	0294	0334	0374	4	8	12	17	21	25	29	33	37
11	0414	0453	0492	0531	0569	0607	0645	0682	0719	0755	4	8	11	15	19	23	26	30	34
12	0792	0828	0864	0899	0934	0969	1004	1038	1072	1106	3	7	10	14	17	21	24	28	31
13	1139	1173	1206	1239	1271	1303	1335	1367	1399	1430	3	6	10	13	16	19	23	26	29
14	1461	1492	1523	1553	1584	1614	1644	1673	1703	1732	3	6	9	12	15	18	21	24	27
15	1761	1790	1818	1847	1875	1903	1931	1959	1987	2014	3	6	8	11	14	17	20	22	25
16	2041	2068	2095	2122	2148	2175	2201	2227	2253	2279	3	5	8	11	13	16	18	21	24
17	2304	2330	2355	2380	2405	2430	2455	2480	2504	2529	2	5	7	10	12	15	17	20	22
18	2533	2577	2601	2625	2648	2672	2695	2718	2742	2765	2	5	7	9	12	14	16	19	21
19	2788	2810	2833	2856	2878	2900	2923	2945	2967	2989	2	4	7	9	11	13	16	18	20
20	3010	3032	3054	3075	3096	3118	3139	3160	3181	3201	2	4	6	8	11	13	15	17	19
21	3222	3243	3263	3284	3304	3324	3345	3365	3385	3404	2	4	6	8	10	12	14	16	18
22	3424	3444	3464	3483	3502	3522	3541	3560	3579	3598	2	4	6	8	10	12	14	15	17
23	3617	3636	3655	3674	3692	3711	3729	3747	3766	3784	2	4	6	7	9	11	13	15	17
24	3802	3820	3838	3856	3874	3892	3909	3927	3945	3962	2	4	5	7	9	11	12	14	16
25	3979	3997	4014	4031	4048	4065	4082	4099	4116	4133	2	3	5	7	9	10	12	14	15
26	4150	4166	4183	4200	4216	4232	4249	4265	4281	4298	2	3	5	7	8	10	11	13	15
27	4314	4330	4346	4362	4378	4393	4409	4425	4440	4456	2	3	5	6	8	9	11	13	14
28	4472	4487	4502	4518	4533	4548	4564	4579	4594	4609	2	3	5	6	8	9	11	12	14
29	4624	4639	4654	4669	4683	4698	4713	4728	4742	4757	1	3	4	6	7	9	10	12	13
30	4771	4786	4800	4814	4829	4843	4857	4871	4886	4900	1	3	4	6	7	9	10	11	13
31	4914	4928	4942	4955	4969	4983	4997	5011	5024	5038	1	3	4	6	7	8	10	11	12
32	5051	5065	5079	5092	5105	5119	5132	5145	5159	5172	1	3	4	5	7	8	9	11	12
33	5185	5198	5211	5224	5237	5250	5263	5276	5289	5302	1	3	4	5	6	8	9	10	12
34	5315	5328	5340	5353	5366	5378	5391	5403	5416	5428	1	3	4	5	6	8	9	10	11
35	5441	5453	5465	5478	5490	5502	5514	5527	5539	5551	1	2	4	5	6	7	9	10	11
36	5563	5575	5587	5599	5611	5623	5635	5647	5658	5670	1	2	4	5	6	7	8	10	11
37	5682	5694	5705	5717	5729	5740	5752	5763	5775	5786	1	2	3	5	6	7	8	9	10
38	5798	5809	5821	5832	5843	5855	5866	5877	5888	5899	1	2	3	5	6	7	8	9	10
39	5911	5922	5933	5944	5955	5966	5977	5988	5999	6010	1	2	3	4	5	7	8	9	10
40	6021	6031	6042	6053	6064	6075	6085	6096	6107	6117	1	2	3	4	5	6	8	9	10
41	6128	6138	6149	6160	6170	6180	6191	6201	6212	6222	1	2	3	4	5	6	7	8	9
42	6232	6243	6253	6263	6274	6284	6294	6304	6314	6325	1	2	3	4	5	6	7	8	9
43	6335	6345	6355	6365	6375	6385	6395	6405	6415	6425	1	2	3	4	5	6	7	8	9
44	6435	6444	6454	6464	6474	6484	6493	6503	6513	6522	1	2	3	4	5	6	7	8	9
45	6532	6542	6551	6561	6571	6580	6590	6599	6609	6618	1	2	3	4	5	6	7	8	9
46	6628	6637	6646	6656	6665	6675	6684	6693	6702	6712	1	2	3	4	5	6	7	7	8
47	6721	6730	6739	6749	6758	6767	6776	6785	6794	6803	1	2	3	4	5	5	6	7	8
48	6812	6821	6830	6839	6848	6857	6866	6875	6884	6893	1	2	3	4	4	5	6	7	8
49	6902	6911	6920	6928	6937	6946	6955	6964	6972	6981	1	2	3	4	4	5	6	7	8
50	6990	6998	7007	7016	7024	7033	7042	7050	7059	7067	1	2	3	3	4	5	6	7	8
51	7076	7084	7093	7101	7110	7118	7126	7135	7143	7152	1	2	3	3	4	5	6	7	8
52	7160	7168	7177	7185	7193	7202	7210	7218	7226	7235	1	2	3	3	4	5	6	7	7
53	7243	7251	7259	7267	7275	7284	7292	7300	7308	7316	1	2	2	3	4	5	6	6	7
54	7324	7332	7340	7348	7356	7364	7372	7380	7388	7396	1	2	2	3	4	5	6	6	7

N	0	1	2	3	4	5	6	7	8	9				Proportional Parts					
											1	2	3	4	5	6	7	8	9
55	7404	7412	7419	7427	7435	7443	7451	7459	7466	7474	1	2	2	3	4	5	5	6	7
56	7482	7490	7497	7505	7513	7520	7528	7536	7543	7551	1	2	2	3	4	5	5	6	7
57	7559	7566	7574	7582	7589	7597	7604	7612	7619	7627	1	2	2	3	4	5	5	6	7
58	7634	7642	7649	7657	7664	7672	7679	7686	7694	7701	1	1	2	3	4	4	5	6	7
59	7709	7716	7723	7731	7738	7745	7752	7760	7767	7774	1	1	2	3	4	4	5	6	7
60	7782	7789	7796	7803	7810	7818	7825	7832	7839	7846	1	1	2	3	4	4	5	6	6
61	7853	7860	7868	7875	7882	7889	7896	7903	7910	7917	1	1	2	3	4	4	5	6	6
62	7924	7931	7938	7945	7952	7959	7966	7973	7980	7987	1	1	2	3	3	4	5	6	6
63	7993	8000	8007	8014	8021	8028	8035	8041	8048	8055	1	1	2	3	3	4	5	5	6
64	8062	8069	8075	8082	8089	8096	8102	8109	8116	8122	1	1	2	3	3	4	5	5	6
65	8129	8136	8142	8149	8156	8162	8169	8176	8182	8189	1	1	2	3	3	4	5	5	6
66	8195	8202	8209	8215	8222	8228	8235	8241	8248	8254	1	1	2	3	3	4	5	5	6
67	8261	8267	8274	8280	8287	8293	8299	8306	8312	8319	1	1	2	3	3	4	5	5	6
68	8325	8331	8338	8344	8351	8357	8363	8370	8376	8382	1	1	2	3	3	4	4	5	6
69	8388	8395	8401	8407	8414	8420	8426	8432	8439	8445	1	1	2	2	3	4	4	5	6
70	8451	8457	8463	8470	8476	8482	8488	8494	8500	8506	1	1	2	2	3	4	4	5	6
71	8513	8519	8525	8531	8537	8543	8549	8555	8561	8567	1	1	2	2	3	4	4	5	5
72	8573	8579	8585	8591	8597	8603	8609	8615	8621	8627	1	1	2	2	3	4	4	5	5
73	8633	8639	8645	8651	8657	8663	8669	8675	8681	8686	1	1	2	2	3	4	4	5	5
74	8692	8698	8704	8710	8716	8722	8727	8733	8739	8745	1	1	2	2	3	4	4	5	5
75	8751	8756	8762	8768	8774	8779	8785	8791	8797	8802	1	1	2	2	3	4	4	5	5
76	8808	8814	8820	8825	8831	8837	8842	8848	8854	8859	1	1	2	2	3	4	4	5	5
77	8865	8871	8876	8882	8887	8893	8899	8904	8910	8915	1	1	2	2	3	3	4	4	5
78	8921	8927	8932	8938	8943	8949	8954	8960	8965	8971	1	1	2	2	3	3	4	4	5
79	8976	8982	8987	8993	8998	9004	9009	9015	9020	9025	1	1	2	2	3	3	4	4	5
80	9031	9036	9042	9047	9053	9058	9063	9069	9074	9079	1	1	2	2	3	3	4	4	5
81	9085	9090	9096	9101	9106	9112	9117	9122	9128	9133	1	1	2	2	3	3	4	4	5
82	9138	9143	9149	9154	9159	9165	9170	9175	9180	9186	1	1	2	2	3	3	4	4	5
83	9191	9196	9201	9206	9212	9217	9222	9227	9232	9238	1	1	2	2	3	3	4	4	5
84	9243	9248	9253	9258	9263	9269	9274	9279	9284	9289	1	1	2	2	3	3	4	4	5
85	9294	9299	9304	9309	9315	9320	9325	9330	9335	9340	1	1	2	2	3	3	4	4	5
86	9345	9350	9355	9360	9365	9370	9375	9380	9385	9390	1	1	2	2	3	3	4	4	5
87	9395	9400	9405	9410	9415	9420	9425	9430	9435	9440	0	1	1	2	2	3	3	4	4
88	9445	9450	9455	9460	9465	9469	9474	9479	9484	9489	0	1	1	2	2	3	3	4	4
89	9494	9499	9504	9509	9513	9518	9523	9528	9533	9538	0	1	1	2	2	3	3	4	4
90	9542	9547	9552	9557	9562	9566	9571	9576	9581	9586	0	1	1	2	2	3	3	4	4
91	9590	9595	9600	9605	9609	9614	9619	9624	9628	9633	0	1	1	2	2	3	3	4	4
92	9638	9643	9647	9652	9657	9661	9666	9671	9675	9680	0	1	1	2	2	3	3	4	4
93	9685	9689	9694	9699	9703	9708	9713	9717	9722	9727	0	1	1	2	2	3	3	4	4
94	9731	9736	9741	9745	9750	9754	9759	9763	9768	9773	0	1	1	2	2	3	3	4	4
95	9777	9782	9786	9791	9795	9800	9805	9809	9814	9818	0	1	1	2	2	3	3	4	4
96	9823	9827	9832	9836	9841	9845	9850	9854	9859	9863	0	1	1	2	2	3	3	4	4
97	9868	9872	9877	9881	9886	9890	9894	9899	9903	9908	0	1	1	2	2	3	3	4	4
98	9912	9917	9921	9926	9930	9934	9939	9943	9948	9952	0	1	1	2	2	3	3	4	4
99	9956	9961	9965	9969	9974	9978	9983	9987	9991	9996	0	1	1	2	2	3	3	3	4

Index